THE SEARCH FOR THE ULTIMATE SINK

TECHNOLOGY
AND THE
ENVIRONMENT

JEFFREY K. STINE AND
WILLIAM MCGUCKEN

SERIES EDITORS

JOEL A. TARR

THE SEARCH FOR THE ULTIMATE SINK

URBAN POLLUTION IN HISTORICAL PERSPECTIVE

THE UNIVERSITY OF AKRON PRESS
AKRON, OHIO

Manufactured in the United States of America
00 99 98 5 4 3 2

LIBRARY OF CONGRESS CATALOGING-IN-PUBLICATION DATA

Tarr, Joel A. (Joel Arthur), 1934–
The Search for the ultimate sink : urban pollution in historical perspective / Joel A. Tarr.—1st ed.
p. cm.—(Technology and the environment)
Includes bibliographical references and index.
ISBN 1-884836-05-4. —ISBN 1-884836-06-2 (pbk.)
1. Urban pollution—United States—History. I. Title. II. Series: Technology and the environment (Akron, Ohio)
TD180.T37 1996
363.73'09173'2—dc20 96-38383

The paper used in this publication meets the minimum requirements of American National Standard for Information Sciences—Permanence of Paper for Printed Library Materials, ANSI Z39.48–1984. ∞

Dedicated to my colleagues in the Departments of Civil and Environmental Engineering and Engineering and Public Policy at Carnegie Mellon University, and to my graduate students in History.

Contents

List of Illustrations

Photographs

Figures

Tables

Series Preface

This book series springs from public awareness of and concern about the effects of technology on the environment. Its purpose is to publish the most informative and provocative work emerging from research and reflection, work that will place these issues in an historical context, define the current nature of the debates, and anticipate the direction of future arguments about the complex relationships between technology and the environment.

The scope of the series is broad, as befits its subject. No single academic discipline embraces all of the knowledge needed to explore the manifold ways in which technology and the environment work with and against each other. Volumes in the series will examine the subject from multiple perspectives based in the natural sciences, the social sciences, and the humanities.

These studies are meant to stimulate, clarify, and influence the debates taking place in the classroom, on the floors of legislatures, and at international conferences. Addressed not only to scholars and policymakers, but also to a wider audience, the books in this series speak to a public that seeks to understand how its world will be changed, for ill and for good, by the impact of technology on the environment.

Acknowledgments

Over the years, many individuals and funding agencies aided in the research and writing that led to the articles included in this publication. My greatest debt is reflected in dedicating this book to my distinguished present and former engineering colleagues at Carnegie Institute of Technology at Carnegie Mellon University and to my graduate students in the Department of History. Carnegie Mellon University has furnished an ideal environment for interdisciplinary research, and I am grateful that I have been able to spend most of my professional career at this stimulating institution. I would especially like to thank engineering faculty members Robert U. Ayres (formerly at CMU and now on the faculty at the Institut Europeen d'Administration des Affaires in France), Cliff Davidson, Steven J. Fenves, Chris T. Hendrickson, Richard G. Luthy, M. Granger Morgan, Harry W. Paxton, Edward S. Rubin, James P. Romouldi (deceased, 1994), and Mitchell J. Small for their generosity with their time and advice. Special thanks are owed to Francis C. McMichael, the Blenko Professor of Civil Engineering, with whom I collaborated on different projects and papers and who, over the years, has explained the complexities of sanitary and environmental engineering to me. My graduate students in environmental history (some of whom were included as co-authors on the original articles) have been a source of constant help and insights. I would particularly like to note the contributions of Nicholas Casner, Charles Jacobson, Kenneth Koons, Bill Lamperes, James McCurley, III, Rachel P. Maines, Todd Shallat, and Terry F. Yosi.

Fellow environmental and urban historians have, over the years, generously listened to my ideas and read and criticized manuscripts. I have especially benefited from conversations about the history of the environment and its relationship to the city with Samuel P. Hays, Martin V. Melosi, Harold L. Platt, Mark H. Rose, Christine M. Rosen, Jeffrey Stine, and Sam Bass Warner, Jr. Other colleagues and friends who have generously shared their comments and criticisms with me include Craig E. Colten, Edward W. Constant, Gabriel Dupuy, David Hounshell, Steven Klepper, Clay McShane, John Modell, and Maurice Shapiro.

This book never would have been produced without the support of Martin Melosi and Jeffrey Stine, both outstanding environmental historians. Marty originally suggested preparing a volume of my essays for inclusion in an environmental history series he was then editing, and he has generously contributed a forward to this volume. Jeff Stine has been an enthusiastic supporter of this collection since I first mentioned its possibility to him. As an editor of The University of Akron Press Series on Technology and the Environment, he has been a source of constant support and informed criticism, pushing consistently for refinement of ideas and clarity in their written expression. Marybeth Mersky of The University of Akron Press has shown patience with my disorganized approach to book preparation, and Elton Glaser of the Press has been a source of good advice on its final production.

The research and writing of these essays was generously supported by grants from various foundations and funding agencies, as well as by my own university. I am grateful for grants from the National Endowment for the Humanities, the National Science Foundation, the National Oceanic and Atmospheric Administration, and the Andrew W. Mellon Foundation.

Article Acknowledgments

For allowing me to make use of previously-published articles, which appear here in substantially unaltered form, except in specifically noted cases, I am obliged to *Agricultural History, American Heritage,* the American Society of Civil Engineers, the *American Journal of Public Health,* Cambridge University Press, *Environmental History Review,* the *Journal of Social History,* Sage Publications, the *Journal of Urban History, Technology and Culture,* the University of Virginia Press, and *Waste Management and Research.*

I. Crossing Environmental Boundaries

"The Search for the Ultimate Sink: Urban Air, Land, and Water Pollution in Historical Perspective" was originally delivered as the 1981 Letitia Woods Brown Memorial Lecture at the Columbia Historical Society of Washington, DC, and was first published in *Records of the Columbia Historical Society of Washington, DC* (Charlottesville, VA: The University Press of Virginia, 1984) 51: 129. I am grateful to Linda Lear, then at George Washington University, for suggesting me as the speaker.

"Land Use and Environmental Change in the Hudson-Raritan River Basins and Estuary, 1700–1980" is a reshaped version of an article co-authored with Robert U. Ayres under the title "Pollution Trends in the Hudson River and Raritan River Basins, 1790–1980" in *The Earth as Transformed by Human Action: Global and Regional Changes in the Biosphere over the Past 300 Years,* edited by B. L. Turner et al. (New York: Cambridge University Press, 1990), 623–640.

"The Pittsburgh Survey as an Environmental Statement" was prepared to appear in *The Pittsburgh Survey Revisited,* edited by Margo Anderson and Maureen Greenwald (Pittsburgh: University of Pittsburgh Press, forthcoming).

II. Water Pollution

"Historical Decisions About Wastewater Technology, 1800–1932" was originally published with Francis C. McMichael as co-author in the *Journal of Water Resources Planning and Management Division, Proceedings of the American Society of Civil Engineers* 103 (May, 1977): 47–61.

"The Separate vs. Combined Sewer Problem: A Case Study in Urban Technology Design Choice" was originally published in the *Journal of Urban History* 5 (May, 1979): 308–339.

"Disputes Over Water Quality Policy: Professional Cultures in Conflict, 1900–1917" was published with Terry F. Yosie and James McCurley, III as co-authors in the *American Journal of Public Health* 70, no. 4 (April, 1980): 427–435.

"Water and Wastes: A Retrospective Assessment of Wastewater Technology in the U.S., 1800–1932" was published with James McCurley, III, Francis C. McMichael, and Terry F. Yosie as co-authors in *Technology and Culture* 25, no.2 (April, 1984): 226–263.

III. Smoke Pollution:

"Changing Fuel Use Behavior and Energy Transitions: The Pittsburgh Smoke Control Movement, 1940–1950, A Case Study in Historical Analogy" was published with Bill Lamperes as co-author in the *Journal of Social History* 14, no. 4 (Summer 1981): 561–588.

"Railroad Smoke Control: A Case Study in the Regulation of a Mobile Pollution Source" was co-authored with Kenneth Koons and appeared in *Energy and Transport: Historical Perspectives on Policy Issues,* edited by George H. Daniels and Mark H. Rose (Beverly Hills, CA: Sage Publications, 1982), 71–92.

IV. Land, Transport, and Environment

"From City to Farm: Urban Wastes and the American Farmer" first appeared in *Agricultural History* 49 (October, 1975): 598–612.

"From City to Suburb: The 'Moral' Influence of Transportation Technology" was published in *American Urban History,* rev. ed., edited by Alexander B. Callow, Jr. (New York: Oxford University Press, 1973), 202–212.

"The Horse-Polluter of the City" was originally published in a slightly different form under the title "Urban Pollution Many Long Years Ago" in *American Heritage* 22 (October, 1971): 65–69, 106.

V. Industrial Wastes as Hazards

"Historical Perspectives on Hazardous Wastes in the United States" appeared in *Waste Management & Research* 3 (1985): 95–102.

"Industrial Wastes and Public Health, 1876–1962" is an expanded and revised version of an article originally published under the title "Industrial Wastes and Public Health: Some Historical Notes, Part 1, 1876–1932" in the *American Journal of Public Health* 75, no. 9 (September 1985): 1059–1067.

"Searching for a Sink for an Industrial Waste" was originally published under the title "Searching for a 'Sink' for an Industrial Waste: Iron-Making Fuels and the Environment" in *Environmental History Review* 18 (Spring, 1994): 9–34.

Foreword

No single person can be credited with advancing the scholarly study of urban ecosystems, but since the 1970s, when urban environmental history began to take clearer form, the leading figure in this field has been Joel A. Tarr. And Tarr's influence has not been limited to the path-breaking studies that he has published on such topics as the history of pollution control and the interdependence of social policy and urban infrastructures. His energy and intelligence have served this interdisciplinary field in many ways. As an editor, he has overseen the completion of essays for special issues of the *Environmental History Review and the Journal of Urban History.* As a conference organizer, most recently for the American Society for Environmental History, he has promoted the study of new topics and the careers of budding historians. And as a mentor and as a bottomless source of information, he has long been someone with whom to share a confidence or a new idea. Now this volume brings together for the first time many of those essays that have helped to shape the field of urban environmental history and, in the process, have served to establish further the scholarly reputation of Joel A. Tarr.

The city is a unique ecosystem—an "open system" really—connected in intricate ways with the outside world.[1] But only within the last hundred years has the growth of the physical city and its confounding problems of air, water, and land pollution been analyzed in any systematic way. Beginning in the 1910s, social scientists sought to develop an ecological approach to studying spatial and social organization based on concepts and principles conceived by nineteenth-century plant and animal ecologists.[2] The swirl of ideas in these various sociological and geographic studies, including location theory and systems analysis, offered suggestive theoretical routes open to those seeking to link the natural history of the city with the history of city building.[3] For their part, historians were latecomers to this discourse, but they ultimately provided a solid empirical base (if not major new theories) to the examination of the urban environment. By the 1960s, historians had broadened the scope of their investigations, producing valuable works on building technology, public works and infrastructure, sanitary services, parks and greenspace, pollution and public health, and energy use and development.[4]

In the last twenty years, no one has contributed more to the expansion and the legitimacy of this field than Joel A. Tarr. For Tarr, urban environmental history is "primarily the story of how human-built, anthropogenic structures ('built environment') and technologies shape and alter the natural environment of the urban site with consequent feedback to the city itself and its populations."[5] This broad view required an interdisciplinary outlook and a willingness to examine the work in existing fields of history from a different vantage point. By exploiting a wide range of historical methodologies and embracing key elements of social science theory, Tarr's resulting in-depth studies of transportation, energy use and development, and waste disposal not only moved beyond the historical works of the 1960s but inspired a wide variety of research among a new crop of urban and environmental historians in the 1970s and thereafter.

This task has not been easy. With some exceptions, urban environmental history has struggled to gain an identity. It began as an unfocused topic existing within several established historical fields rather than as a distinctive area of research with clear ties to the broader tradition of urban ecology. For example, much of the research on infrastructure, public works, and engineering was drawn from the history of technology; the study of building technology from architectural history; interest in public health and disease from medical history; pollution regulation from law; urban reform from political history; and city growth and city services from urban history and city planning history.[6]

Joel Tarr's work, however, has helped to direct attention to three key areas of inquiry that provide a clearer identity for urban environmental history: urban growth, infrastructure, and pollution and health. Understanding how and why cities grow is the first step in grasping the importance of the urban environment and its impact on humans.[7]

Tarr's transportation studies, in particular, speak directly to this issue. Although not represented in this volume, "Transportation Innovation and Changing Spatial Patterns in Pittsburgh, 1859–1934"[8] is probably the best example of Tarr's attempt to grapple with the dynamic impact of outward urban growth in the tradition of Sam Bass Warner's *Streetcar Suburbs,* and it is a model study of its kind.

Because many urban environmental historians have been interested in the internal structure of cities, city building has attracted as much, if not more, attention than outward, linear growth. As a departure point, the study of urban design, building technology, and urban space are essential for exploring the city-building process itself.[9] Attempting to grasp city building in larger environmental terms, however, requires more broadly conceived concepts than "building technology" or "urban landscape."

The relatively recent focus on city "infrastructure" offers a more useful handle. Tarr has explained persuasively that infrastructure provides the vital technological "sinews" of a city: roads and bridges, water and wastewater lines, disposal facilities, power systems, communications networks, and buildings.[10] Thus systems analysis becomes a basic tool for understanding the city-building process in a holistic way.[11]

Growing out of a 1983 conference in Paris sponsored by Centre National de la Recherche Scientifique and the National Science Foundation, *Technology and the Rise of the Networked City in Europe and America,* edited by Joel Tarr and Gabriel Dupuy, presented the work of European (primarily French) and American scholars interested in exploring the growth of cities through scrutiny of various technical networks. In the preface, Tarr and Dupuy made the case for the significance of such networks: "Technological infrastructure makes possible the existence of the modern city and provides the means for its continuing operation, but it also increases the city's vulnerability to catastrophic events such as war or natural disaster. While technology may enhance the urban quality of life, it may also be a force for deterioration and destruction of neighborhoods, as well as a hindrance to humane and rationale planning."[12]

The new "infrastructure" literature of the last decade or so has deepened our knowledge about technical systems and city services, placing the urban environment in a new light. The work of Tarr and others is based on extensive mining of underutilized—but valuable—research materials such as technical journals and tracts, city plans and maps, transactions of engineering societies, and numerous government documents.[13]

Pollution studies examine the role of humans as consumers of the environment, and within the context of the city they also provide a vehicle for exploring the extent and nature of physical degradation caused by population growth, technological change, and industrial production; the formulation and effectiveness of environmental laws and regulation; and the origins and impact of "green" politics and urban environmental reform.[14]

The study of pollution and its ramifications is complemented by the rich literature in the field of the history of medicine and public health. To truly understand the quality of the urban environment, especially from the perspective of consumers, is to explore disease transmission and epidemics, sanitation and health, and the role of doctors, sanitarians and public health officials in combating disease and pollution.

Joel Tarr's work in the areas of pollution and public health may be his most significant empirical contributions to the literature of urban environmental history. This current volume includes many examples of that work from the study of coal smoke to water pollution. Tarr's studies are particularly valuable because they marshal historical data to quantify the extent of pollution problems in the past. As such, they are among the best examples of the utility of history as a policy tool. By going beyond qualitative or anecdotal research methods, several of Tarr's studies provide decision makers with tangible and substantive historical evidence especially useful in trend analysis. "Land Use and Environmental Change in the Hudson-Raritan Estuary Region, 1700–1980" is an exceptionally good example of such a policy-relevant study.

Tarr's own work (scholarly and applied), the studies he inspired, and the application of historical evidence to issues of urban public policy have been instrumental in changing our view of the city itself. Lewis Mumford's "invisible city"—those pipes, conduits, and wires creating a hydraulic, pneumatic, and electrical maze below the streets—and the buildings and bridges standing as concrete forests above the streets are not merely a physical backdrop for human action. They are integral components in a dynamic environmental system constantly in flux.

Joel Tarr has been at the center of these changing perceptions of the historical city for more than two decades. Many excellent samples of his contributions to the study of the urban environment are reprinted in this book. They are carefully researched and thought-provoking. They often carry additional lessons worth heeding: Avoid the sin of "presentism," which does not give the past its due and assumes all-too-simplistic connections to the "now." Do not succumb to a technological determinism that gives too much credit to functionalism and too little to human choice. Remember that "risks," "hazards," and "trade-offs" are relative terms. Look for the not-so-obvious patterns even in the most rational networks and systems. And do not underestimate the power of "values" in the process of decision making.

Tarr's healthy skepticism in not accepting historical events and evidence at face value and his refusal to jump to the easy conclusion have earned his work respect among colleagues in both academic and applied fields. While the essays presented in this book offer a panorama of a career and the evolution in thinking of a key figure in the development of urban environmental history, some of Joel Tarr's contributions are not so obvious in these pages, particularly his role as a teacher and nurturer of scholarly talent.

Several of his essays are coauthored with experts who complement Tarr's knowledge and experience, with colleagues who share his enthusiasm for the subject matter, and with students who have received a rare opportunity to learn by doing. Tarr has never been afraid to share the limelight with others or to admit that others could contribute something useful to a worthwhile project. While he rarely gives quarter to sloppy research, is always willing to debate a controversial point, and demands nothing less than the best product, Joel Tarr never ceases to support the efforts of his colleagues with an enthusiasm often reserved for the freshly minted Ph.D.

For some people, *The Search for the Ultimate Sink* is an opportunity to explore urban environmental history for the first time. For others, it is a chance to follow one scholar's odyssey along a long intellectual path. And for a few more, it is a way to reacquaint themselves with much of the work that inspired the growth and evolution of a significant field of study.

Martin V. Melosi
University of Houston

Notes

1. See Thomas R. Detwyler and Melvin G. Marcus, eds., *Urbanization and Environment: The Physical Geography of the City* (Belmont, Calif.: Duxbury Press, 1972), 10, 21; Ronald J. Johnston, *The American Urban System: A Geographic Perspective* (New York: St. Martin's Press, 1982), 304–5.

2. Urban sociology, born at the University of Chicago during World War I, strongly influenced the development of urban ecology among sociologists and geographers. See Brian J. L. Berry and John D. Kasarda, *Contemporary Urban Ecology* (New York: Macmillan, 1977), 3.

3. For a more thorough discussion on the role of social science in developing urban ecology, see Martin V. Melosi, "The Place of the City in Environmental History," *Environmental History Review* 17 (Spring 1993): 1–23.

4. Pioneering historical research in the 1960s includes Lewis Mumford's sweeping *The City in History* (New York: Harcourt, Brace and World, 1961); Sam Bass Warner, Jr.'s classic case study *Streetcar Suburbs: The Process of Growth in Boston* (Cambridge: Harvard University Press, 1962); Carl Condit's *American Building Art*, 2 vols. (New York: Oxford University Press, 1960–61); Nelson Blake's seminal *Water for the Cities* (Syracuse: Syracuse University Press, 1956); geographer Allan R. Pred's *The Spatial Dynamics of U.S. Urban-Industrial Growth 1800–1914* (Cambridge: MIT Press, 1966); John W. Reps's *The Making of Urban America* (Princeton: Princeton University Press, 1965); Roy Lubove's *Twentieth-Century Pittsburgh* (New York: Wiley, 1969); Charles S. Rosenberg's *The Cholera Years* (Chicago: University of Chicago Press, 1962); and John Duffy's *A History of Public Health in New York City, 1625–1866* (New York: Russell Sage Foundation, 1968).

5. Joel A. Tarr, letter to author, September 6, 1992.

6. For a range of bibliographic references, see Howard Gillette, Jr., and Zane L. Miller, *American Urbanism: A Historiographical Review* (New York: Greenwood Press, 1987); Eugene P. Moehring, "Public Works and Urban History: Recent Trends and New Directions," *Essays in Public Works History* 13 (August 1982): 1–60; Suellen M. Hoy and Michael C. Robinson, eds. and comps., *Public Works History in the United States: A Guide to the Literature* (Nashville: American Association of State and Local History, 1982); Joel A. Tarr, "The Evolution of Urban Infrastructure in the Nineteenth and Twentieth Centuries," in Royce Hanson, ed., *Evolution of the Urban Infrastructure* (Washington, D.C.: National Academy Press, 1984), 4–60; Josef W. Konvitz, Mark H. Rose, and Joel A. Tarr, "Technology and the City," *Technology and Culture* 31 (April 1990): 284–94; Martin V. Melosi, "A Bibliography of Urban Pollution Problems," in Melosi, ed., *Pollution and Reform in American Cities, 1870–1930* (Austin: University of Texas Press, 1980), 199–212; idem, "Urban Pollution: Historical Perspective Needed," *Environmental Review* 3 (Spring 1979):

37–45; idem, "The Urban Physical Environment and the Historian: Prospects for Research, Teaching and Public Policy," *Journal of American Culture* 3 (Fall 1980): 526–40.

Most of these references reflect a major emphasis on North American topics. Part of the reason is that American scholars, until recently, have played a major role in the study of the urban environment. However, the contribution of Europeans to the field has been growing steadily, building on the major contributions of Fernand Braudel and the *Annales* school and the work of various European social scientists and engineers.

7. Eric E. Lampard's work on urban growth is central to the topic: see "American Historians and the Study of Urbanization," *American Historical Review* 67 (October 1961): 49–61; "Historical Aspects of Urbanization," in Philip M. Hauser and Leo F. Schnore, eds., *The Study of Urbanization* (New York: Wiley, 1965), 519–54; and "The Evolving System of Cities in the United States: Urbanization and Economic Development," in Harvey S. Perloff and Lowdon W. Wingo, Jr., eds., *Issues in Urban Economics* (Baltimore: Johns Hopkins University Press, 1968), 81–139.

8. In *Essays in Public Works History,* vol. 6 (Chicago: Public Works Historical Society, 1978).

9. See Carl W. Condit, *The Rise of the Skyscraper* (Chicago: University of Chicago Press, 1952); *American Building Art; American Building* (Chicago: University of Chicago Press, 1967); *Chicago 1910–1929: Building, Planning and Urban Technology* (Chicago: University of Chicago Press, 1973); *Chicago, 1930–1970* (Chicago: University of Chicago Press, 1974); *The Port of New York* (Chicago: University of Chicago Press, 1980); Eugenie Ladner Birch, "Design, Process, and Institutions: Planning in Urban History," 135–54, and Richard Longstreth, "Architecture and the City," 155–94, in Gillette and Miller, eds., *American Urbanism;* David Schuyler, *The New Urban Landscape: The Redefinition of City Form in Nineteenth-Century America* (Baltimore: Johns Hopkins University Press, 1986); Edward Relph, *The Modern Urban Landscape* (Baltimore: Johns Hopkins University Press, 1987); John R. Stilgoe, *Common Landscape of America, 1580 to 1845* (New Haven: Yale University Press, 1982); and *Borderland: Origins of the American Suburb, 1820–1939* (New Haven: Yale University Press, 1988).

10. Joel A. Tarr and Gabriel Dupuy, eds., *Technology and the Rise of the Networked City in Europe and America* (Philadelphia: Temple University Press, 1988), xiii. For more on Tarr's intellectual development, see Bruce M. Stave, "A Conversation with Joel A. Tarr: Urban History and Policy," *Journal of Urban History* 9 (February 1983): 195–232. For infrastructure bibliography, see also Joel A. Tarr and Josef W. Konvitz, "Patterns in the Development of the Urban Infrastructure," in Gillette and Miller, eds., *American Urbanism,* 195–226; Joel A. Tarr, "Infrastructure and Urban Growth in the Nineteenth Century," *Essays in Public Works History* 14 (December 1985): 61–85.

11. Christine Meisner Rosen added an operational dimension to Tarr's definition by arguing that infrastructure development shared the qualities of "capital intensiveness, land extensiveness, and monopolistic production." And a good summary statement is Josef W. Konvitz's notion that "Unlike public works, which it subsumes, the term 'infrastructure' is at once a description of physical assets and of their economic, social, and political role." See Rosen, "Infrastructural Improvement in Nineteenth-Century Cities: A Conceptual Framework and Cases," *Journal of Urban History* 12 (May 1986): 222–23; Konvitz, *The Urban Millennium: The City Building Process from the Early Middle Ages to the Present* (Carbondale: Southern Illinois University Press, 1985), 131.

12. Tarr and Dupuy, eds., *Technology and the Rise of the Networked City in Europe and America,* xiii. See also Eugene P. Moehring, "The Networked City: A Euro-American View: Review Essay," *Journal of Urban History* 17 (November 1990): 88–97. In the spring of 1991, Centre National de la Recherche Scientifique published the first issue of *Flux* under the editorship of Gabriel Dupuy. The journal, published alternately in French and English, expanded on the ideas at the 1983 Paris conference, hoping to display in its pages the most current work dealing with technical networks. "Notions which are at present confused, like 'system effects' and 'network effects' should be clarified by articles on telecommunications or transportation," Dupuy stated. "The role played by the interconnections between networks should appear in the domain of water dis-

tribution as well as electricity." The premier issue of *The Journal of Urban Technology* was published in the fall of 1992 in New York. It may prove to be an important companion to *Flux.*

13. The Public Works Historical Society has played a major role in promoting such studies through publications such as the *Essays in Public Works* series and the recent *Water and the City: The Next Century,* ed. Howard Rosen and Ann Durkin Keating (Chicago: American Public Works Association, 1991), as well as through development of numerous programs at professional meetings. "Public Works History" is often used to describe the interest in the urban infrastructure, but "public works" tends to exclude privately developed structures, technologies, and systems that fit more comfortably under the heading of "infrastructure" when defined broadly to include social, political, and economic factors.

14. On industrialization, see Theodore L. Steinberg, "An Ecological Perspective on the Origins of Industrialization," *Environmental Review* 10 (Winter 1986): 261–76. Early attempts at dealing with urban pollution as an important historical problem can be found in Melosi, ed., *Pollution and Reform in American Cities.* Joel Tarr, with James McCurley and Terry F. Yosie, wrote a key piece for that volume, entitled "The Development and Impact of Urban Wastewater Technology: Changing Concepts of Water Quality Control, 1850–1930."

Introduction

The essays included in this volume were written over a period of more than twenty years, and I am grateful for the opportunity to have them republished as a collection. I have always thought of myself as an urban rather than environmental historian. Much of my interest in the environment stems from my concern with the dynamics of cities and urban technical systems and the manner in which city building interacts with, or modifies, nature to create a special urban environment. These interactions often involve city officials, engineers, and planners making choices about technologies, with these choices having both positive and negative effects. Perhaps predictably, given the propensity of technologies to produce unexpected effects, solutions in one domain have frequently created problems in other domains, condemning cities to future problems.

Decision-making about technology and its interaction with the environment often involves critical trade-offs based on some sort of benefit-cost analysis, whether explicitly called that or not. These analyses historically have rested on different assumptions and forecasts about factors such as population growth, the effectiveness of technology, medical and scientific hypotheses associated with public health, project costs, the expected behavior of nature, and the degree of health or physical risk involved in changing or not changing urban networks. The construction of technological networks in American cities for the transmission of water, wastewater, power, communications, freight, and people dramatically altered the context of city life and the effect that urban centers had upon their surrounding environments. This was especially true in the decades of the late nineteenth and early twentieth centuries, a period of unprecedented population and territorial growth for North American and Western European cities with the consequent development of new urban forms.[1]

I believe that there were four primary ways in which the population and territorial growth of cities, the building of technological systems, and the decisions of urban politicians and developers concerning land use affected and shaped municipalities and their hinterlands: by consuming water, land, and natural resources; by the creation and disposal of vast quantities of human and industrial wastes; by constructing a built environment that sharply modified the natural environment and created miniclimates; and by moving the urban population into larger and larger spatial areas, constantly expanding the urban periphery. In each of these areas, technical factors played a critical role, and issues such as technological choice, the externalizing of environmental cost, or the failure to regulate adequately or measure pollution have had significant effects on environmental quality.

The majority of these essays deal with events that occurred between 1850 and 1950. This was a period of evolving knowledge and discovery in the areas of engineering and public health, of major technological invention and innovation, of rapid national and urban growth, and of an increase in state power. It was also a period of growing industrialization and of great increases in the exploitation of nonrenewable fossil fuels and other natural resources. These powerful forces caused extensive degradation of air, water, and land, some of it so devastating to nature and so destructive to public health that it could not be tolerated. Eventually, regulatory authority shifted from the local to state to federal level, with the latter assuming new and sweeping authority in regard to protecting the environment.

Prior to the full-scale emergence of the environmental movement in the 1960s and 1970s, prominent engineers and public health professionals questioned the costs of rapid and unplanned urban growth, especially as it involved human health. Still others expressed anxiety about the unrestrained exploitation of natural resources, with some emphasizing conservation methods and others preservationist goals. Forces on the local and state levels, including the courts, began slowly to assert authority over unrestricted pollution, first because of damage to property but later because of public health effects. Measures to curtail pollution and curb the despoliation of the environment were limited, however, as progress in some areas was matched by retrogression in others. This problem had geographical dimensions, as one locality's gain became another's loss; legislation or legal action often caused the pollution burden to shift from place to place and from medium to medium (from water pollution to air pollution, for example). It is from this tendency of both municipalities and industries constantly to seek new sites for waste disposal that the title of this book is drawn.

The articles included in this volume deal primarily with issues of urban metabolism: that is, the supply of water and the disposal of wastewater or sewage; the generation and disposal of industrial wastes; and the collection and disposal of solid wastes and garbage from food and consumer products.[2] It is because of my research in these areas, rather than because of any formal training, that I find it possible to accept the designation of "environmental historian." Environmental history is actually one of the newest historical fields, and urban environmental studies as a subfield is even more recent, dating back to the 1970s.[3] Since that time, however, a group of historians has emerged and produced a number of excellent urban environmental studies. Like me, few of these scholars were specifically trained in environmental history; we all largely arrived at the subject through our concern with urban phenomena.[4]

In 1970, after having written about Chicago and Illinois politics for almost a decade, I turned toward an analysis of urban transportation.[5] My earlier research on Chicago had included questions of urban transit and other public utilities, but from a political and policy viewpoint rather than the perspective of urban land use and traffic flows. Relatively recently arrived at Carnegie Mellon University in Pittsburgh, a city that still had an operative streetcar network, I proposed to study the impact of transportation innovation on that city, examining changing spatial and building patterns in response to technological innovation and exploring the development and alterations in residential neighborhoods, the central business district, and industrial locations.

Although I was not particularly attuned to questions involving environmental history, it was impossible to ignore issues of air quality in Pittsburgh, which bore the burden of its reputation as "the Smoky City," even though the days of heavy smoke pollution were long gone. In addition, agitation in 1969–70 about environmental degradation, and particularly the role of automobile emissions and smog (I had lived in the Los Angeles Basin for five years), helped sensitize me to air-quality concerns. (I might mention that, while growing up in industrial Jersey City, we accepted pollution as a matter of course. This included smells from the pig farms feeding on New York City garbage in nearby Secaucus.) In the process of my transportation research, I began to note how often the pollution caused by horses (such as manure, urine, noise, and carcasses) posed problems for both city-dwellers and municipal administrations, while at the same time the city remained heavily dependent on horses for passenger and freight transportation.

The parallel to modern times was intriguing: in the nineteenth century the horse, a primary means of transportation, was a major urban polluter; in 1970, however, the ubiquitous and seemingly indispensable automobile had become a major threat to the urban environment. I began to speculate about the reaction of Americans to the automobile at the time of its introduction and their comparison of it with the horse. My research suggested that rather than viewing the automobile as a dangerous polluter, early-twentieth-century Americans regarded it as an environmental improvement.[6] Believing this perspective worth presentation to a larger audience, I wrote a paper entitled "The Horse: Polluter of the City," and submitted it to *American Heritage Magazine.* The reaction was generally enthusiastic, except for one member of the magazine's editorial board, who believed me to be an apologist for the sins of the automobile! *American Heritage* published my essay under its own title, "Urban Pollution Many Long Years Ago," with some popularization of the text, but in this volume I am presenting it with the original text and title.[7] Perhaps the editors at *American Heritage* had a better sense of the short-term nature of their audience's memories than I did, since several newspapers ran editorials and columns about the article in which they exclaimed surprise at what a bad polluter the horse (not really so long gone from the streets of American cities) had been.[8]

The horse article led to some notoriety (as well as a few bad jokes) and eventually to further environmental work, although my main research interest continued to focus on transportation. In 1973 the public relations director of the firm Envirotech read my horse pollution article and asked me if I could prepare a study dealing with the history of the disposal of human wastes, especially in regard to the land. In the *American Heritage* piece, I had observed that farmers had often collected horse manure from city streets and stables and applied it as crop fertilizer. Envirotech was interested in the history of the land use of wastes because it was involved in a dispute with the U.S. Environmental Protection Agency (EPA): the agency was sponsoring the land disposal of human wastes, while Envirotech was marketing incineration equipment.

My research on the agricultural use of human wastes revealed that American farmers in the middle and latter part of the nineteenth century had frequently applied organic materials collected from urban privy vaults, as well as horse manure, for fertilizer. Furthermore, the research showed that "sewage farming," a method of sewage disposal touted in the 1970s as environmentally beneficial, had been employed by a number of American cities at the turn of the century. It had, however, never been widely adopted as a method of treating human wastes because of its land intensiveness, the difficulties of maintaining a sanitary and controllable treatment process, and the attractiveness of newer and more technology-intensive methods to sanitary engineers. After completing my report to Envirotech, I did further research on the subject and in 1975 published the article included in this volume, "From City to Farm: Urban Wastes and the American Farmer," as well as several articles on the uses of wastes as fertilizer.[9]

The investigations I had conducted into the practices of urban waste collection and disposal surprisingly revealed that this important and timely area of history had been barely studied. Even though Congress had passed the National Environmental Policy Act (1969), created the EPA (1970), and passed stringent legislation against water pollution (the Federal Water Pollution Control Act Amendments, 1974), there was scant literature available to explain the series of decisions that had led to the development of water pollution or its relationship to broad trends, such as urbanization and industrialization. My interest in transportation studies remained strong, but I was developing a larger concern with urban networks. I began to realize that the nation's growing water pollution problems in the late nineteenth century coincided not only with population increases but also with the growth of sewer systems, a phenomena not explained by the existing literature. This knowledge, I came to believe, would also be useful in informing the policy debates about water-quality issues in the 1970s.

In 1974 Robert Dunlap, one of my colleagues in a new program at Carnegie Mellon called Engineering and Public Affairs (today the Department of Engineering and Public Policy, or EPP), brought to my attention that the National Science Foundation (NSF) had issued a request for proposals in an area called Retrospective Technology Assessment (RTA).[10] The technology assessment movement itself had begun in the late 1960s, and in 1972 Congress had established the Office of Technology Assessment (OTA) to study the effects of new technologies, especially their unintended, indirect, and delayed effects, with the goal of helping to devise policy to limit societal and environmental damages. Its purpose was to advise Congress on such matters, allowing it to use this knowledge in its authorizations, appropriations, and oversight. (As of this writing, the Republican-dominated House of Representatives, led by Speaker Newt Gingrich, a Ph.D. in History, has taken the shortsighted step of eliminating the OTA.) How to conduct the future-oriented assessments, however, was unclear. From the perspective of the NSF staff, this is where history entered: Retrospective Technology Assessment was primarily meant to help develop and test technology assessment methodologies by using them in historical situations where the unexpected results of new technologies could be identified.[11]

The appearance of the RTA program was also significant in another regard. Since my arrival at Carnegie Mellon University, I had held a joint appointment in the Department of History and the new School of Urban and Public Affairs (now the H. John Heinz III School of Public Policy and Management). My interactions in the policy school, where I taught courses dealing with historical perspectives on urban problems and participated in problem-oriented project courses, convinced me that history had major relevance to the formation of public policy. Yet it had been largely ignored (the Defense Department was perhaps the leading exception). The RTA program now seemed to suggest that history might be taken more seriously in some Washington circles and that funding would be available to support research on the history of policy problems and to provide for graduate students. With this in mind, in 1974 I launched a campaign with colleagues and administrators to create a graduate program in Applied History at Carnegie Mellon University, a program that successfully came into being in 1975. Peter N. Stearns, our new Heinz Professor of History, was appointed the program director, and I became director of a new universitywide program in Technology and Society funded for five years by the Andrew W. Mellon Foundation.[12]

In addition, in 1975 I joined with Francis C. McMichael, an environmental engineer at the university, and David Wojick, a CMU philosopher of technology, to submit a proposal to the National Science Foundation in the RTA category. We proposed to assess the development of water-carriage technology (or sewerage), examining the decision to adopt the technology; alternative choices not taken; forecasts made about possible impacts; environmental, institutional, and policy effects; and the policies followed to deal with unintended and negative impacts. To our astonishment, our proposal was funded, and I was able to put together a research team that spent the next three years studying the history of sewers. While the chief purpose of the RTA program was to improve technology assessment methodologies, our study was more substantive in orientation, and its contribution was primarily in that direction.[13]

Because sewers or water-carriage networks had far-reaching implications for urban public health, sanitarians had been their most vigorous advocates in the early public health movement. This emphasis helped explain why it had been public health historians who had done most of the writing about the history of wastewater systems and why the technological aspects of the systems and the critical technical decisions had largely been ignored. These critical decisions we identified as the decision to adopt the water-carriage method of waste removal, the decision of large cities to construct combined rather than separate sewers, and the decision to treat water for municipal consumption rather than treat both water and wastewater or sewage.

It may be worth relating here the mental discovery process that led to the answer to the first and the most basic of these three questions (basic because all other decisions followed from it). For some time after I had begun thinking about wastewater technology, I puzzled over the explanation and timing of large-scale municipal adoption of sewerage systems in the late nineteenth century, since the public health model left many questions unanswered. One day in 1974, the members of a seminar I was teaching with my colleague David Wojick were discussing this question. David noted that while he did not know very much about sewers, he did know that you did not have them without water. I realized that this was a possible answer: nineteenth-century cities had built waterworks, thereby greatly increasing the amount of water available for use in the city; but they had to find means to dispose of the fouled wastewater (stormwater as well as domestic waste water), and capital intensive sewerage systems were the means by which they eventually chose to do so. Public health and cultural issues were involved in the decisions but were not the only factors driving municipal policy-making; rather, fouled domestic water and stormwater flooding created nuisances and complications for urban life that were often the definitive element in forcing cities to invest in capital-intensive sewerage systems.*

*I would like to note the publication of an important new work by Maureen Ogle, *All the Modern Conveniences: American Household Plumbing, 1840–1890.* (Baltimore: Johns Hopkins University Press, 1996). This enlightening book argues that "technology is the material embodiment of a people's values and ideas—their culture . . .," and investigates the role of values and culture in the development of domestic plumbing and related waste disposal systems. It greatly enriches our understanding of the period that it examines and challenges the central role that I have placed on exploring the role of technology in urban systems. Without detracting from the great value of Professor Ogle's work, I would like to gently demur from some of her conclusions. While the investigation of the role of technology has driven many of my urban studies, I never intended to suggest that culture and values did not play a definitive role in the choice of systems and alternatives. I would argue, however, that technology, by its very force and character, can shape and change cultural values. The process of change involving technology and society and society and technology, therefore, should be thought of in terms of action and reaction, with constant feedback, rather than as a one-way process.

City after city constructed their sewerage systems *after* they had built waterworks—almost never simultaneously and not before household water usage for domestic and sanitary purposes had vastly increased and overloaded the existing disposal system of privy vaults and cesspools.[14] One of the principal technical devices driving the need for sewers was the flush toilet: waterworks advocates had not predicted its adoption or considered the problems created by its discharges. The other critical decisions we identified—those relating to sewer design and water treatment—were linked directly to the decision to construct water-carriage waste-removal systems of the combined type, the need to rid streets (especially paved streets) of stormwater, and the policy of discharging the wastes mainly in the water bodies from which other downstream cities drew their water supplies.

In addition to studying the major technical decisions in regard to wastewater technology, the effects of sewers and sewage disposal on the public health, and attempts to cope with the effects of water pollution, we also performed an institutional and social analysis. This analysis identified a range of social effects produced by the technology, from the creation of new professions such as sanitary engineering to the formation of governmental institutions such as special district governments. These findings were included in a report to the NSF and also in a number of articles, several of which are included in this volume. Major sections of the report are summarized in "Water and Wastes: An Assessment of Wastewater Technology in the U.S., 1800–1932," while other essays in this volume that derive from the RTA wastewater study include "Historical Decisions About Waste Water Policy, 1800–1932," "The Separate vs. Combined Sewer Problem: A Case Study in Urban Technological Design Choice," and "Disputes Over Water Quality Policy: Professional Cultures in Conflict, 1900–1917."[15]

The wastewater project was extremely significant for my education and involvement in studies of urban environmental technologies. First of all, it brought me into a close working relationship with an exceptional and broad-gauged environmental engineer, Francis C. McMichael, as well as later acquainting me with giants in the field of sanitary engineering, such as Abel Wolman. Because of McMichael, as well as other of my project colleagues, my thinking took on a systems orientation. I began to think more fully about technical decision making and the multiple effects such decisions had throughout society as well as about the validity of forecasts. Finally, because our study stopped before World War II, I had to ponder the validity of analogies and consider how to draw lessons from our historical study for the problems of water pollution policy in the 1970s. While I would like to believe that our study had some effect on environmental policymakers, at least in terms of enlarging their understanding of the problem, this is hard to document. Showing where the history "made a difference," as Edward B. Fiske, education editor of the *New York Times,* once asked me, is difficult given both the nature of history and the nature of politics.[16] Perhaps, somewhat wistfully, I and my colleagues like to believe it did.

By 1978, with the NSF wastewater report completed, and excited about the new perspectives that had been opened up, I began seeking other environmental policy issues that could be informed by the historical record. Part of my motivation in searching for a new area of research was intellectual, but, in addition, I was attempting to secure support for graduate students in our new Ph.D. program in Applied History. In the latter part of the 1970s, the Carter administration was advocating energy conservation measures that had a relation to both energy use and the environment. Living in Pittsburgh, a city that had been famed for its smoke pollution but which had eliminated its dense smoke soon after World War II, I believed that the city's experiences with pollution control provided a suggestive analogy that could be useful in the search for energy conservation policy in the 1970s. I was also weary of historians and policymakers using London's sixteenth- and seventeenth-century fuel crisis and consequent shift from timber to coal as an historical example, as if no more recent examples of fuel substitutions affecting the environment existed.

In Pittsburgh, as in London, fuel type was directly related to air quality, and a major key to improving conditions was to persuade people to change their fuel-consumption habits. Focusing on the question of Pittsburgh smoke control, I began work on a project exploring themes of values, fuel change, economics, technology, and environment. Of the research resulting from this project, two essays are included in this volume: "Changing Fuel Use Behavior: Pittsburgh Smoke Control, 1940–

1950," and "Railroad Smoke Control."[17] Interestingly, while both situations involved fuel shifts, one change, that involving residential fuel use in the city, permitted retrofitting of some existing combustion technology, while the second, that of the railroads, required an entirely new power technology, the diesel-electric engine. The ability to retrofit residential heating plants meant that the costs of shifting to a different fuel were spread over a longer time period for some, thus reducing homeowner opposition to change; however, the costs of the change for those who had to replace their existing heating technology were much higher, generating opposition to change. The shifts in both residential and railroad technologies were accomplished in a generation, causing a vast improvement in Pittsburgh's air quality. In the Pittsburgh case, at least, I believe that the public's support of cleaner air, even at a possible higher cost for heating, as well as the conviction that better conditions were possible, drove the improvement.

In 1981 the Department of History at George Washington University generously invited me to present the Lititia Woods Brown Memorial Lecture in Urban History, cosponsored by the department and the Colombian Historical Society of Washington, D.C. Building on my recent research, I chose as my subject the history of urban pollution, focusing on the relationship between disposal practices and the transfer of pollution problems from one media to another. My talk dealt with all three possible locations for waste disposal—air, water, and land—and interactions among them. As in all my writings on urban pollution, technology played a central role. This essay, drawing on my work on wastewater systems and air pollution, as well as early work on industrial waste disposal, provides the title for this book and its lead essay.[18] Its preparation enabled me to draw some of the connections that existed between different types of waste-disposal methods, policy reactions, and environmental problems. In addition, it provided me with the opportunity to apply to historical data some of the concepts (such as urban metabolism and cross-media pollution) learned from my environmental engineering colleagues that were not yet used in historical analysis.

In the early 1980s, the National Oceanic and Atmospheric Administration (NOAA) invited me to submit a proposal to perform a study of pollution flows over the past century in five East Coast estuaries: the Potomac,

Delaware, Hudson-Raritan, Connecticut, and Narragansett. This invitation was prompted by a March 1981 *New York Times* article that discussed the utility of applying the "lessons of history" to the present and that noted some of my work on the environment.[19] NOAA was particularly interested in the relationship between pollution flows and fish abundance in the estuaries, hoping to identify, through historical pollution studies, the relative role of different types of pollutants in reducing fish populations. The other participant in this project was Martin Marietta Environmental Systems (MMEC), to whom NOAA awarded the contract to study fish abundance over time and correlate their findings with our pollution data.[20]

My work on this project was a humbling experience in regard to the usefulness of the historian's traditional tools in addressing policy questions. My expertise lay in collecting historical information on environmental conditions and in preparing analytical discussions of these conditions from documents. As time went on, it became clear that NOAA's real need in this project was to obtain quantitative pollution-flow data for each estuary and river basin, which could then be correlated with changing fish abundance (a somewhat dubious endeavor, I must admit, in our eyes). Our team was able to collect substantial amounts of data on factors such as population, land use, dredging, miles of sewers, sewage treatment, acres of land in cultivation, changing farm crops, and alterations in pesticide/herbicide use, as well as contemporary reports on pollution conditions for some effluents.[21] However, in regard to flows of industrial effluents, little or no specific information was available for those generated over time. The solution to the difficulty obviously lay outside the use of traditional documentary sources and involved techniques of statistical estimation.

After considerable experimentation, Robert U. Ayres, a faculty member in EPP who had pioneered in the use of the "mass-balance" method of estimating pollution flows, supplied the necessary expertise.[22] (The mass-balance approach is based on the principle of the conservation of mass: matter cannot be destroyed, only altered in form.) Using information we supplied as to the number of workers per plant, the type of processes used, and the kind of raw materials consumed, he was able to prepare estimates for effluent flows of a number of organic and inorganic wastes for the industries of the Hudson-Raritan region. These included heavy metals such as lead, chromium, and copper. In addition, he prepared estimates of flows of other pollutants over time such as nitrogen, phosphorous, carbon, and various pesticides and herbicides. The accuracy of these estimates was checked by reference to core samples and analysis of water-quality data by marine scientists in various parts of the estuary. Thus, although exact numbers were unavailable for industrial and other pollutants that were not measured at the time, we were able to provide order-of-magnitude estimates. While it would have been ideal to have such estimates for all five estuaries, the immense effort required to form them limited our investigation to the Hudson-Raritan.

The mainly textual results of the NOAA study, as opposed to tapes of quantitative data, were submitted to the agency in a report entitled *Retrospective Assessment of Water Quality in Five East Coast Estuaries: The Potomac, the Delaware, the Hudson-Raritan, the Connecticut, and the Narragansett* (1985). A separate three-volume report, authored primarily by Robert U. Ayres and Leslie W. Ayres but with my collaboration, dealt with the estimates for chemicals, heavy metals, and emissions from fossil fuels for the Hudson-Raritan.[23] Finally, the MMEC investigators used our data to produce a report entitled *Assessment of the Relationships among Hydrographic Conditions, Macro Pollution Histories, and Fish and Shellfish Stocks in Major Northeastern Estuaries.* The major conclusion of the latter study was that there was a clear relationship between "low levels of dissolved oxygen in the estuary and declining estuarine stocks . . . [and] a clear link between stock success and water quality." MMEC also maintained that the approach used in the estuary study of linking "stock histories, natural environmental variation, and 'macro pollution histories'" provided a tool that, by linking stock variation to pollution loadings, could strengthen the ecological management of the nation's estuaries.[24]

The material from the two studies of the Hudson-Raritan Estuary was later combined and included as a case study in a volume entitled *The Earth as Transformed by Human Action: Global and Regional Changes in the Biosphere over the Past 300 Years.*[25] This essay has been revised for inclusion in this volume without the quantitative estimates of various effluents. I believe, however, that the ability to draw quantitative estimates of pollution flows over time, even if only order-of-magnitude estimates, is valuable for historians working on pollution and energy questions. It permits them to check the validity of historical pollution estimates as well as make comparisons possible for later periods. And it may help them to identify ecological relationships, such as those between pollution and fish abundance suggested by the MMEC report, that may be of use to policymakers.

While much of my earlier water-related research had involved human wastes and municipal sewerage, the work on the NOAA project emphasized for me the importance of industrial wastes. I was curious as to why industrial wastes, which were a primary focus of concern in the early 1980s, had for many decades ranked lower on the agenda of sanitary engineers and public health officers than had municipal wastes. I also wanted to investigate the problems caused by industrial wastes before the formation of the EPA and examine the response of legislatures and the courts. Building on the NOAA research and results, I began a research project that would explore questions of industrial waste generation and disposal in the Hudson-Raritan, Ohio, and Delaware river basins up to the period of federal involvement in standard setting in the 1970s. This work, still underway, is reflected in this book by three essays: "Industrial Wastes, Water Pollution, and Public Health," "Searching for a Sink for an Industrial Waste," and "Historical Perspectives on Hazardous Wastes in the United States."[26]

In rethinking the issues surrounding the timing of municipal infrastructure construction and subsequent waste disposal, I am no longer surprised that cities did not construct sewer systems to remove dirty water immediately after they built waterworks to supply their populations and industries; that when they did build them, they disposed of the raw sewage in nearby bodies of water, even though these might be the same waterways from which they drew their water supplies; and that they were slow in spending municipal tax dollars for sewage treatment. For them to have done otherwise might have been more ecologically desirable, may have saved considerable lives from various waterborne diseases, and might have preserved rivers and lakes from fish and habitat destruction. But a host of factors militated against such progressive views. Among these were uncertainties as to whether or not running water purified itself; competing hypotheses of disease etiology, especially anticontagionism and the germ theory; competition between various professional groups, such as that between engineers and physicians in the public-health domain; a reluctance of municipalities to spend money for sewage treatment that would have benefited primarily downstream cities; and a municipal disinclination to spend public funds for improvements that did not promise any material return and that might increase the size of budgets or bonded debt.

The situation in regard to industrial wastes poses a somewhat different set of questions. Here the issues relate more to individual firms or groups of firms than to municipal actors. I have argued that less attention was paid to industrial wastes compared to municipal wastes because of the dominance of the bacterial paradigm and the greater threat of epidemic disease from sewage. I have also maintained, as is reflected in the essay "Historical Perspectives on Hazardous Wastes in the United States," that our definitions of hazards have changed over time, depending on shifts in theories of disease etiology, value changes, and the development of new indicators. Although not so obvious in the 1970s, it does seem clear now that the crisis over land disposal of industrial hazardous wastes that erupted after Love Canal can be understood largely in the context of tightening controls over water disposal. Many industries had been disposing of their wastes on the land for decades, so the practice was an old one. But regulation of water pollution also forced a number of firms that had previously used water disposal to turn to the land. With the water sink increasingly unavailable for waste disposal, industries utilized the unregulated land sink.[27]

I am not surprised at this behavior. Economists have clearly established that firms will usually seek a cheaper and less regulated sink for their externalities. This industrial pattern was largely the result of the general regulatory and legal context but is also related to such factors as the availability of information, the type of technology being utilized, and the nature of competition in the industry. As the historical record makes clear, municipalities were also major actors in generating pollution and were slow in attempting to control it, especially if the pollution burden impacted only downstream communities. A focus on economic development, a reluctance to hamper industry with controls, the political power of industrial groups, and the lack of any powerful environmental constituency to compel state action meant that industry often could place its wastes into the air, land, and water with few restrictions.

To a large extent, however, the environmental history of the United States, as well as the rest of the world, is only now being written. Explorations into urban environmental history and the study of the historical relationship between technology and the environment are equally in their formative stages. Excellent and provocative work has recently been done by historians and historical geographers, but many subjects remain unexplored.[28] Only recently have scholars begun conducting historical research that relates the environment to such topics as industrial development and waste disposal; construction of the built environment; the consumption habits of urban populations leading to air, land, and water pollution; the special climatic and natural histories of cities; the differential effects of pollution and regulation on urban inhabitants (environmental equity); the development of environmental indicators; and the formation, implementation, and effects of public policy. There is no doubt that history will make increasingly clear the many ways in which cities and technology affect nature and how in turn nature plays a role critical in urban life.

Notes

1. Joel A. Tarr and Gabriel Dupuy, eds., *Technology and the Rise of the Networked City in Europe and America* (Philadelphia: Temple University Press, 1988).

2. "Metabolism" is defined here as the biological process of building up food into living matter and then its decomposition into waste matter.

3. It should be noted, however, that urban geographers and urban ecologists were somewhat ahead of historians in pioneering this area, just as urban geographers had been producing studies of the city considerably before the field of urban history emerged. See, for instance, Thomas R. Detwyler and Melvin G. Marcus, *Urbanization and Environment: The Physical Geography of the City* (Belmont, Calif.: Duxbury Press, 1972); Spencer W. Havlick, *The Urban Organism: The City's Natural Resources from an Environmental Perspective* (New York: Macmillan, 1974); and Ian Douglas, *The Urban Environment* (London: Edward Arnold, 1983).

4. For recent summaries of the field, see Martin V. Melosi, "The Place of the City in Environmental History," *Environmental History Review* 17 (Spring 1993): 1–23; Christine Meisner Rosen and Joel A. Tarr, "The Importance of an Urban Perspective in Environmental History," *Journal of Urban History* 20 (May 1994): 299–309. Among those scholars who are doing research in urban environmental history, I would note Craig E. Colton, John T. Cumbler, Christopher Hamlin,

Andrew Hurley, Clay McShane, Martin V. Melosi, Christine M. Rosen, and senior scholars Samuel P. Hays and Sam Bass Warner, Jr., as historians whose work I have especially benefited from.

5. See, for example, Joel A. Tarr, *A Study in Boss Politics: William Lorimer of Chicago* (Urbana: University of Illinois Press, 1971).

6. This was true of several technologies, such as the by-product coke oven, that we now consider to be problems for the environment.

7. Joel A. Tarr, "Urban Pollution Many Long Years Ago," *American Heritage* 22(October 1971): 65–69, 106. Because one of my points was that large-scale horse pollution had existed until fairly recently, the title distorted my intent; but since the article was already in print, little could be done.

8. I can remember horse-drawn wagons being used to collect garbage, peddle vegetables, deliver milk, and collect junk while I was growing up in Jersey City in the late 1930s and 1940s. See also Joel A. Tarr, "The Urban Dung Problem in Historical Perspective," *The Public Interest* 25 (Fall 1971): 126–127.

9. Joel A. Tarr, "From City to Farm: Urban Wastes and the American Farmer," *Agricultural History* 49 (October 1975): 499–611. Popular articles that derived from this body of work included "'Out of Sight, Out of Mind': A Brief History of Sewage Disposal in the United States," *American History Illustrated* 10 (January 1976): 40–47; "City Wastes in the 1800s Made It Back to the Farm," *Organic Gardening and Farming* 23 (December 1976): 74–77; and "City Sewage and the American Farmer," *Biocycle* 22 (September–October 1981): 36–38.

10. This RFP was primarily the brainchild of Joseph Coates, an innovative thinker with an acerbic tongue, then at the National Science Foundation. Coates, who had done graduate work in the history of science as well as physics, was concerned with identifying methodologies to be used in the new forecasting area called technology assessment.

11. Joel A. Tarr, "Retrospective Technology Assessment," *Futures: the Journal of Forecasting and Planning* 9 (April 1977): 159; and idem, ed., *Retrospective Technology Assessment* (San Francisco: San Francisco Press, 1977).

12. Peter N. Stearns and Joel A. Tarr, "Applied History: A New/Old Departure," *The History Teacher* 14 (August 1981): 517–531.

13. RTA was funded through a short-lived program called Research Applied to National Needs (RANN). Our grant number was ERP 75-08870. In addition to McMichael and Wojick, the team included Clay McShane, then a visiting assistant professor of history at CMU; two CMU history graduate students, James McCurley III and Terry Yosie; and a research associate, Clifford Hood, then in transit between taking an undergraduate degree at Washington University in St. Louis and a graduate career at Columbia University. The methodology was the brainchild of our philosopher of technology, David Wojick. Basically, it involved performing an analysis of decision making in instances of technology development and resulting social impacts through the construction of decision trees.

14. In her important work on Chicago public works, Robin L. Einhorn notes that the slow growth of the city's sewerage system "probably reflects the less specifically personal benefits of sewage disposal." See, Robin L. Einhorn, *Property Rules: Political Economy in Chicago, 1833–1872* (Chicago: University of Chicago Press, 1991), 137–43.

15. Joel A. Tarr with James McCurley III, Francis C. McMichael, and Terry F. Yosie, "Water and Wastes: A Retrospective Assessment of Wastewater Technology in the U.S., 1800–1932," *Technology and Culture* 25 (April 1984): 226–63; Joel A. Tarr and Francis C. McMichael, "Historical Decisions about Waste Water Technology, 1800–1932," *American Society of Civil Engineers, Journal of Water Resources Planning and Management* 103 (May 1977): 47–61; Joel A. Tarr, "The Separate vs. Combined Sewer Problem: A Case Study in Urban Technology Design Choice," *Journal of Urban History* 5 (May 1979): 308–39; and Joel A. Tarr, Terry Yosie, and James McCurley III, "Disputes Over Water Quality Policy: Professional Cultures in Conflict, 1900–1917," *American Journal of Public Health* 70 (April 1980): 427–35. Other publications not included are Joel A. Tarr and Francis C. McMichael, "The Evolution of Waste Water Technology and the Development of State Regulations," in Tarr, *Retrospective Technology Assessment,* 165–89; Joel A. Tarr and Francis C. McMichael, "Historic Turning Points in Municipal Water Supply and Waste water Disposal,

1850–1932," *Civil Engineering* 47 (October 1977): 82–91; Joel A. Tarr and Francis C. McMichael, "Water and Wastes: A History," *Water Spectrum 10* (Fall 1978): 18–25; Joel A. Tarr, Terry Yosie, and James McCurley III, "The Development of Waste Water Systems," in Martin V. Melosi, ed., *Pollution and Reform in American Cities, 1850–1930* (Austin: University of Texas Press, 1980), 59–82; Joel A. Tarr and Gabriel Dupuy, "Perspectives souterraines," *Les Annales de la Recherche Urbaine* (Paris: Dunod, 1984), 23–24, 65–88; Joel A. Tarr and Gabriel Dupuy, "Sewers and Cities: France and the U.S. Compared," *Journal of Environmental Engineering—Transactions of the American Society of Civil Engineers* 108 (April 1982): 327–38; Gabriel Dupuy and Joel A. Tarr, "Assainissement des Villes: Socio-technique de l'assainissement des villes en France et aux Etats-Unis," *Culture Technique* 2 (February 1981): 215–23; and Joel A. Tarr, "Sewers and Cities in the United States, 1850–1930," in Tarr and Dupuy, *Technology and the Rise of the Networked City*, 159–85.

16. Abel Wolman had been prominent in the field of sanitary engineering since the 1920s and had served as president of both the American Public Health Association and the American Society of Civil Engineers. He was a member of our RTA advisory board. Edward B. Fiske, "Lessons of History Applied to the Present," *New York Times*, March 10, 1981.

17. This project was funded by the Ethics and Values Program of the National Science Foundation, Grant No. SS78-17308. Three graduate students were funded from this grant: Todd Shallat, Gary D. Goodman, and Kenneth Koons. Joel A. Tarr with Bill Lamperes, "Changing Fuel Use Behavior and Energy Transitions: The Pittsburgh Smoke Control Movement, 1940–1950: A Case Study in Historical Analogy," *Journal of Social History 14* (Summer 1981): 561–88. See also "Changing Fuel Use Behavior and Energy Transitions: Policy Lessons from a Historical Case," *Technological Forecasting and Social Change* 20, (1981): 331–46; Joel A. Tarr with Gary D. Goodman and Kenneth Koons, "Coal and Natural Gas: Fuel and Environmental Policy in Pittsburgh and Allegheny County, Pennsylvania, 1940-1960," *Science, Technology, and Human Values* 5 (Summer 1980): 19–21; and Joel A. Tarr with Kenneth Koons, "Railroad Smoke Control: A Case Study in the Regulation of a Mobile Pollution Source," in Mark Rose and George Daniels, eds., *Energy and Transport: Historical Perspectives on Policy Issues* (Los Angeles: Sage Publications, 1982), 71–92.

18. Joel A. Tarr, "The Search for the Ultimate Sink: Urban Air, Land, and Water Pollution in Historical Perspective," *Records of the Columbia Historical Society of Washington, D.C.* vol. 51 (Charlottesville: University Press of Virginia, 1984), 1–29; reprinted in Kendall Bailes, ed., *Critical Issues in Environmental History* (Boston: University Press of America, 1985), 516–52.

19. Edward B. Fiske, "Lessons of History Applied to the Present," *New York Times*, March 10, 1981.

20. Several applied history graduate students were involved in the NOAA project, including James McCurley III, Charles Jacobson, Stuart Shapiro, and Donald Stevens.

21. After collection, these data were provided on magnetic tape to Martin Marietta Environmental Systems.

22. For a discussion of the mass balance approach, see Robert U. Ayres, Francis C. McMichael, and Samuel R. Rod, "Measuring Toxic Chemicals in the Environment: A Materials Balance Approach," in Lester B. Lave and A. C. Upton, eds., *Toxic Chemicals, Health, and the Environment* (Baltimore: Johns Hopkins University Press, 1987).

23. For this report, see Robert U. Ayres, Leslie W. Ayres, Joel A. Tarr, and Rolande C. Widgery, *An Historical Reconstruction of Major Pollutant Levels in the Hudson-Raritan Basin: 1880–1980*, NOAA Technical Memorandum NOS OMA 43, 3 vols. (Rockville, Md.: 1988).

24. See J. K. Summers and T. T. Polgar (co-principal investigators), *Assessment of the Relationships among Hydrographic Conditions, Macropollution Histories, and Fish and Shellfish Stocks in Major Northeastern Estuaries* (Columbia, Md.: Martin Marietta Environmental Systems, 1984), vi–vii. For an earlier statement of these conclusions, see the article published jointly by the Martin Marietta and CMU teams, "Reconstruction of Long-Term Time Series for Commercial Fisheries Abundance and Estuarine Pollution Loadings," *Estuaries* 8 (June 1985): 114–24. See also Joel A. Tarr, Charles Jacobson, Kevin McCauley, James McCurley III, and Donald Stevens, "Histori-

cal Habitat Change in the Hudson-Raritan Estuary: Dredging, Landfill, and Submerged Aquatic Vegetation," in Anthony Pacheco, ed., *Fish and Bricks: Proceedings of the Walford Memorial Convocation, Sandy Hook Laboratory Technical Series Report, No. 85-05*, 1984, 53–59.

25. Joel A. Tarr and Robert U. Ayres, "Pollution Trends in the Hudson River and Raritan River Basins, 1790–1980," in B. L. Turner et al., eds., *The Earth as Transformed by Human Action: Global and Regional Changes in the Biosphere over the Past 300 Years* (New York: Cambridge University Press, 1990), 623–40. For another example of the use of mass balance to estimate historical trends, see Robert U. Ayres, Leslie Ayres, and Joel A. Tarr, "A Historical Reconstruction of Carbon Monoxide and Methane Emissions in the United States, 1880–1980," in R. U. Ayres and Udo E. Simonis, eds., *Industrial Metabolism: Restructuring for Sustainable Development* (Tokyo: United Nations University, 1994), 194–238.

26. This project was funded by the History and Philosophy Section of the National Science Foundation, grant number SS788-17308. Joel A. Tarr, "Industrial Wastes and Public Health: Some Historical Notes, Part 1, 1876–1932," *American Journal of Public Health 75* (September 1985): 1059–67; idem, "Searching for a 'Sink' for an Industrial Waste: Iron-Making Fuels and the Environment," *Environmental History Review 18* (Spring, 1994): 9–34. See also idem, "Industrial Wastes and Municipal Water Supplies in the 1920s," in H. R. and A. D. Keating, eds., *Water and the City: The Next Century* (Chicago: Public Works Historical Society, 1991), 261–274; idem, "Historical Perspectives on Hazardous Wastes in the United States," *Waste Management & Research* 3 (1985): 95–102. See also Joel A. Tarr and Charles Jacobson, "Environmental Risk in Historical Perspective," in Brandon Johnson and Victor Covello, eds., *The Social and Cultural Construction of Risk* (Boston: Reidel Publications, 1987), 317–44.

27. Whether or not firms did know that industrial wastes disposed of in the land sink would migrate to groundwater is an interesting question. Some authors believe that the information was available in the literature. See Craig E. Colten, "A Historical Perspective on Industrial Wastes and Groundwater Contamination," *Geographical Review* 81 (1991): 215–28. My personal belief is that a combination of "careless housekeeping" (Abel Wolman's phrase), a failure of communication, and a deliberate ignoring of possible harmful effects was involved.

28. See, for instance, the essays in Rosen and Tarr, "The Environment and the City," and Joel A. Tarr and Jeffrey K. Stine, eds., *Environmental History Review, Special Issue on Technology, Pollution, and the Environment* 18 (Spring 1994).

PART I

Crossing Environmental Boundaries

Environmental boundaries are often regarded as fixed, whether in terms of the separation of air, land, and water, as the targets of regulatory activity, or even as divisions within the field of environmental history itself. In an attempt to cross these boundaries, the three articles in this section illustrate different themes in the history of the environment. These themes include consideration of how technological fixes for one environmental problem have often produced difficulties in other domains; the nature and scale of regional environmental change; and the reform belief that environmental enhancement would improve human behavior.

While the United States has often prided itself on the reduction of one pollution stream or another, the "solution" itself has frequently generated another set of difficulties. Thus, environmental problem solving—usually involving technological development, policy, and implementation—often produced unpredicted or unanticipated negative effects in other domains or locations. Environmental issues are largely holistic and interrelated, and seldom does the contamination of one media, such as water, avoid having an effect on the other media, air and land.

Over time, the forces of urbanization and industrialization, spurred on by a free-wheeling market economy, caused sweeping changes in the environment, altering the uses of land and polluting air, soil, and water. These trends can clearly be seen in the history of metropolitan areas such as New York City and its hinterland, as they were transformed by agriculture, urbanization, and industrialization from the eighteenth through the late twentieth century. Since the 1920s, but more actively since the 1960s, governments on different levels have used regulations, court orders, cooperative agreements, and technological innovations with some success to reduce environmental degradation. Legislative loopholes and new products, however, have often thwarted these efforts. In addition, advances in science and analytical instrumentation have made it possible to identify new environmental and health threats. By examining the environmental history of an urban region such as New York, it is possible to grasp the immense changes we have made in nature and to confront the limited extent to which we have successfully dealt with society's wastes in the past.

Behind the various campaigns, programs, and pieces of legislation intended to improve the environment have rested sets of beliefs with far-reaching implications. Some reformers have opposed pollution because of its very wastefulness, while others have insisted that nature be restored for its own sake. Still others have believed that human behavior was linked to the quality of the environment and have urged environmental improvement as a means of obtaining a more civic-minded citizenry. Many Progressive Era reformers held these beliefs, and the Pittsburgh Survey—the first major social survey of a large American industrial city, conducted in 1908–9—reflected the belief that environmental quality and human behavior were tightly linked and that improving both the natural and the built environments would result in a greater sense of civic devotion and social order. The Survey writers were naive in their beliefs both about environmental determinism and the ease of accomplishing change, but the power and persistence of their assumptions for today's world cannot be dismissed.

Photo I.1. "Sowing for Diptheria" was the caption on this 1881 etching. Source: *Harper's Weekly,* 1881.

Photo. I.2. "Natural Beauty vs. Industrial Odds." A view down the Monongahela River towards the Pittsburgh Point, 1911. Source: Paul Underwood Kellogg (ed.), *The Pittsburgh District: Civic Frontage, The Pittsburgh Survey* (New York: Survey Associates, 1914).

Photo I.3. "The Valley of Work," by Otto Kubler. Source: *Pittsburgh Record*, 1928.

Photo 1.4. "Flooding in Pittsburgh, 1907." This picture was taken on Pittsburgh's North Side, formerly part of the City of Allegheny. Source: Carnegie Library of Pittsburgh.

CHAPTER I

The Search for the Ultimate Sink

Urban Air, Land, and Water Pollution in Historical Perspective

Introduction

Some years ago, the noted sanitary engineer Abel Wolman wrote an influential essay entitled "The Metabolism of Cities." In the article Wolman described the metabolic requirements of the city as consisting of all the materials and commodities required to sustain the life processes of the city's inhabitants. The metabolic cycle, he said, was not completed "until the wastes and residues of daily life had been removed and disposed of with a minimum of nuisance and hazard."[1] Wolman's model of the city as a metabolic entity has historical as well as contemporary relevance. The processes by which pollutants have been generated have altered over time, but so have the definitions of what pollution actually is. The meaning of the terms "nuisance" and "hazard" are time and culture specific, and their definitions depend on many elements both within the urban container and the larger society.

Urban pollution, therefore, at any time, can be understood as the product of the interaction among technology, scientific knowledge, human culture and values, and the environment. Environmental policy and control technology are further elements that must be added to the model, for at various times they have both reduced and exacerbated pollution problems or resulted in their transfer to different locales or media. The purpose of this paper is to examine three cases of urban air, land, and water pollution in order to explore the interactions among the above variables. More specifically, it will examine three larger and overarching themes or questions:

1. How solutions for one pollution problem often generated new pollution problems in different localities or in different media
2. How both values and scientific knowledge were involved with society's perceptions of the environment and influenced policy to deal with pollution problems
3. How our perceptions of risk and hazard in regard to the urban environment affected our willingness to support policy to deal with these pollution problems (agenda setting)

The cases that will be examined are not necessarily unknown to students of environmental history, but I hope to focus on elements within each that will advance our understanding of the interactive nature of the problems of the urban environment.

The Water: Supply, Waste Disposal, and Pollution

The problems of supplying an adequate and potable supply of water to urban inhabitants and disposing of both human wastes and wastewater are the first situations where American society—in this case, cities—attempted to deal with pollution using a technological solution or fix. These questions of supply and disposal are interrelated, and the solution to one often played a significant role in creating health and sanitary difficulties for other cities. Changing values in regard to the public health and water use have also been important in the society's attempt to deal with these problems over time and have generated new policy initiatives. The search for solutions to the problems of waste disposal and water pollution clearly illustrates the difficulty in finding a sink for wastes without causing further damage to the environment in other locales.

The water supply and human waste and wastewater disposal systems utilized in most cities during the eighteenth and the nineteenth centuries were characterized by a local focus. Water supplies were obtained from local sources such as wells and pumps drawing on groundwater, from nearby ponds and streams, and from rainwater cisterns. Used water (wastewater) and human wastes were usually disposed of in cesspools and privy vaults, although occasionally they were thrown out on the street or in vacant lots. Cesspools and privy vaults were essentially holes in the ground, sometimes lined, from which wastes often leached (deliberately and accidentally) into the surrounding soil. The land thus became the primary sink for both wastewater and for human wastes. In some cities, human wastes were occasionally collected from privy vaults by scavengers (night soil men) or by farmers. These wastes were often recycled on the land as fertilizer or dumped in land depots or nearby waterways. Before the 1850s no city had sewers for human-waste removal, and it was not until after 1880 that most municipalities constructed sewerage systems. Those sewers that existed were largely for stormwater collection, and in some cities ordinances forbade citizens to deposit wastes in them.[2]

This system of local water supply and waste collection could operate without excessive nuisance or sanitary hazard when city populations were small and densities low, but as urban population and density increased in the late eighteenth and early nineteenth century, it became increasingly ineffective. The first part of the system to break down was the water supply. Various studies of city water supplies and public health in the first decades of the nineteenth century document both the growing pollution of the local ponds and wells that served the population of cities such as Boston, New York, and Philadelphia and the problems that developed from inadequate supply. Cleaner and more copious water supplies were needed for normal household functions such as drinking and washing, for fire fighting in crowded urban neighborhoods, for industrial purposes, and for flushing the streets at times of epidemics. Closely associated with the necessity for cleaner water supplies were concerns over the health effects of polluted water and dirty streets and a growing realization that there was a relationship between clean water and freedom from epidemic disease.[3]

Philadelphia was the first city to respond to the inadequacy of local water supplies; it constructed the Fairmount Water Works in 1802 to bring potable water into the city from the Schuylkill River. Cities such as New York, Boston, Detroit, and Cincinnati followed Philadelphia's lead, and by 1860 the sixteen largest cities in the nation had waterworks, with a total of 136 systems in the country; by 1880, this number had increased to 598.[4]

As piped-in water became available, the more affluent urban households installed water-using fixtures. In 1848, for instance, Boston opened the Cochituate Aqueduct, and by 1853 31,750 water-using fixtures of various types were in operation; by 1863 the number had increased to 81,726, of which over 13,000 were water closets. The availability of a constant household supply of water caused a rapid expansion in the number of users and in the volume of use. Chicago, for instance, went from 33 gallons per capita per day in 1856 to 144 in 1882; Cleveland increased from 8 gallons per capita per day in 1857 to 55 in 1872; and Detroit went from 55 gallons per capita per day in 1856 to 149 in 1882. These figures include industrial and other nonhousehold uses, but they are still symbolic of a great increase in water consumption over a relatively short period of time as demand interacted with supply.[5]

But while hundreds of cities and towns installed waterworks in the first three-quarters of the nineteenth century, no city simultaneously constructed a sewer system to remove the water. In most cities with piped-in water, wastewater was initially diverted into existing cesspools or occasionally into stormwater sewers or street gutters. The introduction of large volumes of water into a cesspool system designed to accommodate much smaller amounts unbalanced the system and caused serious flooding and disposal problems. This situation was exacerbated by the widespread adoption of the water closet, which greatly increased the problems of nuisance and sanitary hazard in wastewater disposal by adding "black" water to "grey." Cesspool overflows caused the soil to become saturated, led to cellars that were "flooded with stagnant and offensive fluids," and made cleaning "nearly futile."[6]

Public health officials, especially if they believed in the anticontagionist "filth theory" of disease, viewed overflowing cesspools with water-closet connections as a particularly dangerous threat to a healthful environment. As late as 1894 the secretary of the Pennsylvania State Board of Health, Benjamin Lee, complained that householders persisted in installing water closets in towns without sewers and connecting them to "leaching" cesspools. "Copious water supplies," warned Lee, "constitute a means of distributing fecal pollution over immense areas and no water closet should ever be allowed to be constructed until provision has been made for the disposition of its effluent in such a manner that it shall not constitute a nuisance prejudicial to the public health."[7]

The health and sanitary problems caused in cities by running water and wastewater disposal generated a search for ways to modify the system or for new, alternative methods of disposal. Driving this was the Sanitary Movement, which had begun in Great Britain in the 1840s and 1850s with the work of Sir Edwin Chadwick and his followers and spread to the United States by the 1850s, receiving some impetus from the work of the Sanitary Commission during the Civil War. It was a social and cultural movement that essentially aimed at changing people's ideas about their own personal habits of cleanliness and conditions within the environment around them. At the heart of the Sanitary Movement was the belief in the environmental cause of disease, or what public health specialist Charles V. Chapin called "the filth theory of disease." This anticontagionist theory maintained that disease spread de novo from putrefying organic matter or gave rise to disease-carrying miasmas.[8]

The Sanitary Movement was propagated through a great wave of publicity. Its institutional and organizational embodiments were the American Public Health Association, the National Board of Health, and the multitude of local and state boards of health that appeared in the late nineteenth century. The movement gave a tremendous impetus to organized public health and the physical cleaning of cities. In addition, the value change that caused people to perceive overflowing cesspools and privy vaults as both unpleasant and as a health hazard that could be eliminated, rather than as a natural nuisance to be tolerated, also stimulated a search for technologies to deal with the waste problem.[9]

Among the approaches tried were the pail system, the earth closet, and the "odorless excavator." The pail system and the earth closet were designed as substitutes for the water closet and the privy vault and permitted recycling of human wastes. Although both had their advocates, their inconvenience and labor-intensive qualities, as compared to the water closet, resulted in only limited adoption.[10] The technology that secured the most proponents among engineers, public health professionals, and city officials was the capital-intensive water-carriage system.

Water-carriage technology was designed to utilize the running water to the household as the medium of transport for wastes. Wastes were carried through a system of pipes to a place of disposal outside the immediate locale. The so-called sewerage system, therefore, offered a complete replacement for the previous system of cesspools and privy vaults. The model city for the earliest American sewerage systems was London, which had constructed a system of brick combined sewers (stormwater and household wastewater in the same pipe) in the 1850s.[11] Brooklyn and Chicago both built sewerage systems before the Civil War, and many other municipalities followed in the last quarter of the century. By 1908 American cities with over 10,000 population had 8,199 miles of sewers, and by 1909 cities with more than 30,000 population had 24,972 miles.

The water-carriage system of human waste disposal aroused controversy in many cities because of its costs and concerns over its health impacts. In numerous municipalities, debates over adoption of the system, as well as its design, went on for a period of years. Eventually sewerage systems were constructed in all major American cities (Baltimore was the last, 1911) because their perceived benefits outweighed the costs—the technology promised health and sanitation improvements with minimal costs of disposal. Disposal was accomplished most simply by utilizing adjacent waterways, thereby shifting the sink for the wastes from the land to the water. Warnings by a few chemists and engineers of the potential hazards resulting from the disposal of sewage in streams or lakes were often dismissed with the argument that "running water purifies itself." Up until the 1890s, this hypothesis seemed confirmed by existing methods of chemical analysis of water quality. In 1909, 88 percent of the wastewater produced by the urban sewered population was disposed of untreated in waterways, and the percentage was probably higher a decade before.[12]

Municipal construction of sewerage systems did greatly improve local sanitary conditions and, in many cases, reduced bacterial ailments such as infant diarrhea and typhus. However, soaring morbidity and mortality rates from infectious disease such as typhoid in downstream and lake cities that drew their water supplies from waterways in which other cities disposed of raw sewage raised serious questions about the validity of the disposal hypothesis. The high health costs of sewage disposal in streams spurred research in bacterial science and in the epidemiology of waterborne infectious diseases. Early in the 1890s, chemists and sanitary engineers of the Massachusetts Lawrence Experiment Station identified sewage-polluted water as the carrier for infectious disease and confirmed the dangers of disposal of raw sewage in waterways used for drinking-water supplies. Thus, urban sewerage systems had shifted the sink for human wastes and wastewater from the land to the water and transferred the health and sanitary costs of disposal to downstream cities. Ironically, many of the cities that suffered most severely from sewage-polluted water had themselves spent millions of dollars on sewerage systems to improve local conditions.[13]

Eventually, in the first decades of the twentieth century, researchers in sanitary engineering solved the problem of sewage-polluted drinking water with the development of other retrofits—water filtration and chlorination. These technologies dramatically reduced the incidence of waterborne disease, but they did not improve water quality in terms of other potential waterway uses. Cities with older systems, plus cities with newly constructed sewerage systems, continued the practice of disposing of their untreated wastes in nearby waterways. Except for specific localities with severe nuisance problems from sewage disposal, municipalities resisted installing sewage treatment facilities that promised to provide direct benefits only to downstream cities and instead relied on dilution to dispel the worst concentrations of pollutants. Hence, by 1930 there was a large deficit between that population served by water-treatment facilities and that served by sewage-treatment plants. During the 1930s, a number of sewage treatment plants were constructed by the Works Progress Administration (WPA), but significant progress in improving water quality was not made until the postwar decades. The federal government joined its funds to local resources in this decade in an attempt to control both municipal and industrial pollutants. At the same time, however, new toxic and other industrially based pollutants, new analytic capabilities that made possible the identification of formerly unsuspected health hazards, and problems with controlling more traditional municipal wastes caused increased public and professional concern over water quality.[14]

In the 1960s and 1970s, a widespread public conviction developed that waterways should be limited in the extent to which they should serve as sinks for both municipal and industrial wastes. This belief culminated in the passage of the Clean Water Act in 1972 and the call for zero effluent by 1985. This goal has not been reached, in spite of the development of some innovative treatment systems. Waterways will continue to serve as sinks for the various wastes generated in our urban areas for the foreseeable future, for, as the 1980 Report of the Council on Environmental Quality noted, "cleaning up the nation's water takes a long time."[15]

The Air as a Sink: Coal, Smoke Control, and Acid Rain

The first air-quality problem dealt with by American society involved smoke pollution in industrial cities. The problem of smoke was the result of a conjunction between urbanization, industrialization, and the utilization for fuel of high-volatile bituminous coal. The fuel was the input into the metabolic cycle of the city, and smoke, as well as other air pollutants, such as fly ash, was the output. Like human wastes in cesspools and privy vaults or polluted wells, it presented primarily a local pollution problem. A concern with smoke pollution reached back almost as far as did concerns with water supply, human-waste disposal, and water pollution, but substantive successful action to control smoke came later than it did in regard to water. The lower place of smoke on the environmental agenda can be explained by the fact that smoke pollution presented a somewhat different type of problem than did water supply, human-waste disposal, and water pollution. These differences involved questions of impacts, control technology, and values.

Impacts: Smoke had primarily nuisance impacts on both people and property, causing discomfort and higher cleaning expenses in the city. Physicians suspected that smoke was responsible for many health problems but could not specify health impacts.[16] In contrast, water pollution had more immediate and observable health effects, and bacterial science made it possible to show cause and effect scientifically.

Control Technologies: While there were hundreds of patents issued in the nineteenth century for technologies to control smoke, there was no single technology that had clear cost and efficiency advantages or that demonstrated a consistent series of successes. This contrasted with the record in regard to water supply, waste disposal, and water pollution, which were successively dealt with by the technological fixes and retrofits of water-supply technology, water-carriage systems, and water filtration and chlorination. The substitution of a cleaner manufactured fuel for bituminous coal was suggested in the late nineteenth century, but the idea was rejected as prohibitively expensive. District heating was a technology with some of the characteristics of sewerage systems that could have reduced smoke pollution considerably, but, because of considerations such as lower urban densities, the technology never became as popular in the United States as in Europe.[17]

Values and Perceptions: Smoke had positive value connotations and was often equated with prosperity, production, growth, and jobs.[18] In contrast, dirty or polluted water had no such positive value implications.

Attempts to control smoke in American cities actually began in the middle of the nineteenth century. Some cities banned bituminous coal-burning locomotives from their streets, while Pittsburgh attempted to forbid the construction of beehive coke ovens within the city boundaries in 1869. The Progressive Era saw a rash of municipal attempts to regulate smoke, and by 1912 twenty-three of the twenty-eight cities with a population over 200,000 had smoke-control ordinances. These ordinances were aimed at visible smoke from industrial, commercial, and transport sources, and no city except Los Angeles had regulations controlling smoke from domestic fires. Most cities utilized the Ringelmann Chart, a visual method of measuring smoke density, to determine violations of their ordinances. While these smoke-control efforts had some limited and sporadic successes in reducing dense smoke from industrial and transportation sources, they basically failed to make substantive inroads into the problem.[19]

The 1920s–30s was a period of self-examination and reanalysis of the smoke question by the various professional groups, such as the Smoke Control Association of America and the Fuels Division of the American Society of Mechanical Engineers, involved in control efforts. Smoke-reduction concerns through the 1920s had focused on industries, utilities, and railroads, and professionals generally agreed that these interests had made advances in the elimination of dense smoke because of a desire to economize on fuel. The smoke problem persisted, most professionals agreed, because of a failure to control domestic sources. They considered smoke from household chimneys objectionable because "the amount of black smoke produced by a pound of coal is greatest when fired in a domestic furnace and that domestic smoke is dirtier and far more harmful than industrial smoke."[20]

In order to control domestic smoke, the experts argued, the same approach utilized to solve drinking-water pollution—control at the source before distribution—would have to be followed. This strategy required ordinances that mandated the use of smokeless fuels or of equipment to burn dirty coal smokelessly. Securing support for such legislation, however, required a massive educational effort to change people's fuel-using behavior in cities where cheap bituminous coal was heavily utilized.

Successful efforts to make control of domestic sources of smoke politically acceptable occurred in two cities before World War II—St. Louis and Pittsburgh. In both cities the statutes resulted from intensive media campaigns accompanied by the strong support of organized community, labor, and business groups and important public figures. St. Louis was first to enforce against homeowners, as well as industry and railroads, acting in 1940.[21] The visible signs of its success in reducing smoke inspired Pittsburgh groups to push for a similar ordinance, which was passed in 1941. The Pittsburgh action is especially notable not only because it had a reputation as "the Smoke City" but also because the soft-coal mining industry had a strong base in the region.

The 1941 Pittsburgh Smoke Control Ordinance was the strongest smoke-control law passed by any city up to that time. It had as its policy goal the elimination of dense smoke as well as other components of air pollution, such as fly ash. Consumers would have to burn either smokeless fuel or use smokeless technology if they were using bituminous coal. The ordinance set emission standards for domestic fuel consumers as well as for industrial, commercial, and transportation sources, and it created a Bureau of Smoke Prevention for enforcement.[22]

World War II delayed the ordinance's implementation, and the city suffered extremely bad air-quality conditions during the war because of the use of inferior fuels and old equipment. Convinced that the future of the city depended on smoke control, a coalition of key business leaders (the Allegheny Conference on Community Development) and the newly elected Democratic mayor, David L. Lawrence, united to promote implementation. Beginning in the winter of 1947–48, the Bureau of Smoke Prevention successfully enforced the law by regulating the supply of high volatile coal available to homeowners and forcing them to burn smokeless fuels (including smokeless coal) or use smokeless combustion equipment.[23]

In spite of initial difficulties with fuel supply, Pittsburgh's air improved considerably in the years after the initial implementation. Heavy smoke nearly disappeared from the city's atmosphere. In 1955, for instance, the Bureau of Smoke Prevention reported only 10 hours of "heavy" smoke and 113 hours of "moderate" smoke as compared with 298 hours of "heavy" smoke and 1,005 hours of "moderate" smoke in 1946. Pittsburgh benefited from improved air quality, more sunshine, and improved health as well as saved on cleaning costs, laundry bills, and injury to vegetation. A county law passed in 1949 provided the advantages of smoke-free air to the larger geographical area.[24] The city's decision makers had clearly decided that there were constraints on the extent to which they would allow the air to be utilized as a sink, in spite of the importance of coal to the region.

While the managers of the Pittsburgh Smoke Control movement had originally believed that smoke elimination could occur through the utilization of smokeless coal produced from local bituminous, natural gas soon became the dominant home-heating fuel in the Pittsburgh region. In the post–World War II period, cheap natural gas from the southwest was piped into Pittsburgh and, because of cost and convenience factors, replaced coal. The rates of change for the city are striking. In 1940 80 percent of Pittsburgh households burned coal and 17.4 percent natural gas (from Appalachian fields); by 1950 the figures were 31.6 percent for coal and 65 percent for natural gas.[25] This represented a change in fuel type and combustion equipment by almost half the city's households. Most of the transition took place after 1945 and was accelerated by the smoke-control ordinance.

Throughout the nation in the 1940s and 1950s, natural gas and oil replaced coal not only for domestic heating but also for other industrial, commercial, and transport uses. By 1955 bituminous coal furnished only 27.2 percent of the nation's aggregate energy consumption, compared to 44.8 percent in 1945. During the same years, natural gas increased its percentage from 11.8 percent to 22.1 percent and oil from 29.4 percent to 40 percent. The most significant changes involved the substitution of natural gas for bituminous coal for domestic heating and other household uses and the replacement of the steam locomotive by the diesel electric. The percentage of consumers using natural gas for househeating increased from 35.5 percent in 1949 to 68.4 percent in 1960. During the same period the number of coal-burning steam locomotives decreased from 30,344 to 374. Coal thus lost a number of its traditional markets, and by 1954 the U.S. production of bituminous had reached 391,706,300 tons, the lowest tonnage of any year since 1909 except for the 1931–35 depression years.[26]

The increased substitution of cleaner fuels for coal in the postwar years had a marked impact on air quality in a number of cities, especially in regard to the reduction of visible smoke as a pollutant. Simultaneous with the air quality improvements generated by technology was an increase in the number of communities and states enacting air pollution–control regulations. A 1962 survey by the National Coal Association showed that of 216 urban areas east of the Mississippi River (the largest coal burning area), 140 had ordinances. Most communities (107) would respond to citizen complaint over air pollution, and seven cities with over 200,000 population maintained enforcement bureaus. The ordinances varied in terms of the standards created and the enforcement provided and usually used the simple Ringelmann Chart providing for visual grading of smoke to determine violations. While air pollution specialists warned of problems from other pollutants such as dust, fumes, and sulphur dioxide, the primary criterion of an air pollution nuisance was based on perceptible ground-level effects.[27]

Since air pollution was primarily conceived of as a local problem, a technology that diluted ground-level contaminants—tall stacks—was advocated by a number of air pollution experts in the 1950s and 1960s. These stacks were a method of forcing a pollution plume into the higher levels of the atmosphere, and they spread the pollutants over a much wider downwind area in order to prevent ground-level concentrations. The concept used here resembled that followed in dispersing water pollution in a body of water, that of dilution. The air continued to be a sink, as had the water, but it was believed that natural processes would prevent objectionable or dangerous concentrations of pollutants.[28]

The tall-stack technology was increasingly utilized by the ore smelting and electrical utility industries as a means of avoiding pollutant concentrations that would violate local air pollution ordnances. The electrical power industry, which was located primarily in urban areas, was especially significant. Electrical utilities were expanding rapidly in the postwar generations and had replaced the railroads as the coal industry's largest customer. In 1964, for instance, the electric utility industry consumed 46 percent of the 486,998,000 tons of bituminous coal mined.[29] A "dirty" fuel—bituminous coal—produced a "clean" form of power—electricity—and the tall stacks would use natural processes to dilute the pollutants.

By 1963 some utility stacks had reached the seven-hundred-foot level and predictions were made that they would climb to one thousand feet in the near future. Books on air pollution–control methodology advocated the technology as a means of diluting local concentrations of pollutants, although not without warning of possible "downstream" dangers. In 1965, for instance, the Air Conservation Commission of the American Association for the Advancement of Science reported that the tall-stack approach had "considerable merit." The report also warned, however, that sulfur dioxide emitted in bituminous coal consumption oxidized to form sulfuric acid mist and that, while there appeared to be little danger, the global effects were unknown. The report recommended further study of the question.[30]

During the 1960s and 1970s, as the public became more environmentally conscious and concerned about the health effects and nuisances created by air pollution, stricter legislation appeared on the local, state, and national levels. The Clean Air Act (1970) marked a high point of legislative effort to mandate clean air and brought ambient concentrations of criteria pollutants under the control of national standards within 247 air-quality-control regions. Controls were to be imposed on pollution sources within each region, thus still emphasizing the locality. In order to meet the increasingly stringent standards on local emissions, many utilities and industries accelerated their construction of tall stacks to disperse their pollution. Between 1970 and 1979, for instance, 178 stacks of over five hundred feet were constructed, almost entirely by electric utilities.[31]

In the early 1970s concern surfaced over the phenomenon of acid rain, which had been observed in some localities as early as the mid-1950s. Researchers held that acid rain was largely a product of increased fossil fuel combustion. Tall stacks were critical to the processes resulting in the formation of acid rain because they permitted pollutants (sulfur oxides and nitrogen oxides) to remain aloft. In the upper air levels, photochemistry, water vapor, and trace metals transformed fossil-fuel pollutants into sulfates and nitrates, which then reacted with moisture in the air to form acids. Among the observed effects of acid rain have been the destruction of fish and the release of toxic metals into the environment. Concern over the long-range transport of air pollutants resulted in the inclusion of provisions in the 1977 amendments to the Clean Air Act directing the Environmental Protection Agency (EPA) to propose regulations governing stack heights. The tall stacks, however, still remain, as evidence mounts that the burning of fossil fuels is producing acid rain conditions in parts of the world far distant from the coal-burning facilities.[32] Thus, as in the case of water-carriage systems and water pollution, a technology that had helped deal with a local problem had transferred the adverse effects to a different locality and, in this case, to a different medium.

The Land: Industrial and Municipal Waste Disposal

Of the three media utilized over time for waste disposal, the least attention has been paid—by the public, the government at all levels, and the researchers—to the land. The prime reasons for this lack of notice are that land-deposited wastes neither created the health hazards of water pollution nor had the visibility of smoke pollution. Dumps for garbage and refuse and for industrial wastes of different kinds had always existed on city fringes or vacant lots, but they were largely viewed as a nuisance and an eyesore rather than as a health hazard. Rats, flies, and odors associated with garbage dumps were obviously disagreeable, but they generated neither epidemics nor smoke blankets that blotted out the sun. In addition, until recently land utilized as a waste depository did not appear to possess the transport qualities or cross-media pollution capabilities of air or water.

While land has historically been regarded as an acceptable sink, it has also been utilized more intensively in recent times because surface waters and the air were no longer acceptable sinks for the disposal of some wastes. The 1979 Report of the Council on Environmental Quality noted, for instance, that "the increasing tempo of the cleanup of lakes and streams is literally driving pollution underground." That this would occur, however, could have been predicted from past experience. In the 1940s, for instance, when the Pennsylvania Sanitary Water Board began enforcing the Clean Streams Act, numbers of small industrial plants turned to the use of earthen lagoons on plant property as a means of avoiding controls. These lagoons, many of which were poorly constructed and unlined, ultimately threatened groundwater supplies, created nuisances, and even posed air pollution problems.[33]

Deep-well injection, another method of industrial waste disposal that uses the land, expanded because of regulations restricting disposal in surface waters. The chemical and petroleum industries, facing disposal problems, developed this technique in the 1930s, and its use expanded in the postwar years. In 1960 there were about thirty deep wells throughout the country; within the decade, because of enforcement of water and air pollution statutes by the various levels of government, the number increased to 110. The concept behind a properly designed deep well was that it would take the "effluent out of the human environment and bury it forever," but many firms located the wells in strata that posed a threat to underground aquifers and drinking-water supplies.[34]

Industries have, over the years, punched or dug thousands of holes in the ground, usually on their own property, to dispose of wastes. More closely related to the everyday life of the cities, however, has been the disposal of municipal solid wastes. Solid wastes are defined as refuse of different kinds, such as packaging and food wastes. Up until World War II, ashes made up a large component of the solid wastes of urban dwellers. In 1910, for instance, about three-quarters of a pound of garbage and rubbish and about 4 pounds of ashes were collected from each New York urban dweller. By 1960, with ashes now a negligible ingredient, the average urban dweller generated a little over 2.5 pounds of solid waste a day; by 1979 the figure was up to 3.8 pounds. The largest percentage of these totals was still nonfood materials, such as packaging and glass.[35]

Before World War II, cities disposed of solid wastes using one or a combination of the following methods: open dumps on the city's fringe; pig farms, where garbage was fed to hogs; ocean dumping (coastal cities); incineration; garbage reduction; and composting. The first four methods were used most frequently, with open dumping and pig farms the most common. A fair amount of recycling of urban wastes actually occurred in the late nineteenth and early twentieth centuries. Pig farms provided a way to recycle garbage into pork, composted or milled garbage served as fertilizer, reduction plants provided usable fats, refuse was burned to produce electricity, ashes were used to fill in swamps and low-lying areas, and wastes were sorted to reclaim usable metals, glass, and rags.[36]

All of the above methods, however, whether involving recycling or not, became increasingly unpopular in the interwar years and especially after World War II. Recycling methods disappeared because they required costly source separation or, as in the case of hog farming, transmitted disease (trichinosis) or were wiped out by epidemics (vesicular exantema). Other garbage disposal techniques, such as reduction and incineration, were not only costly but also produced such nuisances as odors and smoke. Incineration plants, for instance, often violated local smoke-control ordinances. Ocean dumping of garbage was banned by the Supreme Court in 1934. Finally, open dumps on the city fringe produced nuisances, fires, and public health hazards that became more noticeable and objectionable as suburbs expanded in the postwar years.[37]

Increasingly, as existing methods of urban solid-waste disposal developed costs and problems, sanitary engineers and public health officials advocated a technique of waste disposal known as sanitary landfill. Sanitary landfill involved the filling of depressions or trenches in the ground with refuse utilizing a technology such as a bulldozer or a bull clam shovel to dig the trench and compact the fill. Each day's deposit was sealed into an individual refuse cell. When the fill reached the desired level, it was covered with earth and again compacted.[38] When the fill was completed, the created land was often used for recreational or even building purposes.

Sanitary landfill bore a resemblance to the technique of garbage burial that had been used by cities such as New York, Seattle, and Boston in the nineteenth century often to fill in areas around their waterfronts. During the first decade of the twentieth century, Champaign, Illinois, and Columbus, Ohio, both compacted their refuse. Great Britain was the pioneer in the use of sanitary landfill, calling the technique "controlled tipping." The British developed the method in World War I, and by 1935 44.5 percent of all English refuse was being deposited in tips. San Francisco constructed a sanitary landfill in 1926 as a replacement for incineration, using the refuse to fill in wetlands along the bay. Two of the most important early sanitary landfill experiments occurred in Fresno, California and New York City in the late 1930s. (It was in Fresno that the Public Works Director, Jean Vincenz, supposedly coined the term "sanitary landfill.") New York City built sanitary fills in Long Island and advocated the technique as both a low-cost method of garbage disposal and a way of reclaiming swampy land.[39]

Public works officials, public health professionals, and municipal engineers responded to sanitary landfill technology as they had greeted other technologies (such as water carriage) touted as providing solutions to urban waste-disposal problems. They were convinced by the technology's advocates that it offered tremendous advantages over the previous techniques, and focused on the benefits while overlooking or down-playing the possible hazards. In the case of sanitary landfill, risk perception was also conditioned by the certification by a panel of "experts" of the technology's acceptability on health grounds.

In 1938, discovering that its incinerators were both expensive and produced nuisances, the New York City Sanitation Department announced plans to build a sanitary landfill on the shores of Jamaica Bay. A group of Queens citizens who lived near the bay, however, protested that the landfill would produce odors, vermin, and gas and cause real estate values to fall. The Borough of Queens sued to halt the building of the landfill, and two New York City commissioners were arrested at the proposed landfill site. Eventually, in an early example of the use of the science court concept, the courts persuaded the citizens and the city to agree to allow a board of public health physicians and sanitary engineers to examine the sanitary landfill technique to determine whether it presented a public health hazard. The committee was headed by Dr. Thomas Parran, surgeon general of the United States.

The Parran commission found the sanitary landfill technique to be free of dangers to public health or safety. Sanitary landfill, it said, was a large health improvement over the open dump, eliminated undesirable marsh and swamp land (today called "wetlands") that harbored rats and mosquitoes, and provided a benefit in terms of filled-in ground. The commission considered several possible landfill hazards, such as fires and low weight-bearing value, but maintained that proper precautions could control them. It did not, however, mention other possible dangers from leachate runoff or groundwater pollution or the possibility of long-term health hazards. Most of the commission's discussion of risk was in terms of nuisances rather than health dangers. The report enumerated fifteen mostly operational rules for the safe conduct of sanitary landfill operations.[40]

New York City officials greeted the Parran report with enthusiasm. The New York commissioner and deputy commissioner of health, for instance, published an article in the *American Journal of Public Health* extolling sanitary landfill as "a program of disposal of rubbish and garbage under sanitary, scientific control, that would be truly economical—cheaper than incineration." The Parran report had vindicated the importance of public health considerations in garbage disposal, said the commissioners. New York City proceeded to incorporate its recommendations for landfill operations directly into the municipal sanitary code.[41]

The Parran report, combined with a favorable experience by the army with sanitary landfills at its camps in the U.S. during World War II (Fresno's Jean Vincenz directed the army operation), gave the technique wide appeal in the postwar period. Public works officials and public health professionals strongly endorsed it as a method of waste disposal and as a preferred replacement for the open dump. Between 1945 and 1960, according to one survey, the number of fills increased from 100 to 1,400, and articles boasting the virtues of the sanitary fill appeared in practitioner journals such as the *American City and Public Works.* The advantages most commonly cited were those listed by the Parran commission: the elimination of the nuisances and health hazards associated with open dumps; the filling in of marshes and swamps and elimination of rats and mosquitoes; and the creation of land for buildings, parks, and recreational areas. In the 1950s, however, a few articles also appeared that noted hazards at operating landfills, such as methane fires, groundwater pollution, and low-weight-bearing capabilities that restricted building. Occasionally these pieces appeared simultaneously with articles describing new landfill operations that ignored the dangers described in the same issue [42]

In 1961 the Sanitary Engineering Division of the American Society of

Civil Engineers (ASCE) published a survey of sanitary landfill practice that examined 250 sites. Of the fills surveyed, 12 percent were less than 250 feet from the nearest dwelling. Completed landfills were most commonly used for recreational and industrial purposes, although some fills were used for homesites and schools. The article noted that while landfill groundwater pollution was a "critical item, in general site planning there has been minimum concern with ground water pollution," with 79 percent of the sample within 20 feet of groundwater and 27 percent at or near groundwater. Only 9.3 percent of the sites reported that operators tested ground boring before fill operations, and only 14 percent had specially engineered drainage devices. The survey also reported that in spite of the purported safety of landfills, citizens often opposed having them located near their residences.[43]

The 1961 survey also noted that over 70 percent of the landfills examined operated under some sort of city or county regulations. During the 1950s, as landfills became more common, cities issued sanitary landfill regulations, states such as California and Illinois suggested operational guidelines, and professional groups, especially the American Public Works Society, the Sanitary Engineering Division of the American Society of Civil Engineers, and the U.S. Public Health Service, conducted investigations on standards to avoid undue risk. By the time of the ASCE survey, professional groups involved in solid-waste questions agreed that, while sanitary landfills reduced disposal costs and were superior to the open dump, they still presented dangers with regard to leachate seepage, groundwater pollution, poor load bearing, methane production, and nuisances such as rats, vermin, and blowing paper. A lack of research on these hazards, however, restricted the availability of technical information on these hazards that could be used to refine practice.[44]

In 1963, in an attempt to generate interest and research in the solid-waste area, the U.S. Public Health Service and the American Public Works Association sponsored the first National Conference on Solid Waste Research. In his keynote address Professor J. E. McKee of the California Institute of Technology commented on the lack of research and offered four explanations for its absence. First, McKee noted, there was no demand for such information from cities, regulatory agencies, or the public. Second, there had been no public health crises involving solid wastes equivalent to those in air and water pollution that would have generated such a demand. Third, there was minimal federal and state involvement. And, fourth, the majority of those concerned with solid-waste disposal considered it an economic and political rather than a scientific or engineering problem.[45] All of these factors applied to the sanitary landfill technique as well as to solid wastes in general. Landfills, however, did not hold an especially prominent place at the conference, and neither the papers on sanitary landfill nor the discussions following their delivery emphasized possible hazards. Concern for the land as a sink for pollutants did not yet have the urgency associated with it that was beginning to characterize the air and the water media in the 1960s.

Conferences such as that sponsored in 1963 by the Public Health Service and the American Public Works Association, however, did highlight the deficiencies in solid-waste research. In a sense they created their own demand for legislation. In addition, solid-waste collection and disposal was growing more expensive, and powerful local politicians, such as Chicago's mayor Richard Daley, pushed for a federal role to lighten the burden on cities. In 1965, after President Lyndon B. Johnson had spoken of the need for "better solutions to the disposal of solid waste" and called for federal legislation, Congress passed the Solid Waste Disposal Act. This act created the Office of Solid Wastes and provided the federal government with a more formal role in regard to municipal wastes.[46]

The Solid Waste Disposal Act provided funds for research, investigations, and demonstrations in the area of solid waste and for technical and financial assistance to state and local governments and interstate agencies in "the planning, development and conduct" of disposal programs. The most important impacts of the program were on stimulating research and inspiring state government activity in the solid-waste area. In 1965, for instance, there was no state-level solid-waste agency in the country, but by 1970 forty-four states had developed programs. During the 1970s, however, the focus of federal legislation turned away from research into conventional methods of solid-waste disposal and instead focused on the reuse and recycling of resources. This was reflected in the passage of the Resource Conservation Act of 1970 in the form of amendments to the 1965 legislation.[47]

Section 212 of the 1970 Solid Waste Act required that the EPA undertake a comprehensive investigation of the storage and disposal of hazardous wastes. This led to a report to Congress in 1974 on the disposal of hazardous wastes and eventually, in 1976, in the passage of the Resource Conservation and Recovery Act (RCRA). The act attempted to fill the regulatory gaps in the disposal of hazardous wastes left by the states, and early in 1980, acting under the requirements of RCRA, the EPA announced new regulations implementing cradle-to-grave controls for handling hazardous wastes.[48] The use of the land as a sink was now to be severely curtailed.

The various acts passed from 1965 on, and investigations conducted under their authority, caused a convergence of the different streams of research concerning municipal wastes on the one hand and industrial hazardous wastes on the other. The point of convergence was landfill-type operations, with special concern over site construction and hazards and groundwater pollution. The exact number of active landfills in existence today is uncertain, although surveys estimate approximately 75,000 industrial and 16,000 municipal sanitary landfills, with no sound information on the number that have been abandoned or closed. Many of these landfills, both active and inactive and municipal and industrial, contain potentially hazardous wastes.[49]

These landfills pose a serious threat to groundwater supplies. Groundwater furnishes drinking water for about half of all U.S. residents and constitutes about 25 percent of all freshwater used in the country. The degree of threat to groundwater depends on the material underlying the surface site and existing hydrological and geologic conditions. Groundwater moves very slowly, and it may take decades for a source polluted in one location to contaminate a water supply elsewhere, but the transport possibilities often exist.[50] As in the cases of surface waters and the air, the use of the land as a sink has transferred pollution problems to different locales and to a different medium.

A widespread awareness of the danger posed by landfill leachate to groundwater purity is very recent, although isolated warnings about this hazard occurred as far back as the 1930s, if not before. In the postwar years several state departments of health issued statements about potential chemical pollution of groundwater from sanitary fills; but in the general enthusiasm for the approach, the warnings tended to be ignored.[51] When landfill research accelerated during the 1960s, so did warnings about possible groundwater pollution. By 1970 many states had regulations requiring field investigations of groundwater location in the siting of new municipal and industrial landfills. Problems, however, usually centered around older sites that had been developed without adequate investigations of the risk of possible groundwater contamination.

There are several different reasons why the potential of groundwater contamination from landfills was ignored. One is the lack of research in the area of solid-waste disposal in general and sanitary landfills in particular, and limited knowledge about underground processes. A 1972 article on landfill leachate, for instance, noted that "before 1965 very few people were aware of the fact that water passing through refuse in a landfill would become highly contaminated . . . few cases were noted where leachate had caused harm to someone." In addition, there was a lack of analytical instrumentation necessary to trace certain contaminants from landfills or to detect extremely low levels of potentially hazardous substances. Before 1965 (the Solid Waste Act) there was no incentive system to spur research in either analytical chemistry in regard to groundwater processes or groundwater-leachate-soil exchanges.[52]

Another important factor is the absence of a clear hazard or crisis in regard to groundwater pollution from solid waste. As one sanitary engineer noted in 1968, a "major obstacle to the solution of solid waste problems is the lack of awareness on the part of governmental decision-makers that the problem even exists. This lack of awareness exists at all levels." Up to 1970 few incidents of the pollution of groundwater drinking supplies by landfill leachate had been reported, and municipalities ignored the potential problem. Rather than spend money on expensive testing and monitoring, municipalities put their dollars in areas where need appeared more immediate.[53] But the caveats about the limitations of technical knowledge and crises notwithstanding, some municipal governments and their consulting sanitary engineers in the postwar decades carried on landfill siting and operations with a disregard for the state-of-the-art techniques and the knowledge available at the time.

In the late 1970s public agencies and environmental groups began to direct their attention toward the pollution of drinking-water supplies drawn from groundwater as new analytic techniques revealed many incidents of groundwater contamination from toxic organic chemicals. By 1980 every state had one or more laws pertaining to groundwater pollution, and both the Environmental Protection Agency and many state agencies had projects underway to identify especially hazardous situations. In 1980 and 1981, the EPA moved to suggest guidelines for preserving groundwater purity under the authority given it by the various environmental control acts passed during the 1970s, such as the Clean Water Act (1972), the Safe Drinking Water Act (1974), and the Toxic Substances Control Act (1976). The proposed new regulations for groundwater, when taken in combination with the restrictions on disposal of hazardous wastes in landfills imposed in the Resource Conservation and Recovery Act (1976), meant that another sink—the land—would in all likelihood be severely restricted as a disposal site for society's wastes.[54]

Conclusion

As society's use of the air, land, and water as sinks for wastes illustrates, we have used technology to improve the local environment and to provide for growth without fully considering the possible problems created by effluents for downstream or downwind cities. Thus, such technologies as sewers, tall stacks, and sanitary landfills helped reduce pollution problems in one city only to transfer those same problems to another locale. In addition, policies developed to deal with pollution in one medium often resulted in the transfer of contaminants to another, less regulated medium. To deal with the different negative effects, we have usually utilized other technologies, thereby involving the society in loops of retrofits and technological fixes.

Cities, as well as the state and federal governments, have responded to environmental problems with policies utilizing a rough benefit-cost calculus based on health and nuisance considerations. Thus, the environmental agenda featured action against dirty water and human waste pollution first, dirty air second, and land and groundwater pollution last. Whatever action taken was conditioned by the costs involved, contemporary social values, and existing levels of analytical instrumentation. Thus, for example, in the early twentieth century Pittsburgh could have reduced its smoke problem by building (as Andrew Carnegie suggested) a huge coal gasification plant; or, as some public health authorities maintained, water quality could have been improved by forcing cities to construct sewage-treatment plants. The costs, however, would have been large compared to benefits, and there was no incentive to generate such actions. From the perspective of the time, it was cheaper to burn smoky bituminous coal or to dispose of raw sewage in streams than to invest in control technologies. In the economists' terminology, air and water were regarded as free goods available to absorb the externalities of the industrial city.[55]

Problem transfer, or the ignoring of ill effects, was not always a willful act on the part of the producer of the effluents. The environmental hypothesis "running water purifies itself" gave sanction to cities that wanted to dispose of their sewage in nearby streams, while chemical analysis seemingly provided a "scientific" stamp of approval. Ringelmann charts supplied a method of grading smoke but did not identify other insidious but largely invisible air pollutants. Sanitary and industrial landfills produced leachates that contaminated groundwater used for drinking-water supplies, but monitoring and detection capabilities were limited. Research in these areas often only developed after the occurrence of crisis situations and as a result of specific public policies, not before. But even after research had pinpointed the mechanisms by which capital technologies such as sewers or tall stacks or landfills produced negative effects, it was difficult to persuade the operators of these technologies, be they private or public, to stop using the polluting technology or to stop building new systems having the same results.[56]

A new era of environmental consciousness generated by changing values has resulted in a series of laws that seek to close off the traditional sinks for pollutants and to force cities and industries to think in terms of recycling and conservation. In an historical reversal society has begun focusing on the costs of new technologies rather than only on the benefits. Public attention is now occupied with the risks and hazards associated with technology rather than its potential for progress. This concern with technological risk, however, should not obscure the important role that technology must play if we are to cope with our environmental problems.

In the past technologies helped us achieve short-term objectives in coping with environmental difficulties, but the problems were often only displaced or delayed in their effects. The question facing our society today is how to protect the environment most effectively on a long-term basis given a range of uncertainties. We need mechanisms and institutions for environmental priorities that will prevent us from using new sinks for effluents that may offer temporary solutions but in the end create long-term and disastrous hazards.[57]

Notes

Grants from the National Science Foundation and from the Andrew W. Mellon Foundation supported various sections of the research from which this chapter was drawn. As author, however, I am solely responsible for the opinions expressed. I would like to thank my colleagues Steven Klepper, Francis C. McMichael, and M. Granger Morgan for their useful comments.

1. Abel Wolman, "The Metabolism of Cities," in *Cities: A Scientific American Book* (New York: Knopf, 1968), 156–74. See also Brian J. L. Berry et al., *Land Use, Urban Form and Environmental Quality,* Department of Geography Research Paper No. 155 (Chicago: University of Chicago, 1974).

2. Joel A. Tarr and Francis C. McMichael, "The Evolution of Wastewater Technology and the Development of State Regulation: A Retrospective Assessment," in Joel A. Tarr, ed, *Retrospective Technology Assessment* (San Francisco: San Francisco Press, 1977), 165–90; B. A. Segur, "Privy-Vaults and Cesspools," *Papers and Reports of the American Public Health Association* 3 (1876): 185–87; Mansfield Merriman, *Elements of Sanitary Engineering* (New York: Wiley and Sons, 1906), 139–42; Joel A. Tarr, "From City to Farm: Urban Wastes and the American Farmer," *Agricultural History* 49 (October 1975): 598–612; and Jon A. Peterson, "The Impact of Sanitary Reform upon American Urban Planning, 1840–1890," *Journal of Social History* 13 (Fall 1979): 83–101.

3. Nelson M. Blake, *Water for the Cities* (Syracuse: University of Syracuse Press, 1956), 3–27; John Duffy, *A History of Public Health in New York City, 1625–1866* (New York: Russell Sage Foundation, 1968). There is a good description of the New York situation in Eugene P. Moehring, "Public Works and the Patterns of Urban Real Estate Growth in Manhattan, 1835–1894" (Ph.D. diss., City University of New York, 1976), 42–85.

4. "Ownership of American Water Works," *Engineering News* 27 (January, 1892): 83–86.

5. Tarr and McMichael, "The Evolution of Wastewater Technology," 170.

6. Ibid., 170–73; Moehring, "Public Works and the Patterns of Urban Real Estate Growth in Manhatten," 139.

7. Benjamin Lee, "The Cart before the Horse," *Papers and Reports of the American Public Health Association* 20 (1895): 34–36.

8. Francis Sheppard, *London 1808–1870: The Infernal Wen* (Berkeley: University of California Press, 1971), 253–358; George M. Fredrickson, "The Sanitary Elite: The Organized Response to Suffering," *The Inner Civil War* (New York: Harper and Row, 1965), 104–5; Richard L. Schoenwald, "Training Urban Man: A Hypothesis about the Sanitary Movement," in H. J. Dyos and Michael Wolff, eds., *The Victorian City: Images and Realities,* 2 vols. (London: Routledge and Kegan Paul, 1973), 2:675–82; and C. V. Chapin, "The End of the Filth Theory of Disease," *Popular Science Monthly* 60 (1902): 234–39.

9. Howard D. Kramer, "Agitation for Public Health Reform in the 1870's," pts. 1–11 *Journal of the History of Medicine* 3 and 4 (Autumn 1948, Winter 1949): 388–473, 75–89; Joel A. Tarr et al., "Values and the Technology," *Retrospective Assessment of Wastewater Technology in the United*

States 1800–1972, A Report to the National Science Foundation (Washington, D.C.: National Technical Informations Services, 1978, PB 275 272/3/s), chap. 6.

10. Tarr and McMichael, "Evolution of Wastewater Technology," 171–73.

11. Sheppard, *London 1808–1870,* 253–58.

12. For a discussion of these debates, see Tarr and McMichael, "Evolution of Wastewater Technology," 173–79; and Joel A. Tarr, "The Separate vs. Combined Sewer Problem: A Case Study in Urban Technology Design Choice," *Journal of Urban History* 5 (May 1979): 308–39. William T. Sedgwick, *Principles of Sanitary Science and the Public Health* (New York: Macmillan, 1918), 231–37; and Jay Slater, "The Self-Purification of Rivers and Streams," *Synthesis* 2 (Autumn 1974): 41–45.

13. Tarr, "Separate vs. Combined Sewer Problem," 330–32. For a comparison between sewer construction and typhoid fever mortality rates in fifteen cities in the 1888–1915 period, see Tarr et al., *Retrospective Assessment of Wastewater Technology,* 1–26a.

14. Allen Hazen, *Clean Water and How to Get It* (New York: J. Wiley and Sons, 1907), 8–75; George C. Whipple, "Fifty Years of Water Purification," in Mazyck P. Ravenal, ed., *A Half-Century of Public Health* (New York: American Public Health Association, 1921), 161–80. The policy issue of deciding between water filtration and sewage treatment is dealt with in Joel A. Tarr et al., "Disputes over Water Quality Policy: Professional Cultures in Conflict, 1900–1917," *American Journal of Public Health* 70 (April 1980): 427–35; Tarr et al., "The Development of a Federal Role in Water Pollution Control," *Retrospective Assessment of Wastewater Technology,* chap. 7.

15. U.S. Council on Environmental Quality (CEQ), *Environmental Quality—1980: The Eleventh Annual Report of the Council on Environmental Quality* (Washington, D.C.: Government Printing Office, 1980), 81–138.

16. W. C. White, "What Is Known about the Effect of Smoke on Health," *Mechanical Engineering* 49 (June 1927): 655–57; Dr. I. H. Alexander, "Smoke and Health," 35th Annual Convention, Smoke Prevention Association of America (1941), 66–75.

17. Carlos Flick, "The Movement for Smoke Abatement in Nineteenth-Century Britain," *Technology and Culture* 21 (January 1980): 29–50. See *District Heating in American Housing, National Building Studies* (London: His Majesty's Stationary Office, 1949).

18. Joel A. Tarr and Bill C. Lamperes, "Changing Fuel Use Behavior and Energy Transitions: The Pittsburgh Smoke Control Movement, 1940–1950," *Journal of Social History* 14 (Summer 1981): 563.

19. Samuel B. Flagg, *City Smoke Ordinances and Smoke Abatement,* Bureau of Mines, Bulletin 49 (Washington, D.C.: Governmental Printing Office, 1912); Victor J. Azbe, "Rationalizing Smoke Abatement," *Proceedings of the Third International Conference on Bituminous Coal* 2 (Pittsburgh, 1931): 593–645.

20. Azbe, "Rationalizing Smoke Abatement" 2:603. See the papers on smoke abatement presented at the fuels division session of the 1926 American Society of Mechanical Engineers annual convention and printed in *Mechanical Engineering* 48 (mid-November 1926): 1193.

21. "Cities Fight Smoke," *Business Week,* April 6, 1940, 33–34; Oscar H. Allison, "Raymond R. Tucker: The Smoke Elimination Years, 1934–1950" (Ph.D.diss., St. Louis University, 1978).

22. See Tarr and Lamperes, "Changing Fuel Use Behavior and Energy Transitions," 563–70. One of the keys to the ordinance's passage was that the campaign managers were able to convince most Pittsburghers that the benefits of smoke control would outweigh the costs and that the policy would have minimal distributional effects. Two groups whose combined opposition might have blocked or delayed the law's passage—organized labor and the coal industry—were coopted by involving them in the decision-making process and by the argument that smokeless fuel could be produced from local bituminous coal, thereby expanding coal production and jobs for miners.

23. Ibid., 570–75.

24. Ibid., 575.

25. Ibid., 576.

26. For energy consumption see Sam H. Schurr and Bruce C. Netschert, *Energy in the American Economy, 1850–1875: An Economic Study of Its History and Prospects* (Baltimore: Johns Hopkins University Press, 1960), 36. For the use of natural gas for househeating, see American Gas Association (AGA), *Gas Facts 1958* (New York: AGA, 1958), 140; *Historical Statistics of the Gas Industry* (New York: AGA, 1956), 239. U.S. Bureau of the Census, *Historical Statistics of the United States, Colonial Times to 1970* (Washington, D.C.: Government Printing Office, 1975), pt. 2:728–29; and National Coal Association (NCA), *Bituminous Coal Facts 1962* (Washington: NCA, 1963), 61.

27. W. C. L. Hemeon, "Review of Latest Development in Outdoor Air Pollution," American Industrial Hygiene Foundation, Bulletin No. 22 (1952): 60–64; Harry C. Ballman and Thomas J. Fitzmorris, "Local Air Pollution Control Programs—a Survey and Analysis," *Journal of the Air Pollution Control Association* 13 (November 1963): 386–491; Samuel M. Rogers, "A Review and Appraisal of Air Pollution Legislation in the United States," ibid. 7 (February 1957): 308–15; W. C. L. Hemeon, "Current Conceptions on Air Pollution," American Iron and Steel Institute, 1954, 10; and Arthur C. Stern, ed., *Air Pollution,* 2 vols. (New York: Academic Press, 1962).

28. See, for instance, Arthur C. Stern and Leonard Greenburg, "Air Pollution—the Status Today," *American Journal of Public Health* 41 (January 1951): 27–37. The authors noted, "As we learn more and more about micrometeorology, it serves to emphasize the importance of high stacks as a primary means for achieving ground level dilution of contaminants to within acceptable safe limits" (31). Air Pollution Control Association (APCA) *The Electric Utility Industry,* Technical Manual No. 4 (Pittsburgh: APCA, 1968), 13.

29. National Coal Association, *Bituminous Coal Facts 1966* (Washington: NCA, 1966), 85.

30. American Association for the Advancement of Science (AAAS), *Air Conservation* (Washington: AAAS, 1965), 37, 67. "There is a time and place for everything. The time and place to pollute air is downwind" (294).

31. CEQ, *Environmental Quality—1980,* 173–77.

32. Ibid.; CEQ, *Environmental Quality—1979* (Washington, D.C.: Government Printing Office, 1980), 70–71; U.S. Environmental Protection Agency (EPA), *Acid Rain* (Washington, D.C.: EPA, July 1980); National Research Council, *Atmosphere-Biosphere Interactions: Toward a Better Understanding of the Ecological Consequences of Fossil Fuel Combustion* (Washington, D.C.: National Academy Press, 1981); CEQ, *Environmental Quality—1980,* 175–77; Richard A. Kerr, "Pollution of the Arctic Atmosphere Confirmed," *Science* 212 (May 1981): 1013–14; and Bette Hileman, "Acid Precipitation," *Environmental Science and Technology* 15 (October 1981): 1119–24.

33. CEQ, *Environmental Quality—1979,* 110; Donald A. Lazarchik, "Pennsylvania's Pollution Incident Prevention Program," *Proceedings of the 25th Annual Purdue Industrial Waste Conference (1970),* 528. See also Sheppard T. Powell, "Industrial Wastes," *Industrial and Engineering Chemistry* 46 (September 1954): 95A–96A, 98A.

34. "Speed-up for Deep-Well Waste Disposal," *Chemical Engineering* 75 (January 1968): 88–90; William R. Walker and Ronald C. Stewart, "Deep-Well Disposal of Wastes," *Journal of the Sanitary Engineering Division, Proceedings of the American Society of Civil Engineers* 94 (October 1968): 945–1068.

35. Martin V. Melosi, *Garbage in the Cities: Refuse, Reform, and the Environment, 1880–1980* (College Station: Texas A&M University Press, 1981), 192–94. Melosi's volume is the best survey of methods of refuse disposal before World War II.

36. Ibid. See also American Public Works Association (APWA), *Municipal Refuse Disposal* (Chicago: Public Administration Service, 1960).

37. Melosi, *Garbage in the Cities,* 189–231.

38. APWA, *Municipal Refuse Disposal,* 2d ed. (Chicago: Public Administration Service, 1966), chap. 4.

39. Rudolph Hering and Samuel A. Greeley, *Collection and Disposal of Municipal Refuse* (New York: McGraw Hill, 1921), 243–57; A. L. Thomson, "What European Cities Are Doing in Handling Refuse," *Municipal Sanitation* (December 1935): 365–68; "Sanitary-Fill Refuse Disposal at San Francisco," *Engineering News-Record (ENR),* February 27, 1936, 314–16; APWA, *Municipal Refuse*

Disposal, 89–90; and E. J. Cleary, "Land Fills for Refuse Disposal"; *ENR,* September 1, 1938, 270–71. But also see Harrison P. Eddy, Jr., "Caution Regarding Landfill Disposal," ibid., December 15, 1938, 766–67.

40. "Health Experts Endorse Landfills and Recommend Best Practice," *ENR,* March 28, 1940, 54–55; "Board Reports on New York Landfills," *American City (AC),* 55 (July 1940): 11; and Rolf Eliassen and Albert J. Lizee, "Sanitary Land Fills in New York City," *Civil Engineering* 12 (September 1942): 483–86.

41. John L. Rice and Sol Pincus, "Health Aspects of Land-Fills," *American Journal of Public Health* 30 (December 1940): 1991–98; Richard Fenton, "Landfills and Inspection," Bureau of Sanitary Engineering, Department of Health, New York City, May 1947.

42. Rolf Eliassen, "War Conditions Favor Landfill Refuse Disposal," *ENR,* June 4, 1942, 912–14; "Where and How to Use the Sanitary Fill: Based on Experience Gained from Army and U.S. Public Health Sources," *AC* (January 1944): 47–56. Booster articles for the sanitary landfill can be found in the following issues of *AC*: 62 (December 1947); 63 (May 1948); 63 (September 1948); 64 (March 1949); 64 (December 1949); 66 (October 1951); 67 (December 1952); 68 (October 1953); 68 (November 1953); 71 (May 1956); 71 (June 1956); 76 (May 1961); 77 (June 1962). For *AC* articles containing warnings, see 66 (February 1951); 66 (May 1951); 71 (April 1956).

43. Committee on Sanitary Engineering Research, American Society of Civil Engineers, "A Survey of Sanitary Landfill Practices," *Journal of the Sanitary Engineering Division, Proceedings of the ASCE* 87 (July 1961): 65–83.

44. "Sanitary Fill Standards," *AC* 66 (February 1951): 104–5; William S. Foster, "The Elements of Refuse Collection and Disposal," *AC* 66 (May 1951): 125; Wilbur C. Webb, "Limitations in the Use of Sanitary Landfill as a Method of Solid Trash Disposal," *Ninth Annual Purdue Conference on Industrial Wastes (1954),* 138-46; Robert C. Merz, et al., *Leaching of a Sanitary Landfill,* California State Water Pollution Control Board Publication No. 10 (1954); Committee on Sanitary Landfill Practice of the Sanitary Engineering Division, "Sanitary Landfill," *American Society of Civil Engineers, Manual of Engineering Practice,* No. 30 (New York, ASCE, 1959); U.S. Public Health Service, *Recommended Standards for Sanitary Landfill Operations* (Washington, D.C.: GPO, 1961); and APWA, *Solid Wastes Research Needs* (Chicago: APWA, 1962).

45. APWA, *National Conference on Solid Waste Research* (Chicago: APWA, 1963), 1–7.

46. "Solid Waste Disposal Gets Federal Effort," *Chemical and Engineering News,* December 12, 1966, 50–54; Melosi, *Garbage in the Cities,* 199–200.

47. Melosi, *Garbage in the Cities,* 201–3.

48. CEQ, *Environmental Quality—1979,* 174–88; *Environmental Quality—1980,* 214–22. See *Federal Register,* 45, June 30, 1980 (Solid Waste), no. 127, p. 44, 126, and 46, April 27, 1981 (Hazardous Wastes, RCRA), no. 80, 23, 722.

49. CEQ, *Environmental Quality—1980,* 88–89. In 1979 the CEQ reported an EPA estimate of 32,000–50,000 disposal sites containing hazardous wastes. See *Environmental Quality—1979,* 1974. Estimates of the number of active and inactive dumps vary widely.

50. CEQ, *Environmental Quality—1980,* 83–85.

51. See, for instance, "Sanitary-fill Refuse Disposal at San Francisco," *ENR,* February 27, 1936, 314–17; Eddy, "Cautions Regarding Land-Fill Disposal," 766–67. See also David Keith Todd and Daniel E. Orren McNulty, *Polluted Groundwater: A Review of the Significant Literature* (Huntington, N.Y.: Water Information Center, 1976); G. W. McDermott, *Pollution Characteristics of Landfill Drainage,* Activity Report no. 3, R. A. Taft Sanitary Engineering Center, U.S. Public Health Services, January-March, 1950; C. W. Klassen, "Sanitary Fill Standards," *AC* 66 (February 1951): 104–5; Mertz et al., *Leaching of a Sanitary Landfill,* 123–27; and California State Water Pollution Control Board, *Effects of Refuse Dumps on Ground Water Quality, Publication 24,* 1961.

52. W. C. Boyle and R. K. Ham, "Treatability of Leachate from Sanitary Landfills," *Proceedings of the 27th Annual Purdue Conference on Industrial Wastes (1972),* 687–91; Harvey Brooks, "Science Indicators and Science Priorities," *Science, Technology & Human Values,* 38 (Winter 1982): 17.

53. Harvey F. Ludwig and Ralph J. Black, "Report on the Solid Waste Problem," *Journal of the*

Sanitary Engineering Division, Proceedings of the American Society of Civil Engineers (April 1968): 368; "Sanitary Landfills: The Latest Thinking," *Civil Engineering* (March 1973), 69.

54. CEQ, *Environmental Quality—1980*, 81–92; EPA, *Proposed Ground Water Protection Strategy* and *Groundwater Protection* (Washington, D.C.: EPA, 1980).

55. See, for instance, Allen V. Kneese and Charles L. Schultze, *Pollution, Prices, and Public Policy* (Washington: Brookings Institution, 1975).

56. In the 1920s many new sewer systems were constructed, for instance, that discharged untreated sewage into waterways utilized by downstream cities as sources of water supply. The same sort of adherence to a technology, even after serious questions have been raised about its production of negative externalities, appears to be present in regard to tall stacks and acid rain, See R. Jeffrey Smith, "Administration Views on Acid Rain Assailed," *Science* 214 (October 1981): 38.

57. See Brooks, "Science Indicators and Science Priorities," 17, 28–39; Robert C. Harris, Christopher Hohenemser, and Robert W. Kates, "Our Hazardous Environment," *Environment* 20 (September 1978): 38–39; and Aaron Wildavsky, "No Risk Is the Highest Risk of All," *American Scientist* 67 (November 1979): 32–37. See William C. Clark, "Witches, Floods, and Wonder Drugs: Historical Perspectives on Risk Management," in Richard C. Schwing and Walter A. Albers, Jr., *Societal Risk Assessment: How Safe Is Safe Enough?* (New York: Plenum, 1981), 287–318, for perceptive analysis and suggestions.

CHAPTER II

Land Use and Environmental Change in the Hudson-Raritan Estuary Region, 1700–1980

Human beings have shaped and changed natural areas for thousands of years in North America. Some of the longest-term changes have occurred along the East Coast of the continent, the scene of the most extensive early European settlement. This essay will examine approximately three centuries of change in the Hudson-Raritan estuary and its associated river basins from approximately 1700 through 1980 and explore the pollution history of the region and patterns in the major effluent flows over time.[1] The Hudson-Raritan estuary region is one of the most heavily urbanized and industrialized areas on the coast, and it consists of 12,487 square miles, including twenty-one New York and ten New Jersey counties.

Three major rivers flow into the estuary—the Hudson, the Raritan, and the Passaic. The lower Hudson, which will be the primary focus of this article, stretches 154 miles from Manhattan to Troy. The Federal Dam at Troy, located about two miles downstream of the confluence of the Mohawk with the Hudson, marks the end of tidal influence. At its lower end the river discharges into the Upper New York Bay and subsequently through the Verrazano Narrows into Lower New York Bay, an arm of the Atlantic Ocean, draining an area of 4,940 square miles.

Because of its depth and steep mountain borders, the river valley itself resembles a fiord. The Hudson was formed at the conclusion of the last ice recession, approximately 15,000 years ago, when glacial melting caused a rise in sea levels and drowned the old river mouth, producing the present estuarien environment and the unusually deep river conditions. From Manhattan to Troy, however, the river's surface elevation remains the same, and a powerful flood tide keeps a substantial stretch saline or brackish. Because the Hudson has no gravity flow beneath its confluence with the Mohawk River, it has not built up a floodplain.[2]

The Raritan River is the second major river draining into the estuary and, with the exception of the Delaware River, is the largest stream in New Jersey. Formed by the junction of the North and South branches and their tributaries, it flows in a southeasterly direction to Raritan Bay and is tidal until about four miles above the city of New Brunswick. The river drains an area of 1,105 square miles that is hilly near its upper part but flat and low lying as it approaches the Raritan Bay, cutting through low and marshy land.[3]

The Passaic River is the third major river draining into the estuary, flowing approximately ninety miles from Mendham, New Jersey, south into Newark Bay. It drains an area of 947 square miles, including large portions of eleven northern New Jersey counties and part of southern New York. The upper limit of tidewater is located at Passaic, where the river was drained in order to create a water power. From about three miles below the dam the river flows through an open, level plain, much of it consisting of wetlands, including the Newark meadows near the mouth of the river at Newark Bay and the Hackensack meadowlands between the Passaic and the Hackensack rivers.[4]

During the past three centuries, as a result of human settlement and development, the region has experienced many alterations. Forests have been destroyed, rivers damned and dredged, wetlands filled, and railroads, highways, bridges, and other structures constructed. Equally significant in terms of environmental effects have been the pollution flows generated by the human activities of agriculture, urbanization, industrialization, and consumption. This essay will analyze these activities over time and discuss pollution emissions into the estuarine environment.

Records of emission flows that are today considered as priority pollutants, however, are limited or nonexistent for most of the period under consideration. In addition, systematic and comparable water-quality data was not collected until the beginning of this century. For almost fifty years, the data focused on indicators such as coliform counts (bacteria), biochemical oxygen demand (BOD), dissolved oxygen (DO), and suspended solids (SS).[5] Only within the last two decades or so have indicators and measurements been taken of many pollutants of current concern, such as heavy metals and toxic chemicals. In order to provide a better account of what pollutants were actually entering the environment, the appendix of this essay discusses the sources of some of these emissions.

Agriculture and Agricultural Land Use

From the perspective of land forms, the Hudson River Basin provides a convenient route between the heavily settled New York metropolitan area and the interior of the country. It is composed primarily of soft sedimentary rocks and overlying glacial deposits that have been eroded in a manner to provide several different types of terrain. Originally, forest covered nearly the entire land area of the basin. The trees found in the area's forests are primarily oak, which occupy the warmer regions and thinner soils of the area, and the northern hardwoods, which flourish in deeper soils but also somewhat cooler highlands. Both agriculture and commercial lumbering, however, resulted in extensive deforestation; in 1850, New York State led the nation in terms of lumber production with 30 percent of the total U.S. cut. For the next forty years, the annual cut was over one billion board-feet, and New York State remained among the nation's first ten lumber producers. Pine, spruce, and hemlock from virgin timber stands were most frequently harvested. During the period of greatest timber harvesting, New York was also an important producer of wood pulp, providing about 30 percent of the national total.[6]

In spite of relatively dense settlement patterns, forests again cover large areas of the Hudson-Raritan region, and there are now actually more trees than there were fifty years ago. Between 1950 and 1968, "commercial" forestland, as defined by the U.S. Forest Service, increased by 14 percent in New York State to more than 14 million acres. Within the Hudson River basin, however, the forest resources have declined in quality, and much of the woodland exists largely for amenity functions in spread suburbs and parks rather than for productive purposes.[7]

While the greatest amount of land transformation (through agriculture, lumbering, and urbanization) occurred after European settlement began, the previous human inhabitants—the Amerindians—also shaped the environment. Before Europeans arrived in the Hudson-Raritan area, an Indian population of approximately 40,000 to 60,000—the Algonquins and the Iroquoians—had resided there for centuries. Both the Algonquins and the Iroquois were primarily agriculturists at the time of European arrival. They followed a slash-and-burn method of clearing the forest, primarily to produce a crop of maize or corn. Village and planting sites were subjected to intense human use for agriculture, waste disposal, and firewood gathering. By moving their habitat according to the season of the year, Indian tribes held their ecosystem impacts to a minimum, as selective burning created a mosaic forest environment in many different states of succession.[8]

The ecological relationships of permanent European settlers in the New World were also cyclical in terms of resources, but, unlike the migratory Indian pattern, they constructed permanent settlements. These settlements required improvements such as cleared fields, pastures, and various structures, and these became a fixed part of the landscape. Since the environment appeared so rich and plentiful, settlers followed labor-saving rather than resource-preserving exploitive techniques, and settler deforestation constituted one of the "most sweeping" forms of change the land underwent.[9]

Both colonists and Indians became ruthless exploiters of fur-bearing mammals and other animals for trade purposes, and by 1800 much of the animal population that had formerly abounded in these areas had disappeared. Domesticated animals such as hogs, cows, sheep and horses took the place of wild animals and constituted a larger burden on plants and soils than had the native species. Colonial farmers exploited the land extensively through the practice of monoculture (raising corn and letting livestock eat unharvested material) and seldom fertilized the land, leading to early evidence of soil exhaustion.[10]

Well into the nineteenth century much of the land in the Hudson River Valley was divided into large estates established during the colonial period. These were either held for speculation or rented to tenants. Settlement in the Hudson Valley was slow and usually remained close to the river banks. Except for the east bank of the lower Hudson, the hill country in eastern New York was still composed of almost unbroken forest at the time of the American Revolution. After the Revolution, however, the number of independent farmers increased, and the era of the large landed manor was eventually ended. Even on the large estates, however, land was usually operated by tenant farmers whose agricultural practices were similar to those of the independent farmers. That is, they cleared the forests ruthlessly, cropped the land until exhausted, and ignored the principles of rotation and fertilizer use.[11]

In the years after the Revolution, land ownership and agriculture spread rapidly throughout the Hudson and Mohawk river bottoms, as transport improvements and urban markets opened commercial opportunities. For the first decades of the nineteenth century, agriculture in eastern New York focused on the ruthless exploitation of timber and soil resources. When land was exhausted, the farmer turned to new lands and let exploited fields lie fallow for a few years. Farmers close to cities and towns often attempted to use street refuse, dung, ashes, and other fertilizers for their fields, but they were in a small minority in regard to general agricultural practice.[12]

In the decades following 1820, western competition drove eastern New York farmers away from wheat as a cash crop, and they gradually substituted specialized products such as vegetables and wool and dairy products. Commercial opportunities often brought with them extreme cropping practices, and by 1850 almost every eastern New York county reported serious soil exhaustion, especially in the bottom lands of the Hudson and Mohawk river valleys. The rise of an urban market for garden products, however, as well as for dairy products, spurred farmers to begin to improve their farming methods and to adopt fertilizers, rotate crops, and follow selective breeding practices.[13]

By 1850 a regional agricultural pattern had been established. Dairy was the chief source of farm income in the Hudson and Mohawk valleys, while fodder crops like corn and oats occupied the most acreage. From 1850 through 1900, the amount of farmland in the Hudson-Raritan area stabilized at about 6 million acres. In the years after 1900, however, land in farms diminished as farmers abandoned sites with poor soil. Growing urbanization and land-use changes due to transportation innovation, par-

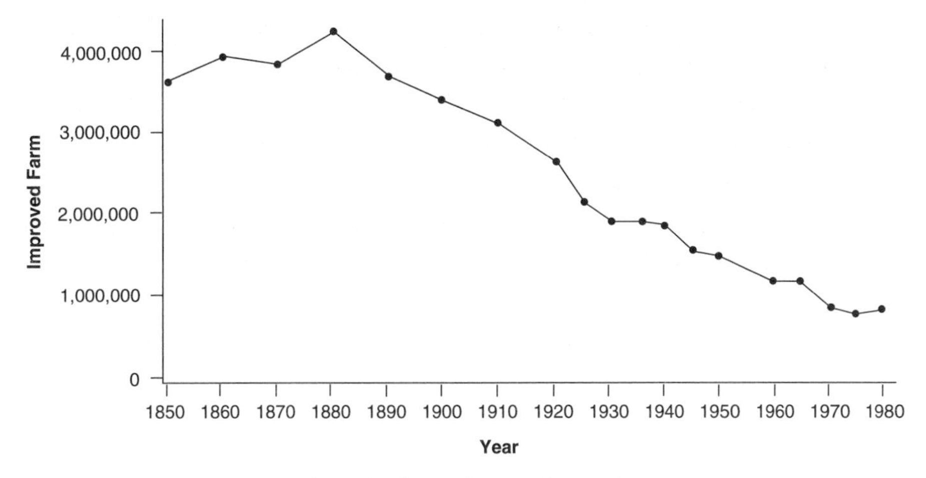

Fig. 2.1. Total Acres in Improved Farmland—Hudson–Raritan Region

ticularly the automobile, also played a role. From 1900 to 1920, over a million acres of farms disappeared; another million acres was lost in the 1920s, and half a million acres from 1930 to 1949, leaving a little over 3 million acres in farmland.[14]

In the decades after World War II, rapid suburbanization, as well as the unprofitability of agriculture, increased withdrawals, and in the five-year period from 1949–54 land was taken out of farming at twice the rates of the previous twenty-year period. By 1969, only 1,795,017 acres remained in farmland, with little change in the next decade. One major result was a large increase in the proportion of land covered by forest. By the 1970s, high-value commercial crops, such as dairy, fruit, and vegetables, dominated Hudson Valley agricultural production, especially in lowland areas. Forage crops occupied large acreage to serve the dairying interests; thousands of acres of fruit orchards were concentrated in the upper Hudson Valley; and vegetable, dairy, and poultry dominated the farms close to New York City. In terms of acreage, agriculture was in decline, but production per acre and per farm was increasing.[15]

Population Patterns: Urbanization, Centralization, and Decentralization

The spread of urbanization in the nineteenth and twentieth centuries was a major force for regional change. The Hudson-Raritan area includes two major metropolitan complexes that are intimately related to its river systems. The largest is the New York Metropolitan Zone. Historically, the most extensive development in the zone has been in the sections of New York and New Jersey adjacent to New York port. The port area is divided into five main divisions: the Lower Bay, the Narrows, the Upper Bay, the Hudson River, and the East River. The Lower Bay is open to the sea for six miles, with the Upper Bay forming the main harbor. Manhattan Island divides the upper end of the harbor in half. The bay complex forms part of one of the world's busiest seaports and the great New York metropolitan area has developed in the flat areas of the coastal plain around it, as urban uses have steadily encroached on surrounding wetlands and on lands formerly devoted to agriculture.[16]

The second urban complex in the region is the Upper Hudson Metropolitan Zone, centered on the urban centers of Albany and Troy, near the confluence of the Hudson and Mohawk rivers. The growth of this urban cluster has historically been strongly related to transportation and trade, with the Hudson, the Erie Canal, the New York State Barge Canal, the various railroads, and the New York State Thruway successively playing a critical role.[17]

Because the Bureau of the Census has altered its definition of an urban area several times, this essay uses county aggregates to describe changing regional population patterns over the years from 1790 to 1980. While counties have changed in size and name, they are more stable than other politically determined units of land. The discussion will use three time periods: 1790–1880, marked primarily by the growth of agriculture and the rise of industrial cities; 1880–1940, marked by the decline of agriculture and the development of metropolitan areas and suburbs; and 1940–80, which witnessed the decline of the central cities, the spread of low-density suburbs, and an increase in nonfarming rural land use.

While the use of county aggregates is convenient, the use of a unit as large as a county to plot urbanization can be misleading. It can obscure city growth within the county and exaggerate the amount of urbanized land. The two major metropolitan complexes in the basin today—the New York and the Upper Hudson metropolitan zones—have been the location of the basin's major cities since the late eighteenth century. Before the transportation improvements of the nineteenth century—initially, omnibuses, horsecars, and commuter railroads, followed by electric streetcars, rapid transit, and the automobile—urban population was densely concentrated in relatively limited spatial areas. Cities occupied a small amount of the land area of most counties, including those in the New York Metropolitan Zone. Since the last part of the nineteenth century, however, growth in urban centers has been accompanied by movement from the urban core, by declines in population densities, and by a flattening of the density curve between central and outlying areas. For over a hundred years, the basic population movement has been toward an ever-receding urban periphery.[18] Thus, the proportion of the land devoted to urban uses (defined as "residential, commercial, industrial, institutional and transportation") has steadily increased due to transportation and communication innovations, even though urban densities have declined.

Growth on the periphery of major cities has been most marked since large-scale automobile and truck adoption in the 1920s. After World War II, the rate of urban population deconcentration increased, spurred by widespread automobile ownership, the development of express highway networks, and the availability of improved financing for housing. Population growth concentrated outside the political boundaries of central cities in a progressively expanding zone of suburban and exurban development. The abrupt transitions between "city" and "country" that had characterized the landscape throughout the eighteenth and nineteenth centuries were replaced by more gradual population and development slopes as a gradual symbiosis of urban and rural land uses occurred in the basin. Urbanization, on the periphery, however, often advanced in hopscotch fashion rather than like a wave, in a process that one major study of the Hudson River basin described as "scatteration of uses." This intermingling of urban and rural has occurred in a context of declining agricultural land use, as farming has decreased in regional importance.[19]

Population, 1790–1880

Between 1790 and 1880, the total population of the study area grew from 403,098 to 3,977,877. The rate of increase was highest in the 1840s, 1850s, and 1870s, with advances ranging from 25 to 30 percent per decade. While

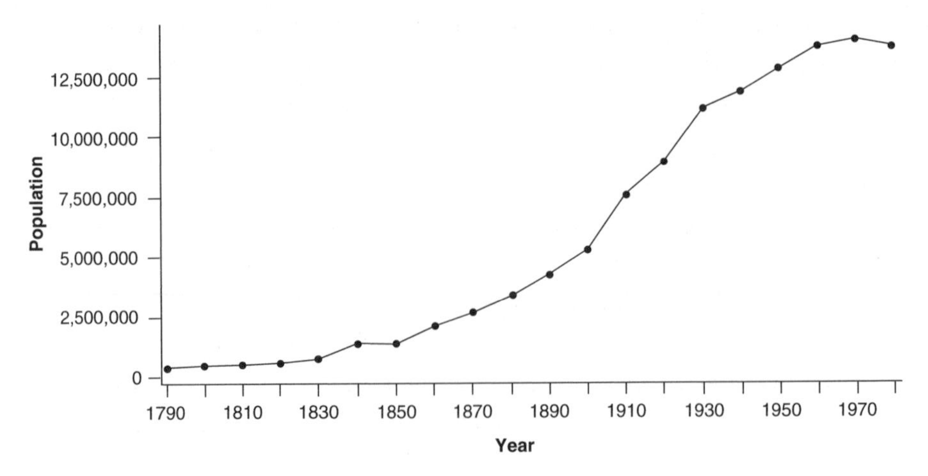

Fig. 2.2. Human Population—Hudson–Raritan Region

all counties grew, growth was much more rapid in those with major cities. The seven counties not included in either the Upper Hudson or the New York metropolitan zones grew modestly from 155,790 inhabitants in 1800 to 332,982 in 1880. In contrast, the three counties in the Upper Hudson Metropolitan Zone grew from 64,454 inhabitants in 1800 to 293,756 in 1880. The largest city in the zone, Albany, increased from 3,498 inhabitants in 1790 to 90,758 in 1880. Most striking, however, was the increase in the New York Metropolitan Zone, which grew from 200,308 inhabitants in 1800 to 3,144,128 in 1880. New York was the largest city in this zone, which grew from 33,131 in 1790 to 1,911,698 in 1880. By 1880, 3,437,884 people, or 86.4 percent of the region's population, lived in one of the two urban zones.[20]

The changes in the demography of the New York Metropolitan Zone can be effectively visualized using a framework of counties designated as the core, the inner ring, and the outer ring. In 1800, the core counties—those of New York, Hudson, Kings, Queens, and Westchester—had a population of 110,544, but by 1860 they had reached a population of 1,312,436 and by 1880 a remarkable 2,193,300 (until 1897 the Bronx was part of Westchester County, and Hudson County, New Jersey, did not appear in the census as a separate county until 1840). The core counties contained 55.2 percent of the zone's population in 1800, 72.4 percent in 1860, and 66.2

percent in 1880. The percentage peak in 1860 reflects continued rapid population growth without commensurate expansion of the city's spatial area. These were the counties that had the highest degree of land development, with extremely high densities in the key cities of New York, Brooklyn, and Jersey City.[21]

The inner-ring counties of Bergen, Essex, Passaic, and Union in New Jersey and Richmond and Nassau counties in New York also experienced striking growth, especially after 1850. Population grew from 41,989 in 1800 to 184,568 in 1850 and then to 723,147 in 1880. In 1800 the inner ring was the place of residence for 21 percent of the zone's population, but by 1860 this had shrunk to 11.2 percent, as population poured into the core cities. By 1880, however, with transportation improvements, the inner ring's percentage of the metropolitan population had reached 21.8 percent, as manufacturing activities and population spilled over from the core. Many of these counties were in New Jersey and contained rapidly growing industrial cities such as Newark (the largest in the state), Paterson, Passaic, and Elizabeth.

Finally, in this period the outer-ring counties of Rockland, Orange, Putnam, Dutchess, and Suffolk in New York and Middlesex, Morris, Monmouth, and Somerset in New Jersey experienced modest growth, as they remained predominantly rural and agricultural. The transportation improvements of the 1840s and 1850s, however, had begun to transform them into sources of garden crops and dairy products for the core cities.[22] Outer-ring population increased from 47,775 inhabitants in 1800 to 296,139 in 1860 and to 396,122 in 1880, but its share of the zone's population dropped steadily from 23.8 percent in 1800 to 16.3 percent in 1860 and to 11.9 percent in 1880 as population in the core counties grew at a rapid rate.

Population, 1880–1940

Between 1880 and 1940, Hudson-Raritan area population grew from 3,977,877 to 10,966,828. In 1880, 3,606,325, or almost 91 percent of its inhabitants, lived in counties of the New York Metropolitan Zone or the Upper Hudson Metropolitan Zone. The population of the latter increased from 293,756 in 1880 to 384,177 in 1910 and then to 465,643 in 1940, with its largest city, Albany, growing from 90,758 to 130,577. The population of the

New York Metropolitan Zone increased from 3,144,128 in 1880 to 7,548,825 in 1910 and to 11,901,962 in 1940. New York City grew from 1,911,698 to 7,454,995, largely by annexing adjacent counties and neighboring cities such as Brooklyn and Queens. From 1880 to 1910, the population of the non–Metropolitan Zone counties remained stable, increasing from 332,973 in 1880 to 365,125 in 1910 and then decreasing to 363,716 in 1940. Hudson-Raritan area rural population grew from 847,390 in 1900 to 1,165,294 in 1940, although each decade from the turn of the century witnessed a shifting pattern.

In the New York Metropolitan Zone, the core grew from 2,193,300 inhabitants in 1880 to 5,501,000 in 1910 and then to 8,506,182 in 1940, or from 66.2 percent of the zone's population in 1880 to 69.1 percent in 1910 and down to 65.8 percent in 1940.[23] Inner-ring counties increased their population to 1,459,941 in 1910 but dropped in share from 21.8 percent in 1880 to 19.3 in 1910. By 1940, however, as their population advanced to 3,039,460, they increased their share to 25.5 percent. The outer ring of counties increased their population from 396,122 in 1880 to 587,884 in 1910 and then to 929,908 in 1940, but still dropped from 11.9 percent in 1880 to 7.8 percent of the metropolitan-area population in both 1910 and 1940.

Thus, in the period from 1880 through 1940, although suburbanization increased in the New York Metropolitan Zone, the bulk of the population remained concentrated in New York City and surrounding urban employment centers such as Yonkers, Jersey City, and Newark.[24] Central cities actually absorbed more than half of the basin's increase in urban population between 1920 and 1940. Nevertheless, a trend toward decreased urban densities was evident after 1920, as transit lines and the automobile provided access to outlying areas. Density changes in New York City exemplified these trends. Between 1920 and 1940, the average population density in Manhattan declined from 103,823 persons per square mile to 85,906; simultaneously, densities in the outlying borough of Queens nearly tripled, increasing from 4,343 persons per square mile to 12,015 per square mile. In Brooklyn, densities increased from 28,427 to 34,017 persons per square mile, and in the Bronx from 17,854 to 34,017.

Outside the core and the older cities such as New York City, Newark, and Jersey City, in the inner-ring counties such as Nassau on Long Island

and Bergen in New Jersey, densities increased, but from a much smaller base. Density in Nassau County increased from 460 persons per square mile in 1920 to 1,356 in 1940, while density in Bergen County increased from 889 persons per square mile to 1,758. Reflecting the tendency for growth to move toward the fringe and for increased urban land uses, between 1920 and 1940 the Bureau of the Census expanded the boundaries of the New York, Northeastern New Jersey Metropolitan District and increased its land area from 1,174 to 2,561 square miles. By 1940, average population densities had declined from 6,733 persons per square mile to 4,565.

Population, 1940–1980

From 1940 to 1980, the population of the Hudson-Raritan area counties expanded from 12,768,257 to 16,266,713. Urbanized population increased from 11,602,963 to 15,188,490, while rural population dropped slightly from 1,165,294 to 1,078,223. Upper Hudson Metropolitan Zone population grew moderately, from 465,643 to 587,038, while that of the New York Metropolitan Zone increased from 11,901,962 to 14,973,972. The non–Metropolitan Zone counties experienced considerable growth after World War II as rural population expanded. In 1940, non-Metropolitan counties had a population of 363,716, but by 1960 this had expanded to 436,310 and by 1980 to 583,980, making them the region's fastest-growing counties. By 1980, the so-called rural population included a substantial nonfarm population.

Within the New York Metropolitan Zone there were dramatic shifts in population redistribution between cities and suburbs and rural areas. After World War II, spectacular growth occurred in the region's suburbs and modest growth in its rural areas. Central city population, however, stagnated, and by 1960 most had begun to lose decline. From 1960 to 1980, New York Metropolitan Zone population remained relatively stable, only increasing by approximately 160,000. But between 1950 and 1970, the Bureau of the Census nearly doubled its estimate of the land it defined as urbanized, from 1,325.5 to 2,576 square miles, while average density declined from 9,583 to 6,480 persons per square mile.

Population of core counties, however, stagnated. Between 1940 and

1960, these counties grew from 7,932,594 to 8,165,827, but their share of the zone's population dropped from 65.8 percent to 55.1 percent. New York City had a net population loss, as did older New Jersey cities such as Jersey City and Newark. Manhattan population densities declined from 89,096 persons per square mile in 1950 to 66,923 per square mile in 1960; Queens's densities increased from 13,724 persons per square mile to 18,400 in 1960, and Brooklyn from 36,029 persons per square mile to 37,172. From 1960 to 1980, core county population dropped to 7,242,719, while the core's share of the Metropolitan Zone population slid to 48.4 percent. Again, the heaviest loses were in New York City and the older industrial cities of New Jersey.

While the core was stable, the inner ring of counties experienced rapid growth in the immediate postwar decades. Population increased from 3,039,460 in 1940 to 4,945,726 in 1960, and the inner ring's share of the zone grew from 25.5 to 33.4 percent. Suburbanization in counties with undeveloped land such as Essex and Union in New Jersey and Nassau in Long Island accounted for much of the increase. Older New Jersey industrial cities like Newark, Passaic, and Paterson either showed losses or no change. From 1960 to 1980, however, population growth in the inner ring nearly stopped, increasing only slightly to 5,181,623 as suburbanization spread to more peripheral areas; population share remained steady at 34.6 percent.

The outer ring of counties experienced the greatest growth in the New York Metropolitan Zone, with population advancing to 2,549,630 from 1940 to 1980. From 1940 to 1960, the outer ring increased its share from 7.8 percent to 11.5 percent, and by 1980 it held over 17 percent of the zone's population. A preference for low-density living, the availability of mortgages at affordable rates, increased automobile ownership, new expressways and interstate highways, and the out-movement of jobs drove this movement toward the periphery.[25]

The Region as a Whole

From 1970–80, the region's population declined from 17,189,311 to 16,266,713, the first population loss in its history. The earlier pattern of central-city losses continued, accompanied now by inner suburban counties, thus outweighing continued growth in outlying counties. According to the Bureau of the Census, urbanized land in the region increased from 2,576 to 3,006 square miles, but urbanized population declined from 16,148,898 to 15,188,490. Metropolitan population also dropped, from 12,098,534 to 12,098,553. Rural population, however, advanced from 1,040,413 to 1,078,223, reflecting the continued preference for low-density living.

Urban decline accounted for the bulk of the loss. New York City population dropped by more than 800,000, from 7,895,563 in 1970 to 7,071,639 in 1980. With the exception of Staten Island (Richmond County), population density in every borough declined in the decade, with the Bronx experiencing the most massive loss. Several major cities shared in the decline, with Newark population dropping from 381,930 to 329,248, Jersey City from 260,350 to 223,532, and Albany from 115,781 to 101,727. Inner suburban counties joined the cities, as Bergen County densities dropped from 3,839 to 3,567 persons per square mile and Nassau County densities dropped from 4,944 to 4,598. Peripheral counties still showed some population gains but at a more modest rate than in the previous post–World War II decades. Morris County, New Jersey, for example, gained 47 inhabitants per square mile between 1970 and 1980, compared to an increase of 259 persons per square mile during the previous decade.

Industrial and Manufacturing Patterns, 1790–1980

In the middle of the nineteenth century, New York was the most heavily industrialized state and New York City the most heavily industrialized city in the Union. By the twentieth century, however, much of this dominance had been lost. Tracing industrial patterns over time is more difficult than tracing population. At times the census collected material on counties, at other times on cities, and occasionally on both. Information on firm size is often unavailable because the census presents its material in ranges of firm sizes or withholds information on certain industries because of privacy restrictions. This section, therefore, will present a general picture of manufacturing trends, with special attention to those industries with highly polluting characteristics.

During the period from the Revolution through the 1820s, manufacturing in the Hudson-Raritan region—except for a few exceptions, such as the textile manufacturing complex at Paterson, New Jersey (established to take advantage of the water power at the fall line), or the Troy iron works—was decentralized. Its main elements were the gristmill, sawmill, and country forge, all of which utilized the water power supplied by the area's abundant streams and served localized areas. By 1820, however, near major waterfalls, large mills had replaced many small gristmills and tanneries. Within the cities, craftsmen and artisans who worked in small shops produced the preponderance of goods, but in industries such as leather tanning, shipbuilding, sugar refining, printing, and construction larger operations were emerging.[26]

Between 1825 and 1850, New York State and City reached their peak as the nation's leading manufacturing centers. The opening of the Erie Canal in 1825, as well as the construction of other important canals, such as the Delaware and Hudson, the Chaplain, and the Oswego, greatly strengthened the Hudson-Mohawk axis of traffic and development, mainly channeling western goods east. Railroad construction began after 1836, and by 1855 a well-developed network provided a direct link between Lake Erie and New York City. Manufacturing tended increasingly to cluster in urban locations at important canal and railroad nodes, and the number of small mills scattered at rural locations diminished yearly.[27]

The nation's backbone of heavy industrial and metals production extended from Albany and Troy down the Hudson River Valley to New York City and shifted westward into New Jersey and then into the Delaware River Valley. Iron and coal were the basis of the growing industrialism, with large ironworks located in the Albany-Troy and New York City regions. In addition, industries developed in other manufacturing sectors such as textiles, clothing, shoes, and furniture as well as food and printing. Heavy industry developed in sectors such as iron molding and casting, machine and tool making, brewing, distilling, sugar refining, and manufactured gas production but only employed about 5 percent of New York manufacturing workers in the 1850s.[28] In New York City, in 1850, there were 588 establishments with more than twenty workers each, with a total of approximately 90,000 workers engaged in crafts and manufacturing.

Most factories and manufacturing establishments were still relatively small. The light handicraft industries, especially in building and consumer finishing trade, remained the most important sectors of the economy. During the 1850s, the area surrounding New York Harbor, especially the cities, became the center of one of the world's great industrial regions, producing in value 14 percent of all U.S. manufactures. The Albany-Troy-Choose-Schenectady area was the other major center of manufacturing, with concentrations of cotton and iron-making and metal-working firms and facilities for processing local products such as lumber and barley. In other Hudson Valley counties, factories and mills employed 8,497 workers in 1850, but workplaces were scattered throughout the countryside.[29]

During the eighty years from 1850–1930, industries based on the processing of local resources such as lumber, flour, brewing, meat packing, and tanning declined, while those involved in producing products such as iron and steel, shoes, machinery, and chemicals increased. Flour milling, for instance, constituted New York State's leading manufacture in 1860, but by 1900 it had slipped to twelfth place. Men's clothing was first in 1880, with women's clothing second, foundry and machine-shop products fourth, and textiles fifth. By 1900, the state produced almost half of the men's clothing in the United States and two-thirds of the women's clothing, most of which was manufactured in New York City. The major manufacturing center was in the New York metropolitan area, while a much smaller center existed in the Upper Hudson metropolitan area. In addition, a variety of small plants were still scattered up and down the main Hudson-Mohawk axis, producing products such as cotton goods, paper, and agricultural equipment.[30]

In the post–Civil War decades, the economy completed its transition from an agricultural and commercial-mercantilist base to an industrial-urban one. Manufacturing output grew at a faster pace than urban population as innovation interacted with growing demand. Completion of an elaborate national rail network cheapened the costs of transport and reduced the importance of river cities such as Albany. New industrial employment and its multipliers produced a shift from mercantile-type occupations toward manufacturing employment, with the consequent generation of new markets. Urban expansion also produced jobs as the city-building process accelerated.[31]

The Hudson-Raritan region experienced very rapid industrial growth up to the late-nineteenth century, then slowing after this time. The core counties of the New York metropolitan zone set the pace. Manufacturing employment grew at a somewhat slower rate than population up to 1869, but from 1869 to 1889 it outdistanced population. In these two decades, the zone's share of the nation's manufacturing workers rose from 12 to 16 percent, a larger increase than the rise in its share of the nation's population. From 1889 to 1929, however, the region's growth in manufacturing workers again sank beneath its rate of population rise.[32]

The declining industries in the post-1880 period were those where the costs of moving raw materials to plant and/or finished product to market were of prime importance. Transport-sensitive industries such as iron and steel, lumber, glass, pottery, and locomotive manufacture relocated closer to sources of raw materials in order to reduce transportation costs.[33]

Simultaneously, the New York metropolitan area became the manufacturing center for many industries for whom transport costs were relatively unimportant. Between 1860 and 1910, the apparel industry increased its employment from 30,000 to 236,000, and its plants from approximately 600 to 10,000. By 1910, New York City possessed over 60 percent of the nation's jobs in the men's and women's apparel industry. In addition, it became the nation's most important center of printing and book and periodical publishing. In 1919, 20 percent of the nation's printing industry was concentrated in the city, employing approximately 91,000 people.[34]

Metal products was the single most important industry in the New York region from 1880 through 1922. In 1880, New York City, Brooklyn, Jersey City, and Newark had 483 firms with almost 16,000 workers producing foundry and machine-shop products, making it the nation's largest concentration of metalworking activity. Between 1900 and 1922, the number of metal products plants increased from 3,216 to 9,426, and the number of employees increased from 150,461 to 269,079. During this period, the region made a major shift from the manufacture of predominantly heavy-metal products to light-metal products. In 1900 more than half the industry's workers were employed in its heavy- and bulky-products branch, but by 1922 only two-fifths of metalworkers were so employed. Many of the plants producing heavy, bulky products moved out of lower Manhattan, some to Queens (Long Island City) and some to New Jersey.[35]

From an environmental perspective, there were two important factors involved in this movement. The first was the disappearance of sites in Manhattan for the foundries—as well as for other industries such as sugar refining and gas manufacturing—to dispose of their wastes, while the second involved complaints and nuisance suits because of externalities such as smoke, dust, and noise. The availability of waste-disposal sites in New Jersey was an important factor in their movement to that state.[36]

In the late nineteenth century, the New York metropolitan zone was also an important center for copper refining, with some associated copper smelting. Copper refineries found it advantageous to locate closer to markets because of the limited weight-reducing character of the refining process. In the period before 1910, four major copper smelting and refining complexes, plus two smaller refineries, were built around New York Harbor. In 1899, these refineries produced 17,000 tons or 39 percent of the U.S. total of refined copper ingots, bars, and wire products as well as 7,400 tons of copper sulfate (blue vitriol), or 54 percent of the national total. By 1918, 63 percent of U.S. copper refinery output was found in the New York region. While the region's percentage share dropped after this date, as late as 1970 the New York zone still accounted for 25 percent of U.S. copper refining.[37] Copper smelting, as compared to refining, was relatively limited in the New York region, although most of the large facilities had blast or reverberatory furnaces. In the 1920s, three of the nation's smelting plants, producing about 4 percent of U.S. smelter output, were located there. The last smelter to operate in the region was closed in 1963.

Copper is a heavy metal and, from an environmental point of view, its emissions are of serious consequence. Gaseous waste streams are generated from copper smelters and converters, while slag is the major solid waste (3 tons of slag produced per ton of blister copper, but as high as 10 to 1 for the concentrates used in the New Jersey–New York smelters). Slag is mostly inert oxides, but it does contain some copper and other trace metals. Arsenic is a by-product of the processing of copper ores and is emitted both in gaseous form and as a component of the slag. The slag was normally disposed of in landfills, often on refinery sites located in low-lying marshy areas. A major source of water pollution from copper refineries located in the Hudson-Raritan basin could be leaching from old slag piles.

During the period from 1900 to 1922, industries producing heavy-metals products declined, but plants manufacturing technical instruments and assembly, as well as light-metal products including welding and electroplating, grew. Some of these activities, such as electroplating, were conducted in widely scattered small shops but had a large polluting potential.[38]

The chemical industry was a second major industry centered in the New York Metropolitan Zone, and by the turn of the century the region had become the nation's largest chemical producer. Between 1900 and 1922, the number of chemical plants in the region increased from 600 to 1,351, while the number of employees increased from approximately 28,000 to over 70,000.[39] In 1922, between 15 and 20 percent of the nation's chemical production was located in the New York Metropolitan Zone.

The branches of the industry that expanded the most were heavy chemicals, explosives, and toilet preparations. The basic pattern for chemical plant location was similar to that for the metals industry. That is, large chemical firms manufacturing products like soap and fertilizer moved out of Manhattan to be replaced by smaller, specialty firms making products such as perfume and cosmetics. The most rapid growth patterns were on the region's periphery, especially near waterways, while the core areas became much more diverse. In the period before World War I, a number of chemical firms manufacturing paint and varnish and pharmaceuticals, as well as petroleum refiners, located in Brooklyn and Queens along Newtown Creek. Here, the emissions they produced and their waste-disposal practices created severe pollution problems, leading to demands for their curtailment. Faced with these pressures and with limited land available for expansion, firms sought desirable sites elsewhere. Increasingly in the twentieth century, the preferred site for large chemical firms was New Jersey, especially on the shores of New York Bay and on the banks of the Passaic River near the wetlands area known as the meadows.[40]

The production of paints, dyes, inks, and varnishes made up another important branch of the chemical industry. Paint and varnish manufacturers initially concentrated in Brooklyn, which in 1880 had thirty-four paint and varnish plants. Newer plants were constructed in Long Island City, Richmond on Staten Island, and in New Jersey. The production of dyes from coal-tar was given a large impetus by World War I, with several large plants located in the New York metropolitan area. Total employment in this manufacturing sector nearly doubled between 1900 and 1922, increasing from approximately 6,000 to 12,000 workers.[41]

One of the fastest-growing industries in the New York Metropolitan Zone was petroleum refining. Between 1900 and 1922, the refining industry expanded from approximately 7,000 employees to almost 15,000 at twelve major refineries (primarily Standard Oil, but also Tide Water and Getty). In 1922, the zone was the world's largest refining center, possessing 11.9 percent of the nation's refining capacity, or 258,000 barrels a day. Like many other chemical firms, the pattern of movement for the refineries in these years was from the core to the inner ring or the outer ring, to sites located primarily in New Jersey (Bayonne and Linden) along the shores of the estuary, where oil pollution became an increasingly larger problem. While refineries existed at Long Island City and in Brooklyn (Newtown Creek, especially), they only constituted a small fraction of the region's total refining capacity, and the New York refineries constituted an ever-smaller amount of the nation's petroleum refining capacity.[42]

During the 1947–77 period, the Hudson-Raritan basin as a whole lost approximately 17 percent of its manufacturing jobs, as the manufacturing work force dropped from 1,750,200 to 1,452,700. New York City, the older suburbs of the Bronx and Brooklyn, and the older New Jersey cities were especially hard hit. The largest growth sectors in the region became the service sectors (finance and business) rather than manufacturing. In 1982, approximately one-fifth of the jobs in the New York region were goods-producing, with 79 percent in the service and information-producing sectors.[43]

The manufacturing sector experienced a large amount of diversity and change during these years. In the New York Metropolitan Zone heavy industry continued to decline and new industry moved toward the region's periphery. New plant construction took place overwhelmingly in the outer ring of counties, drawn by access to transportation and cheap land, and its share of the zone's manufacturing jobs more than doubled. Employment in chemicals in New Jersey and New York, for instance, almost all of it located within the New York Metropolitan Zone, dropped from approximately 95,000 employees in 1977 to about 84,000 in 1985. Employment in fabricated metals, historically one of the zone's strongest industries, dropped from 58,000 to 50,000 in the same period.[44]

The manufacturing decline reduced pollutant flows into the river, but within the core areas and in the inner zone remaining heavy industries (many of which were nuisance industries) created a substantial share of the zone's environmental problems. One recent study (1985) of point-source pollution in the Hudson found 147 industries and utilities that discharged at least one toxic chemical into the Hudson from 1978 through 1983, with many firms discharging more than one toxic stream. The most commonly discharged chemicals were oil and grease, chromium-hexavalent, lead, toluene, cyanide, chloroform, mercury, cadmium, methylene chloride, and trichloroethylene. Of the non–point sources, PCBs were the greatest source of violation of water-quality standards, followed by lead, mercury, and cadmium.[45]

In addition, many of the heavy industries that had moved to the inner- and outer-ring counties (especially in New Jersey) in the 1920s and 1930s because of the ease of waste disposal left a legacy of environmental problems in terms of existing and abandoned waste dumps. These dumps were concentrated in the New Jersey meadowlands and in the areas around the Lower New York Bay and Raritan Bay. Industries seeking to relocate out of New York City in the earlier part of the century had found these locations attractive because of the "swamps" (wetlands) available for waste disposal.[46]

History of Water Pollution in the Hudson-Raritan Estuary

Up until the late-nineteenth century, most studies of the region's environmental conditions focused on conditions on the land and on the edges between the land and the water, rather than on the water itself, except in the cases of obvious nuisances. Sanitary conditions and their relationship to the public health, rather than water quality per se, were the concern of investigators. Investigations also noted problems with nuisance trades such as slaughterhouses, rendering factories, and gas manufacturers, and municipal ordinances often relegated these highly polluting industries to the urban fringes, such as the river banks. Many accounts recorded a great abundance of fin fish and shellfish in the waters of the estuary throughout the century, a sign of sound-water quality conditions.[47]

In the late nineteenth century many municipalities constructed sewerage systems that discharged their raw sewage directly into the rivers and the bays, greatly increasing water pollution, causing obvious nuisances, and raising concerns over threats to the public health. The waste-disposal practices of the growing industrial sector also added to water-quality deterioration. The legal response in regard to water pollution as opposed to sanitary conditions, however, was minimal. Two laws—one state, the Sludge Acid Act (1886), and one federal, the Refuse Act of 1899—were passed. The former related to the injurious effects of industrial wastes on shellfish, while the other concerned navigation impediments.[48] These laws were of limited effectiveness but constituted a clear sign of environmental degradation.

By the beginning of the twentieth century, population and industrial growth had increased the pollution load carried by the basin rivers and the estuary. This was largely related to the rapid construction of sewerage systems by area municipalities and their disposal of raw sewage into adjacent waterways. Dilution in the rivers and bays was relied on to disperse the sewage; but as loads increased, water-quality conditions deteriorated, as evidenced by visible nuisances and bacterial and chemical measures. In addition, by 1900, industries producing oxygen-consuming and toxic wastes had greatly expanded, further increasing the effluent load. In 1900, there were 660 "chemical plants" in the New York metropolitan area, including 37 petroleum establishments and 73 "heavy"-chemical plants with over 28,000 employees, all producing significant amounts of externalities. In the same year, there were 1,283 heavy-machinery and medical products firms located in the district, with 83,508 employees.[49]

In 1902, the U.S. Geological Survey conducted the first significant examination of sewage pollution in the major rivers of the Hudson-Raritan basins, although it did not examine the New York, Raritan, or Newark bays. The study noted that pollution was worse in the Passaic River Basin than in either the Hudson or the Raritan rivers. The report maintained that while there were specific local nuisances in the Hudson below the larger cities such as Albany or Poughkeepsie, the river's dilution capacities sufficed to prevent "material damage" that would require sewage treatment. The report found similar conditions in the Raritan River, with pollution limited to an area near the mouth of the stream. The water of the Raritan's tributaries was potable without treatment and "fisheries have been preserved intact." The Passaic, however, had been badly polluted by both sewage and industrial wastes, with fishing destroyed in the lower river, the natural ice industry damaged, and the water rendered unsuitable for drinking or manufacturing purposes.[50]

Concerns over increased contamination of New York Harbor produced a series of detailed pollution studies in the decades before World War I. The most extensive examination was that of the Metropolitan Sewerage Commission (MSC), a joint project of New York and New Jersey (1907–14).[51] The MSC found the Upper Bay to be heavily contaminated, with floating sewage solids, large greasy slicks, and turbid water conditions. The Lower Bay and the waters of the Lower Hudson above New York City, however, were not "badly polluted by sewage." In contrast, the East and Harlem rivers were turbid and foul smelling and filled with floating sewage solids. The waters of Newark Bay were visibly polluted, with oil slicks and sewage pollution near the mouths of the Passaic and Hackensack rivers.

Dissolved oxygen measurements (DO) showed that, while the inner harbor was heavily polluted, the Upper New York Bay served as "a great equalizer so far as oxygen is concerned." Bacterial studies found the highest counts in the Upper Bay, the Harlem River, and the Passaic River, with a large part of the harbor's bottom covered with sewage sludge deposits. A retrospective study conducted in 1981 for the National Oceanic and Atmospheric Administration (NOAA) applied present water-quality standards to the DO conditions reported by the MSC in 1912. The NOAA report found that the Upper New York Bay, Jamaica Bay, the Passaic River, and the East River below Hell's Gate would not have met the 1981 standards for dissolved oxygen.[52]

While the New York State Conservation Commission and the State

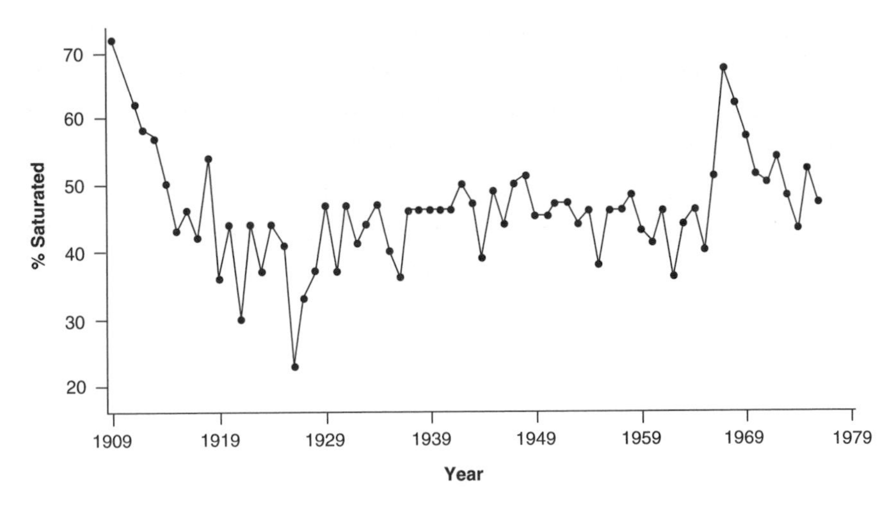

Fig. 2.3. Average Dissolved Oxygen during Summer Months—Hudson River

Department of Health conducted additional water-quality investigations of the Hudson River Basin during the 1920s, there were no published studies comparable to the MSC report. The Corps of Engineers, however, conducted an intensive investigation of pollution conditions in the Hudson River Basin and estuary based on its mandate under the Oil Pollution Act of 1924. The Corps report, which was never published, noted that since the investigations of the MSC more than ten years previously almost no "corrective measures" had been carried out in the estuary and river basins and that water quality conditions had further deteriorated. The DO average in the harbor, for instance, had declined from 62 percent in 1911 to 43 percent in 1925. Conditions were so bad in the fifty-square-mile area between the Narrows, Sputeen Duyvil, and Throgs Neck that there was not "enough oxygen during one third of the year to permit active fish to live and the remainder of the time it is impossible for even sluggish varieties to thrive." The amount of untreated sewage discharged by municipalities in the New York metropolitan area had increased from 621.7 mgd (million gallons daily) in 1910 to 739.1 mgd in 1920 and emptied through 375 outlets. Of this, the East River received 50 percent, the Hudson River 20 percent, and the Harlem River and Upper Bay about 10 percent, with the remaining 10 percent discharged into smaller waterways. Examinations of bottom conditions showed that sewage sludge deposits now covered more bottom areas than in 1910, and "an overlying stratum of semi-fluid scum" on the bottom produced gases that made dredging difficult and damaged the paint on the dredges.

In addition to the sewage, the Corps reported that industry discharged 48,494,250 gallons of trade wastes into the district waters in 1920, with 44 percent originating from factories in Brooklyn, 24 percent from Manhattan, 13 percent from Queens, 11 percent from New Jersey, 7 percent from the Bronx, and 1 percent from Richmond. Most of the trade wastes originated from factories located in a narrow strip along the waterfront. Laundering and cleaning and dyeing establishments discharged the largest volume of wastes, consisting largely of soap, soda, bleach, inorganic acids and bases, and organic dyes as well as dirt removed from textiles and clothes.

Other types of industrial pollution included oil and gas house wastes. The Corps noted that oil pollution was widespread throughout the estuary, although considerably reduced compared to the 1922–25 period. The primary sources of oily refuse were ships, shipyards, refineries, transfer stations, and broken oil lines. The Corps investigators estimated that automobile-owning inhabitants and garages dumped 7 million gallons of crankcase oil yearly into the harbor through the sewers. The report also made special mention of wastes from gas manufacturing, an object of concern in both New York and in New Jersey, warning of their polluting character.[53]

The Corps report concluded its section on industries by noting that water pollution was driving the fishing industry from the waters near New York City and causing the rapid decline of the culture of oysters. It quoted with approval a paragraph from the 1924 report of the New York Conservation Commission that noted sharp alterations in conditions in the Hudson River since the turn of the century.

> Conditions have changed entirely from what they were in the Hudson twenty or thirty years ago. Thirty years ago the Hudson was not the polluted stream it is today. There was not nearly the footage of nets in the river that there is today. There were plenty of natural spawning beds in the river that have since been ruined by the operation of dredges. . . . There were not the number of fast propelled boats plying the water which wash the spawn in shallow beds nor were the carp, the greatest of all menaces to fish reproduction, as firmly established as they are now.[54]

In spite of the MSC and Corps of Engineers investigations, little official attention was paid to improving water-quality conditions in the Hudson-Raritan estuary from the mid-1920s through the mid-1930s, although records were kept of standard indicators. Several New York City departments performed regular analysis of chemical, bacterial, and oxygen water-quality conditions; the New York Department of Conservation kept track of conditions in the Upper Hudson; and several New Jersey groups recorded the deterioration of the Raritan River and Bay. All the reports noted the deterioration of conditions since earlier in the century. The major improvements during this decade involved the opening of the Bronx Valley Sewer and the Passaic Valley Trunk Sewer, both of which improved water-quality conditions in the rivers they served. New York City, however, had attempted to block the projects on the grounds that their effluent (subjected to primary treatment) would further pollute the harbor.[55]

Altered Pollution Flows and Measurement Changes

During the 1930s, due primarily to the funding supplied by the federal government through the Public Works Administration (PWA), cities in the estuary area constructed a number of sewage-treatment plants. The most important of these were in New York City and included the Coney Island (1935, 70 mgd), Wards Island (1937, 180 mgd), Tallmans Island (1939, 40 mgd), and Bowery Bay (1939, 40 mgd) sewage-treatment plants. In 1939, in the Hudson River Valley (Hudson and Mohawk rivers), 561,575 people were served by treatment, while a population of 287,277 still discharged raw sewage into the rivers. Of the thirty-two treatment plants in operation, twenty-five had been constructed during the years from 1930–39. In addition, a number of treatment plants were built along the Raritan River in New Jersey, and the capacity of the primary treatment plant of the Passaic Valley Sewerage Commission was enlarged.[56]

In 1936, New York and New Jersey signed a compact creating the Interstate Sanitation Commission (ISC) for the purposes of abating existing water pollution conditions and controlling future pollution of the tidal waters of the New York metropolitan area. Connecticut joined the compact in 1941. One of the commission's first tasks was, with the help of the Federal Works Agency (FWA), to monitor district water-quality conditions using standard parameters such as DO and coliform indicators. The commission's 1938 survey showed that the district's maximum pollution was in the East and Harlem rivers, closely followed by the Upper Bay, the Narrows, the Kills, and Newark Bay. Conditions in the Lower Bay, the Raritan Bay, and Sandy Hook Bay were found to be satisfactory (FWA, 1939). By 1940, several New York sewage-treatment plants had begun operation, and improvements were observed in the DO conditions of the East and Harlem rivers, as well as in Jamaica Bay, the Kill Van Kull, and the Arthur Kill. A separate study of the Raritan River and Bay compared the results of water-quality surveys taken in 1927–28, 1937–38, and 1940–41 and also found beneficial effects of sewage and industrial waste-treatment facilities installed during the 1930s. "It is difficult to visualize," observed the report, "what the condition of the river would be if municipal sewage plants and several of the larger industrial treatment works had not been built and operated."[57]

World War II brought deteriorating environmental conditions in a number of areas but particularly in regard to increases in industrial pollution. The New York Conservation Department reported an increase in the number of fish kills occurring because of industrial wastes. This marked a shift from the traditional pattern of pollution. Throughout the twentieth century, the principle industries producing oxygen-consuming wastes were canneries and dairies. In 1948, however, the Department reported that firms using chemical methods in manufacturing, especially cyanide processes, had become the worst offenders. Studies of the Raritan River in 1951 showed a similar increase in industrial waste loadings, which had doubled the biochemical oxygen demand (BOD) of municipal sewage.[58]

Alarmed by the rise in pollution from industrial wastes, in 1951 the ISC published its first Industrial Waste Inventory. The commission studied a sample of approximately 1,500 of the district's 29,000 industrial plants located along 1,500 miles of coastline. Of the 1,500, 306 discharged 515 mgd of industrial wastes directly into district waters. The largest concentrations of these plants were in Brooklyn, Queens, the Arthur Kill, Kill Van Kull, and in Bridgeport, Connecticut. The heaviest polluter was the petroleum industry, which discharged 65 percent of the total industrial-waste stream, even though petroleum plants only constituted 5 percent of the plants surveyed. Utilities were next with 14.4 percent of the wastes, while the chemical industry contributed 10.1 percent. The ISC estimated that the oxygen demand (BOD) of the industrial wastes was equivalent to that of the raw sewage discharged by 2,100,000 people, or a city the size of Philadelphia. In addition to using the oxygen-demand parameter, the commission utilized a new set of parameters such as toxicity, grease and oil, acidity, suspended solids, and color to measure pollution conditions.[59]

Concern over water pollution increased in the 1950s as evidenced by statutes passed by federal and state legislatures strengthening water quality standards and providing matching grants for sewage treatment plant construction. From 1952 to 1963, nine major treatment plants (six in New Jersey) providing primary treatment were constructed, and by 1962 about 75 percent of New York City's sewage was being treated. During the same period, in response to legislation requiring the classification of the state's water resources, the New York State Health Department studied conditions in the major drainage basins, exploring the hydrology, uses, and pollution conditions of the state's waters and classifying them for best usage. Various parameters for both municipal and industrial wastes were explored, and point-source dischargers were identified and classified by waste type and treatment. Aside from the Arthur Kill and the Kill Van Kull, heavily polluted by industrial wastes, municipal sewage was identified as the primary determinant of water quality in the Hudson opposite New York City and in the Upper Bay. The East River, as in the past, was a prime offender, dumping about 100,000 pounds of BOD each day from untreated sewage into the Upper Bay.[60]

In 1964, as a result of these water-quality surveys, the New York State Water Resources Commission adopted the following official classifications: Class I for fishing; Class II for water not primarily for recreational purposes, shellfish culture, or development of fish life.[61] Standards utilized in the classification scheme were based on traditional parameters such as floating solids and DO as well as evidence of various types of toxic wastes that might affect "edible fish and shellfish." The waters of the Hudson River below the New York–New Jersey state line, including those of the Upper New York Bay, were classified as primarily Class I, while the lower East River, most of the Harlem River, and the Kill Van Kull were Class II.

In 1965, the U.S. Public Health Service (USPH) reported that the total oxygen-demanding load discharged by municipal sources to the Hudson River from Troy to the Narrows was equivalent to a population of 10 million. Of the total of 74 municipal dischargers on the main stem and its tributaries, 6 provided secondary treatment, discharging an effluent with a PE (Population Equivalent) of 1.959 mgd, 44 provided primary treatment and discharged an effluent with a PE of 3.76 mgd, while 24 discharged raw untreated sewage, contributing a PE of 4.297 mgd. The largest flows of raw sewage were in the upper Hudson and in the New York City area. In terms of contributions to the pollution of the Bays, .487 mgd PEs came from the Newark Bay–Kill Van Kull area and 3.69 mgd from the East River and Harlem River. Raw sewage amounting to 3.92 mgd of PE were discharged from New York City outlets.

One-third of the PEs discharged to the river and bays came from New Jersey dischargers, with the Passaic Valley Sewerage Commission being the leading offender, a confirmation of New York's original concerns over the outlet. The remaining two-thirds came from New York sources. In terms of DO, the USPHS report noted that depressed oxygen conditions existed for a distance of sixty miles downstream from the Troy dam, while the river was almost devoid of dissolved oxygen for twenty miles from Albany downstream (survey taken in August). According to the Health Service report, the river "never returned to a state of relatively complete oxygen saturation until it reached the ocean in the New York Bight." The maximum levels reached were near the Tappan Zee Bridge, while the low point was in the area between the Battery and 42d Street. The report also noted the presence of "vast quantities of sludge deposits" on the bottoms of the East River, Newark Bay, and parts of the Hudson.[62]

Reductions in Traditional Parameters and the Development of New Pollution Concerns

The 1965 totals for the traditional DO and BOD parameters, however, represented a peak, and they declined sharply in the following decade. Studies in the early 1970s showed that while wastewater flows into the river had increased along with population, pounds of organic material (measured in BOD) were reduced by about 35 percent over the previous ten years. These improvements had undoubtedly occurred because of the construction of new sewage-treatment plants and the upgrading of existing facilities. By 1972, for instance, New York City was treating 1,410 mpg of the 1,610 mpg of sewage it was discharging. In the decade of the seventies, in response to the 1972 Federal Pure Water Act (Federal Law 92:500), further upgrading of municipal facilities took place. In 1981 the ISC reported that DO and coliform counts for the lower river and the estuary had improved considerably between 1974–75 and 1981. Heavy-metal concentrations, however, had increased in the East River.[63]

But while traditional waste loads had been reduced in the estuary and its rivers, new pollutants had been identified and water-quality standards had become more stringent. In the 1980s, the attention of regulatory agencies and environmental groups focused on pollutants having significant environmental as well as health effects, such as chlorinated hydrocarbons and heavy metals. The General Electric Company had released into the river large amounts of polychlorinated biphenyls (PCBs), and these raised special concerns because of their concentration in bottom sediments. However, the largest source of contaminants in the estuary for pollutants of both traditional and recent concern in the 1980s continued to be wastewater discharges.[64] Thus, in terms of pollution of our waterways, society is still dealing with the basic decision made over a century ago to transport our wastes via water carriage systems and to discharge in the most convenient water locations.

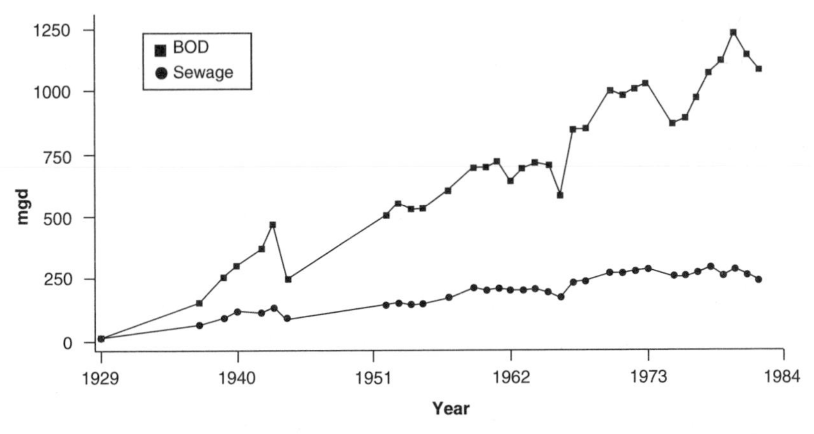

Fig. 2.4. Sewage and BOD Loading (mgd) from treatment plants discharging into the Hudson River downriver of the Bear Mountain Bridge, East River, and Upper New York Bay. 1929 is used as the base date when sewage treatment of some portion of the discharges into these water bodies began.

Conclusion

Major changes have occurred in patterns of population and economy in the Hudson-Raritan Basin over the past three centuries. These changes have, in turn, had important consequences for the types and magnitude of pollution flows into the environment. The record suggests that the environment in the Hudson-Raritan Basin was most heavily abused in the decades from about 1880 through 1970. These were decades of intense agricultural land use, of rapid industrialization, of increasing consumption, and of extremely rapid city growth.

More recently, there has been a considerable environmental restoration and regulation of pollution aided by the withdrawal from the region of a substantial share of manufacturing activity, but severe burdens are still placed on the environment. Industrial pollution remains a problem. One recent study of point-source pollution in the Hudson found 147 industries and utilities that had discharged at least one toxic chemical into the river from 1978 through 1983, and many firms discharged more than one toxic stream. In addition, many of the plants that once operated in the inner- and outer-ring counties, especially in New Jersey because of the ease of waste disposal there, have left a heritage that persists today in the form of existing and abandoned waste dumps.

Current pollution burdens are the product of several factors, including continued leaching from wastes discarded in the past, the unprecedented large sprawl of population and settlement into areas not formerly urbanized, and the unanticipated effects of new products. Success in controlling pollution from processes of production has not been matched by control of pollution from now widely dispersed processes of consumption that have become the principal source of heavy-metals emissions (see appendix). The regulation of leaded gasoline, however, has led to noticeable improvements in the lead burden. In addition, we now have not only higher standards of environmental quality but also the ability to measure injurious substances in much smaller quantities than was possible in the past. The future will almost undoubtedly bring with it both more stringent regulation of traditional pollutants and the discovery of injurious effects from unsuspected sources.

Appendix: Reconstruction of Selected Anthropogenic Pollutant Emissions in the Hudson-Raritan Estuary and Associated River Basins

The previous pages of this article presented primarily a narrative history of anthropogenic changes in the Hudson-Raritan estuary and associated river basins. This history was intended to supply perspective on the dynamics of population growth, agricultural and industrial change, and pollution impacts on regional waterways in the Hudson-Raritan Basin. Data, however, is uneven in regard to these variables. Relatively full statistics are available for population changes, and satisfactory information is available for compiling a picture of agricultural and industrial developments. The study of water pollution, however, is much more difficult. The previous discussion of pollution is based on information reported in various contemporary studies and is therefore limited by the adequacy of contemporary observations and measurements. Thus for many years water-quality indicators measured only BOD, DO, and coliforms, focusing primarily on the hazards posed by human wastes in sewage.[65]

The absence of indicators or the relative lack of observations or measurements in contemporary reports for other pollutants such as heavy metals or chemical wastes does not mean that such substances were not present. To the contrary, they were often present in large amounts, but unless they created a noticeable nuisance or interfered with water or sewage treatment methods they were likely to be ignored. Industrial pollutants often had a devasting effect on fish and shellfish in the rivers and estuary, and industries such as the oystermen or groups such as the Isaac Walton League pushed for regulations. But, given the relatively low environmental consciousness of the time and government's reluctance to burden industry with environmental controls, strong action was seldom taken. Yet, such materials as heavy metals, in contrast to organic human wastes in sewage, are often persistent in the environment.

Forming a more quantitative estimate of what today are priority pollutants has value in order to understand long-term ecological trends, to provide environmental policymakers with an estimate of the magnitude of pollution flows and a sense of their origin, and to supply benchmarks. In an attempt to serve these purposes, in this study emission estimates were made over time for heavy metals, nutrients and biocides, and total organic carbon (TOC) from historical data such as firm and work-force size, materials used in processes, the character of process technologies, and quantities of fuel used. The method followed was that of calculating mass balances, operating on the basis of the law of the conservation of mass: that matter cannot be destroyed but only altered in form. In addition, estimates of flows of carbon, nitrogen, and phosphorus emissions were calculated over time from human and animal population figures, and pesticides and PCBs were calculated from crop and industrial data.

This appendix will discuss the origins of these effluent streams without providing statistical estimates. Those interested in statistical emissions estimates as well as a fuller discussion of the methodology used will find them in the cited works by Robert U. Ayres, Francis C. McMichael, and Samuel R. Rod.[66]

Pesticides and Herbicides—Agriculturally Related Pollution Flows

Before World War II, arsenic was the most widely utilized insecticide in the United States. Arsenic sprays were especially widely used in fruit orchards. Mercury-based fungicides for agricultural purposes were also quite important from the 1930s through the 1950s, with rapid decline after 1968. Most of these pesticides, fungicides, and herbicides are immobilized by the soil or biologically degraded and volatilized although a small amount is lost via runoff. Chlorinated pesticides were used from 1946 through the early 1970s, with restrictions beginning in the early 1960s. Their major farm uses were to control pests of cotton and corn, with lesser amounts used on soy beans, tobacco, potatoes, apples, citrus fruits, and other crops.[67] In the Hudson-Raritan Basin as a whole, deciduous fruit is today the region's most important crop, followed by hay. The use of chlorinated hydrocarbon pesticides in the basin appears to be quite small, suggesting that nonagricultural pesticide uses involving lawns and road borders have been most significant in accounting for pesticides found in the estuary's waters.

Metallurgical Refining Processes

The Hudson-Raritan Basin was an important producer of iron in the middle of the nineteenth century; but while its leadership continued in the metal-products industry, in the years after 1880 its share of the U.S. iron/steel industry declined. In contrast, the copper- and lead-refining industries were highly concentrated in the region from the 1890s to the 1970s. The convenience of New York Harbor as a site for processing imported copper ores from Chile for reexport as finished products and the early dominance of New York as an electro-metallurgical center accounted for this clustering.[68] Most of the important copper-using industries were also located in the area.

The metallurgy of copper is complex, with each smelter and refinery having a different pattern. Gaseous waste streams are generated from copper smelters and converters, while slag is the major solid waste (3 tons of slag produced per ton of blister copper, but as high as 10 to 1 for the concentrates used in the New Jersey–New York smelters). Slag is mostly inert oxides, but it does contain some copper and other trace metals. Arsenic is a by-product of the processing of copper ores and is emitted both in gaseous form and as a component of the slag. The slag was normally disposed of in landfills, often on refinery sites located in low-lying marshy areas.[69] A major source of water pollution from copper refineries located in the Hudson-Raritan Basin could be leaching from old slag piles.

Other heavy metals such as arsenic and cadmium are associated with copper and zinc ores. Some is lost in mining and ore-processing operations. Until recently, there was one zinc mine operating in New Jersey on a tributary of the Hudson River; it produced 29,000 tons of recoverable zinc in 1980. In addition, there was one plant manufacturing zinc oxide and a secondary zinc recovery unit in New Jersey.[70]

There are no lead mines in the Hudson-Raritan Basin, but three major lead refineries existed there in the past and conducted lead refining and desilverizing from imported ore or bullion. In 1899, two of these plants accounted for 32.5 percent of the total of 222,000 metric tons of refined lead in the U.S. Lead-refinery output in New Jersey probably peaked sometime before World War I, although one refinery operated until 1945 and a second until 1961.[71]

Control of emissions from copper and lead smelters came early in the twentieth century, driven by damages to plants and animals near copper smelters in Tennessee, Arizona, Montana, and Ontario. The initial control efforts involved mechanical techniques, such as baffles, expansion chambers, and "bag houses." In 1907, F. G. Cottrell of the University of California developed the first successful electrostatic precipitator, which was quickly applied in the copper-smelting industry. By the end of the 1920s most copper smelters had Cottrell Treators (ESPs), usually operating at 90 percent efficiency or better.[72] Averaging over all nonferrous smelters and related activities in the Hudson-Raritan Basin, we estimate roughly 50 percent dust recovery by 1900 (entirely mechanical), 80 percent by 1920 (mostly mechanical), 90 percent by 1930 (ESPs), 93 percent by 1940, 95 percent by 1950, 97 percent by 1960, and 98 percent by 1970.

Heavy Metals Released by Fossil Fuel Combustion

Up until the 1820s and 1830s, wood was the primary fuel used by both urban and rural dwellers in the Hudson-Raritan region. At this time, however, anthracite coal from eastern Pennsylvania entered the urban markets and by the middle of the century it was the dominant domestic and industrial fuel. Bituminous coal from western Pennsylvania began to capture industrial markets in the late nineteenth century, and by 1900, although anthracite was still the primary fuel used in New York State and New York City, bituminous was increasingly used in industry. In that year, 13,572,000 tons of anthracite were consumed for all uses, while 5,721,000 tons of bituminous were used for industrial purposes. In addition, over 16 million tons of bituminous were used by New York State's railroads. Anthracite use reached a peak in 1920 but by 1925 had declined to the 1900 level of use. Bituminous was the preferred fuel for industry (12,285,000 tons) and especially in the growing electrical utility industry (5,354,000 tons). In addition, by the 1920s both residual and distillate oil had become important fuels. In 1940, bituminous and anthracite were used about equally in the state, although bituminous was the preferred fuel for industrial and electrical-utility use, and anthracite was used primarily for domestic purposes. The use of fuel oil also accelerated, about equally divided between electric utilities, industry, and domestic use. In 1940, about 69 percent of households in the New York metropolitan area burned coal for fuel, about 25 percent used oil, and the remainder used gas or wood.[73]

By 1950, bituminous coal was being used in the state at a ratio of about 2 to 1 over anthracite, with electrical utilities and industry continuing as the dominant users. Fuel oil had made sharp inroads into the New York metropolitan area domestic markets and was now about even with coal in terms of percent of users (coal, 38 percent of households; oil, 37 percent; gas, 18 percent). By 1960, anthracite use in the state was down sharply to a total use of 1,577,000 tons, while bituminous had soared to 23,316,000 tons. Fuel oil was also up sharply in all areas of use. In 1960, in the New York metropolitan area, coal was only used by 434,817 households, while oil was the fuel for over 2.5 million users. Another 700,000 households used gas for fuel.[74]

In the 1970s, the energy crisis and rising oil prices caused utility and household oil use in the state to diminish sharply, but industrial use continued to be high. Bituminous coal use was also down considerably, while anthracite use was almost nonexistent. Natural gas use by electric utilities and by commercial and residential users, however, soared, with increases also by industry.

There is evidence that minor trace metals may be mobilized by coal combustion to an extent comparable with natural processes or mining. The quantity of trace metals actually released to the environment depends on the fraction that escapes as vapor and the fraction that condenses on very small particulates. There are actually three distinct phases in the history of coal-ash control. In the first phase before 1920, a large fraction (about 50 percent of stoker-fired coal combustion) of the ash was collected and removed in solid form and usually deposited in landfills or dumps. In the second period, after 1915, mechanical baffles, filters, and cyclones were used, although pulverized fuel increased the fly ash generated. In 1929, however, N.Y. Edison Company and other utilities adopted electro-static precipitators with about 90 percent recovery efficiency.[75] Households and industries, however, continued to burn coal without fly-ash controls until 1950. Averaging all users, we estimate 50 percent control for 1930, 60 percent for 1940, 70 percent for 1950, 90 percent for 1960, and 97 percent by 1970.

Heavy Metals Released by Product Dissipation

During the past thirty to forty years, the major source of heavy-metals emissions into the estuary have shifted from production-related to consumption-related emissions. Copper prior to 1920, chromium prior to 1930, cadmium prior to 1940, and lead prior to 1950 were primarily derived from productive sources. Arsenic, mercury, silver, and zinc in the environment, however, were always derived more from consumptive than productive processes. There are now various consumption uses that release heavy metals into the environment in ways that contribute to water pollution via surface runoff. These more recent sources of heavy metals vary in their degrees of dissipation in use and their modes of release to the environment.

Since 1950, tetraethyl lead (TEL) used as an "anti-knock" additive for gasoline, has been the principal source of lead emissions to the environment and the most significant consumption-related source of any one heavy metal. (Before the 1950s, pigments for paints—"white lead" and "red lead"—were the major dissipative use of lead.) Consumption of tetraethyl lead in any region can be assumed to be roughly in proportion to the region's gasoline consumption. The gasoline consumption of the Hudson-Raritan watershed area in 1970, for instance, was 6.65 percent of the U.S. total, and on this basis it can be assumed that close to 6.65 percent of the 246,000 tons of lead consumed in gasoline in 1970, or 16,400 tons, was consumed in the Hudson-Raritan Basin. Corresponding ratios for other years can be estimated.

Of the total consumption of lead as TEL in 1970, about 70 percent, or 11,500 tons, must have been emitted directly to the basin's air, the rest being retained in auto exhaust systems, oil filters, and waste motor oil. Approximately 32 percent of all motor oil is burned or lost by leakage onto city streets and washed into the sewers. Some waste motor oil is dumped "privately" and often commingled with refuse. In urbanized areas, much of it is ultimately incinerated. Some oil is collected and used later as fuel. This may account for the high lead content in some incinerator emissions. In any case, most of the lead emitted by motor vehicles is deposited locally, and most of it finds its way into watercourses via runoff.

For heavy metals as a group, we have identified ten consumption categories that are readily distinguishable in terms of their different degrees of dissipation in use and different modes of release into the environment.

1. Metallic uses, such as in alloys, incur environmental losses mainly in the production stage and as a result of corrosion or discharge to landfills.

2. Plating and anodyzing (excluding paints and pigments) generate some losses in the plating or treatment process and some corrosion loss.

3. Paints and pigments generate losses at the point of application and from weathering and wear. Some are ultimately disposed of in landfills along with discarded objects or building materials.

4. Batteries and electronic devices have relatively short useful lives of one to ten years. Production losses can be significant. Most are discarded to landfills.

5. Other electrical equipment as above but may be longer lived.

6. Industrial chemicals and reagents (such as catalysts and solvents) that are not embodied in products have short useful lives. Catalysts and solvents are partially recycled; others are lost directly to air or water.

7. Chemical additives to consumer products, including fuel additives and rubber vulcanizing agents and pigments, detergents, plasticizers, and photographic film are disposed of mainly to landfills or incinerators. There is no recycling.

8. Agricultural pesticides, fungicides, and herbicides are used dissipatively on farms and nurseries. Most are immobilized by soil or are biologically degraded and volatilized. There is some uptake into the food chain and a small amount of loss via runoff.

9. Nonagricultural biocides, including the above, are used in homes and gardens. These uses are dissipative, but most biocides again are immobilized by soil.

10. Pharmaceuticals and germicides are used in the home or in health-service facilities and are largely discharged via sewage or to incinerators.

Almost no data have been published on emissions coefficients for consumption activities, and many analysts have not considered such activities to be "sources" of pollutants. Yet, in the total metal emissions for the Hudson-Raritan

Basin, in recent decades consumption-related emissions have dominated production related emissions by a large factor.

PCBs

The emissions of PCBs into the Hudson River are largely due to a single point source. Two General Electric capacitor plants at Fort Edward and Hudson Falls in the upper basin purchased about 35,000 tons of PCBs during the nine years from 1966–74. This constituted about 15 percent of the total U.S. consumption during that period and 25 percent of U.S. consumption for electrical equipment. Assuming similar patterns of U.S. consumption during the earlier period from 1948 to 1965, for which GE data are lacking, the total amount used at these locations was about 75,000 tons. Given the amounts known to have been removed by dredging and that remaining in sediments, there must have been a total accumulation of at least 500 tons in the upper basin corresponding to a loss rate from the capacitor plants of 0.67 percent. Roughly 50–60 percent of this is now immobilized on dump sites from dredging in 1978–80. Around 64 tons remained in the riverbed of the upper basin as of 1978.

Notes

The research for this chapter was originally conducted under grants from the National Oceanic and Atmospheric Administration (NA82RAD00010 and NA83AA-D00059) and the National Science Foundation (SES-8420478). For the full reports on this material, see Joel A. Tarr and James McCurley III, *Historical Assessment of Pollution Impacts in Estuaries: The Potomac, Delaware, Hudson, Connecticut, and Narragansett. A Report to the National Oceanic and Atmospheric Administration* (Pittsburgh: NOAA, 1985); and Robert U. Ayers (principle investigator), Leslie W. Ayers, Joel A. Tarr, and Rolande C. Widgery, *An Historical Reconstruction of Major Pollution Levels in the Hudson-Raritan Basin: 1880–1980*, NOAA Technical Memorandum NOS OMAIA 43 (Rockville, Md., 1988). The authors would like to thank James McCurley III, Mitchell Small, Samuel R. Rod, and Martin Marietta Environmental System for their help in collecting and analyzing the data.

1. This essay was originally published in somewhat different form in B. L. Turner II, William C. Clark, Robert W. Kates, John F. Richards, Jessica T. Mathews, and William B. Meyer, eds., *The Earth as Transformed by Human Action: Global and Regional Changes in the Biosphere over the Past 300 Years* (New York: Cambridge University Press, 1990).

2. For descriptions of the Hudson River, see Robert H. Boyle, *The Hudson River* (New York: W. W. Norton Co., 1979); Oswald A. Roels, ed., *Hudson River Colloquium, Annals of the New York Academy of Sciences*, Vol. 250. For the New York Bight, see Charles G. Gunnerson, "The New York Bight Ecosystem," in R. A. Geyer, ed., *Marine Environmental Pollution*, Vol. 2: *Dumping and Mining* (Amersterdam: Elsevier Scientific Publishing Co., 1981), 313–77.

3. W. Rudolfs and H. Heukelekian, "Raritan River Pollution Studies: Comparison of Results Obtained in 1927–28, 1937–38. 1940–41," *Sewage Works Journal* 14 (1942): 839–65.

4. N. F. Brydon, *The Passaic River: Past, Present, Future* (New Brunswick, N.J.: Rutgers University Press, 1974), 3–23.

5. Joel A. Tarr and James McCurley lll, *Historical Assessment of Pollution Impacts in Estuaries: The Potomac, Delaware, Hudson, Connecticut, and Narragansett. A Report to the National Oceanic and Atmospheric Administration* (Pittsburgh: NOAA, 1985).

6. John E. Sanders, "Geomorphology of the Hudson Estuary," in Roels, ed., *Hudson River Col-*

loquium, 5–39. W. Cronon, *Changes in the Land: Indians, Colonists, and the Ecology of New England* (New York: Hill and Wang, 1983), 27; Michael Williams, *Americans and their Forest: A Historical Geography* (New York: Cambridge University Press, 1989), 111–189, 331–390; and Michael Williams, "Clearing the United States Forests: Pivotal years 1810–1860." *Journal of Historical Geography,* 8 (1982): 12–28.

7. J. Gottmann, *Megalopolis: The Urbanized Eastern Seaboard of the United States* (Cambridge, Mass.: MIT Press, 1961), 232–41, 341–83; R. W. Richardson and G. Tauber, eds., *The Hudson River Basin: Environmental Problems and Institutional Response* 2 vols. (New York: Academic Press, 1979), I: 273–74.

8. Historians William Cronon and Timothy Silver have described the alternative effects of Indian occupation of the land with that of colonial Americans for the New England and for the South Atlantic regions, and their analysis also applies to the Mid-Atlantic region and will be followed here. See Cronon, *Changes in the Land,* 46–53, 147; and Timothy Silver, *A New Face on the Countryside: Indians, Colonists, and Slaves in South Atlantic Forests, 1500–1800* (New York: Cambridge University Press, 1990). Also see Ulysses Prentiss Hedrick, *A History of Agriculture in the State of New York* (New York: Hill and Wang, 1966), 20–39.

9. Cronon, *Changes in the Land,* 53, 109–113; D. M. Ellis, J. A. Frost, H. C. Syrett, and H. J. Carman, *A History of New York State* (Ithaca, N.Y.: Cornell University Press, 1968), 66–69; and Williams, "Clearing the United States Forests," 12–28.

10. Cronon, *Changes in the Land,* 98–102, 127–156; Ellis et al. *A History of New York State,* 79.

11. Ellis et al., *A History of New York State,* 5–15, 72–75; Hedrick, *A History of Agriculture,* 40–63.

12. Ellis et al., *History of New York State,* 16–65, 71–73, 91–93; Richard A. Wines, *Fertilizer in America: From Waste Recycling to Resource Exploitation* (Philadelphia: Temple University Press, 1985), 6–15.

13. Ellis et al., *A History of New York State,* 159, 186-187, 211-216; D. W. Meinig, "Geography of Expansion," in J. H. Thompson, ed., *Geography of New York State* (Syracuse: Syracuse University Press, 1966), 166.

14. D. W. Meinig, "Elaboration and Change," in Thompson, ed., *Geography of New York State,* 178.

15. Jean Gottmann, *Megalopolis,* 32; Richardson and Tauber, *The Hudson River Basin* I: 113–19; and Gottmann, *Megalopolis,* 261.

16. G. W. Carey, *A Vignette of the Metropolitan Region* (Cambridge, Mass.: Ballinger Publishing Co., 1976), 1–6.

17. Richardson and Tauber, *The Hudson River Basin* I: 113–19.

18. Kenneth T. Jackson, *Crabgrass Frontier: The Suburbanization of the United States* (New York: Oxford, 1985); Peter O. Muller, *Contemporary Suburban America* (Englewood Cliffs, N.J.: Prentice Hall, 1981).

19. Jackson, *Crabgrass Frontier,* 231–245. For urban landscape changes, see Edward Relph, *The Modern Urban Landscape* (Baltimore: Johns Hopkins University Press, 1987); Richardson and Tauber, *The Hudson River Basin,* I: 113–22; and J. Gottmann, *Megalopolis,* 263.

20. The seven counties not included were Columbia, Greene, Montgomery, Saratoga, Ulster, Warren (1880 only), and Washington counties; the three counties in the upper Hudson Metropolitan Zone were Albany, Renselaer, and Schenectedy counties. Annexations accounted for some of the increase.

21. This follows the schema used by the Regional Plan Association in its 1950s study of the New York metropolitan Region, except for deletion of Suffolk County from the outer ring. The rationale for its removal is that its runoff patterns are not toward the estuary. See E. M. Hoover and R. Vernon, *Anatomy of a Metropolis: The Changing Distribution of People and Jobs within the New York Metropolitan Region* (Cambridge, Mass.: Harvard University Press, 1959), 8. In 1968 the Regional Plan Association increased the number of counties it designated as part of the region to thirty-one. Allan R. Pred, *The Spatial Dynamics of U.S. Urban-Industrial Growth, 1800–1914: Interpretive and Theoretical Essays* (Cambridge, Mass.: MIT Press, 1966), 1–83.

22. Ellis et al., *A History of New York State*, 205–06, 223–24.

23. Bronx County was formed early in the century and is added to the core totals for 1920. Westchester County was added to the totals for the inner ring for that time.

24. Richard Harris, "Industry and Residence: The Decentralization of New York City, 1900–1940," *Journal of Historical Geography* 19 (1993): 169–76.

25. Richard Harris, "The Geography of Employment and Residence in New York since 1950," in John H. Mollenkopf and Manuel Castells, eds., *Dual City: Restructuring New York* (New York: Russell Sage Foundation, 1991), 129–52.

26. Ellis et al., *A History of New York State*, 183; Sean Wilentz, *Chants Democratic* (New York: Oxford University Press, 1984), 31.

27. Meinig, "Geography of Expansion," 140–71.

28. Thomas C. Cochran, *Frontier of Change: Early Industrialism in America* (New York: Oxford University Press, 1981), 90; Kenneth Warren, *The American Steel Industry 1850–1970: A Geographical Interpretation* (London: Clarendon Press, 1973), 15–18; and E. K. Spann, *The New Metropolis: New York City, 1840–1857* (New York: Columbia University Press, 1981), 405–6.

29. Wilentz, *Chants Democratic*, 108–12; Spann, *The New Metropolis*, 402–3; and Meinig, "Geography of Expansion," 168.

30. Ellis et. al., *A History of New York State*, 512; Meinig, "Elaboration and Change," 178–79.

31. Pred, *The Spatial Dynamics of U.S. Urban Industrial Growth*, 24–32.

32. R. M. Lichtenberg, *One-Tenth of a Nation: National Forces in the Economic Growth of the New York Region* (Cambridge, Mass.: Harvard University Press, 1960), 6–14.

33. Ibid., 13–17.

34. A. F. Hindrichs, *The Printing Industry, Monograph No. 6, Regional Plan of New York and Its Environs* (1928; reprinted, New York: Arnold Press, 1974), 13.

35. U.S. Bureau of the Census, 1880; Vincent W. Lanfear, *The Metal Industry, Monograph No. 2. Regional Plan of New York and Its Environs* (1928; reprinted, New York: Arnold Press, 1974), 16–18.

36. Ibid., 26–27.

37. U.S. Bureau of Mines, *Mineral Facts and Problems* (Washington, D.C.: Government Printing Office, 1975); T. R. Navin, *Copper Mining and Management* (Tucson: University of Arizona Press, 1978), 57, 65–69; and Lanfear, *The Metal Industry*, 27.

38. Lanfear, *The Metal Industry*, 29–34.

39. Mabel Newcomer, *The Chemical Industry, Monograph No. 1, Regional Plan of New York an its Environs* (1922; reprinted, New York: Arnold Press, 1974), 12.

40. Andrew Hurley, "Creating Ecological Wastelands: Oil Pollution in New York City, 1870–1900," *Journal of Urban History* 20 (May 1994): 340–64; Newcomer, *The Chemical Industry*, 17–23.

41. Newcomer, *The Chemical Industry*, 48–49.

42. Hurley, "Creating Ecological Wastelands," 357–60; Joseph Pratt, "The Corps of Engineers and the Oil Pollution Act of 1924" (unpublished manuscript, 1983); and H. M. Larson, E. H. Knowlton, and C. S. Popple, *History of Standard Oil Company (New Jersey), Vol. 3: New Horizons 1927–1950* (New York: Harper and Row, 1971), 771–72.

43. Regional Plan Association, *Outlook for the Tri-State Region Thru 2000* (New York: Regional Plan Association, 1986); Harris, "The Geography of Employment and Residence in New York Since 1950," 130–32.

44. Hoover and Vernon, *Anatomy of a Metropolis;* Harris, "The Geography of Employment," 132–34; U.S. Bureau of the Census, *Census of Manufactures* (Washington, D.C.: Government Printing Office, 1985).

45. Steven O. Rohmann et al., *Tracing A River's Toxic Pollution: A Case Study of the Hudson: An Inform Report* (New York: Inform, Inc., 1985);` Ronald J. Breteler, ed., *Chemical Pollution of the Hudson-Raritan Estuary, NOAA Technical Memorandum NOS OMA 7* (Rockville, Md.: NOAA, 1984).

46. Newcomer, *The Chemical Industry*, 18–20, 22; Michael R. Greenberg and Richard F.

Anderson, *Hazardous Wastes Sites: The Credibility Gap* (New Brunswick, N.J.: Rutgers University Press, 1984), 130–63.

47. Citizens' Association of New York, *Report upon the Sanitary Condition of the City* (New York: D. Appleton and Co., 1866); John Duffy, *A History of Public Health in New York City, 1625–1966* (New York: Russell Sage, 1974), 1–142; D. R. Franz, "An Historical Perspective on Molluscs in Lower New York Harbor with Emphasis on Oysters," and S. C. Esser, "Long-term Changes in Some Finfishes of the Hudson-Raritan Estuary," both in G. F. Mayer, ed., *Ecological Stress and the New York Bight: Science and Management* (Columbia, S.C.: Estuarine Research Federation, l982), 181–98, 299–314.

48. M. J. Klawonn, *A History of the New York District U.S. Army Corps of Engineers, 1775–1975* (New York: U.S. Army Engineer Corps–New York District, 1977); J. W. Mersereau, *1887 Report of J. W. Mersereau State Oyster Inspector, Report of Oyster Investigation and Shellfish Commissioner,* New York State Assembly Document No. 37, 1888.

49. Newcomer, *The Chemical Industry,* 18; Lanfear, *The Metal Industry,* 16. See also, Hurley, "Creating Ecological Wastelands," 342–56.

50. M. O. Leighton, *Sewage Pollution in the Metropolitan Area near New York City and Its Effect on Inland Water Resources, USGS Water-Supply and Irrigation Paper No. 72* (Washington, D.C.: Government Printing Office, 1902).

51 Other important studies include *Reports of the New York Bay Pollution Commission to the Governor,* March 31, 1905, and April 30, 1906; and William M. Black and Earle B. Phelps, *Report Concerning the Location of Sewer Outlets and the Discharge of Sewage in New York Harbor to the New York Board of Estimate and Apportionment,* March 23, 1911.

52. J. A. Mytelka, M. Wendell, P. L. Sattler, and H. Golub, "Water Quality of the Hudson-Raritan Estuary" (New York: Interstate Sanitation Commission, 1981), 152–79.

53. In 1908 the New Jersey State Sewerage Commission sponsored an investigation into the generation and polluting characteristics of gas house wastes. G. C. Whipple, "Gas Wastes," *Report of the New Jersey State Sewerage Commission* (Somerville, N.J., 1908). The annual reports of the New York Department of Conservation and the New Jersey Sewerage Commission throughout the 1920s noted consistent problems with gas house wastes.

54. U.S. Engineer Office, First District, NYC, "Report on Investigation of Pollution of Navigable Waterways and Their Tributaries," n.d., in U.S. Army Corps of Engineers Records, National Archives, Suitland, Md.; New York Conservation Commission, *Annual Report* (Albany: 1924), 55.

55. Mytelka et al., "Water Quality of the Hudson-Raritan Estuary," 152–79; New York Conservation Commission, *Annual Reports* for 1923, 1930, 1933; New York State Department of Health, *Annual Report,* for 1925, 1926, 1929; and Remington, Vosbury, and Goff, *Methods for Abatement of the Pollution of the Raritan River, Report to the Port Raritan District Commission* (New Brunswick, N.J., 1930); New Jersey Passaic Valley Sewerage Commission, *Report* (Newark, N.J., 1926); and New York City Sanitation Department, *Preliminary Reports on the General Plans for Sewage Disposal for the City of New York* (New York, 1931).

56. A. S. Loop, *History and Development of Sewage Treatment in New York City* (New York: Department of Health, 1964), 52–70; New York State Legislature, *Report of the Hudson Valley Survey Commission to the New York Legislature, Legislative Document No. 71* (1939); and Rudolfs and Heukelekian, "Raritan River Pollution Studies," 839–65.

57. Mytelka et al., "Water Quality of the Hudson-Raritan Estuary," 152–79; Rudolfs and Heukelekian, "Raritan River Pollution Studies," 865.

58. New York Department of Conservation, *Annual Reports for 1946, 1948;* Leonard Metcalf, Harrison P. Eddy, and Elson T. Killiam, *Summary Report to the Middlesex County Sewerage Authority upon Abatement of Water Pollution in the Raritan River, Its Tributaries, and Raritan Bay* (New Brunswick, N.J.: 1951).

59. *Industrial Waste Inventory* (New York: Interstate Sanitation Commission, 1951).

60. Water Pollution Control Board, *New York City Waters Survey Series, Reports 1–5* (Albany: New York State Department of Health, 1960).

61. The use of such classification schemes had been pioneered by Massachusetts early in the century and had also been used by Pennsylvania in the 1920s. See, Joel A. Tarr, "Industrial Wastes and Public Health: Some Historical Notes," *American Journal of Public Health* 75 (September 1985): 1061–62.

62. U.S. Department of Health, Education, and Welfare, *Report on Pollution of the Hudson River and its Tributaries* (Washington, D.C.: Public Health Service, 1965).

63. C. G. Gunnerson, "New York City: Costs, Financing, and Benefits of Conventional Sewerage," in *Project Monitoring and Reappraisal in the International Drinking Water: Supply and Sanitation Decade* (New York: American Society of Civil Engineers, 1981); Leo Hetling, *An Analysis of Past, Present, and Future Hudson River Wastewater Loadings Technical Paper 37* (Albany: New York State Department of Environmental Conservation, 1974); and Mytelka, "Water Quality of the Hudson-Raritan Estuary," 32–150.

64. A useful survey of pollution conditions in the New York Bight is Donald F. Squires, *The Ocean Dumping Quandary: Waste Disposal in the New York Bight* (Albany: State University of New York Press, 1983); James A. Mueller, Theresa A. Gerrish, and Maureen C. Casey, *Contaminant Inputs to the Hudson-Raritan Estuary. NOAA Technical Memorandum OMPA 21* (Boulder: NOAA Office of Marine Pollution Assessment, 1982); and Breteler, *Chemical Pollutants of the Hudson-Raritan Estuary*, 7.

65. Tarr, "Industrial Wastes and Public Health," 1064–66.

66. Ayres et al., *An Historical Reconstruction of Major Pollutant Levels in the Hudson-Raritan Basin*; Robert U. Ayres, *Resources, Environment and Economics: Applicants of the Materials/Energy Balance Principle* (New York: John Wiley and Sons, 1978); Robert U. Ayres and Samuel R. Rod, "Patterns of Pollution in the Hudson-Raritan Basin: Reconstructing an Environmental History," *Environment* 28 (May 1986): 4, 14–20, 39–43; Robert U. Ayres, Francis C. McMichael, and Samuel R. Rod, "Measuring Toxic Chemicals in the Environment: A Materials Balance Approach," in Lester B. Lave and A. C. Upton, eds., *Toxic Chemicals, Health, and the Environment* (Baltimore: Johns Hopkins University Press, 1987); and Samuel R. Rod, "Estimation of Historical Pollution Trends Using Mass Balance Principles: Selected Metals and Pesticides in the Hudson-Raritan Basin, 1880–1980," (Ph.D. diss., Carnegie Mellon University, 1989).

67. J. Whorton, *Before Silent Spring: Pesticides and Public Health in Pre-DDT America* (Princeton: Princeton University Press, 1974); John H. Perkins, *Insects, Experts, and the Insecticide Crisis: The Quest for New Pest Management Strategies* (New York: Plenum Press, 1982), 3–48; and Christopher J. Bosso, *Pesticides and Politics* (Pittsburgh: University of Pittsburgh Press, 1987), 109–42.

68. M. M. Trescott, *The Rise of The American Electrochemical Industry 1880–1910* (Westport, CT: Greenwood Press, 1981); T. R. Navin, *Copper Mining and Management* (Tucson: University of Arizona Press, 1978).

69. Robert U. Ayres, et. al., *An Historical Reconstruction of Major Pollutant Levels in the Hudson-Raritan Basin: 1880–1980, Volume II: Heavy Metals and Fossil Fuels, NOAA Technical Memorandum NOS OMA 43* (Rockville, Md.: NOAA, 1988), 41–67.

70. Ibid., 82–91.

71. Ibid., 68–81.

72. Ibid., 102–6.

73. F. M. Binder, *Coal Age Empire: Pennsylvania Coal and Its Utilization to 1860* (Harrisburg: Pennsylvania Historical and Museum Commission, 1974), 53, 92, 141–48; Alfred D. Chandler, Jr., *The Visible Hand: The Managerial Revolution in American Business* (Cambridge, Mass.: Harvard University Press, 1977), 76–77; and Bureau of the Census, 1940.

74. Bureau of the Census, 1950, 1960.

75. R. Hering and S. A. Greeley, *Collection and Disposal of Municipal Refuse* (New York: McGraw-Hill, 1921); Carleton-Jones, "Electrostatic Precipitation of Fly-ash Reaches Fiftieth Anniversary," *Journal of the American Pollution Control Association* 24 (1974): 11.

CHAPTER III

The Pittsburgh Survey as an Environmental Statement

Historians have frequently explored the Pittsburgh Survey for insights into the problems of the industrial city. The Survey highlighted the discrepancy between industry's use of extensive planning and expertise in the name of production and profit, and the limited attention paid to housing, social, and sanitary conditions in Pittsburgh working-class neighborhoods. In addition, it dramatically illustrated the unsafe working conditions existing in the city's and region's factories and workshops. But while utilizing the Survey for its depiction of social and physical conditions in the industrial city, historians have neglected to view it as an environmental statement that provided a graphic picture of Pittsburgh's degraded environment and landscapes. They also have not fully assessed the relationship that the Survey investigators perceived between the city's environment and the "moral" behavior of its working-class and immigrant inhabitants.[1]

Other American industrial cities also possessed befouled environments, although not necessarily as extreme as Pittsburgh's. Progressive reformers viewed such urban conditions as a challenge to the possibility of establishing a moral social order among the city's masses. Rather than despair, however, many reformers concluded that environmental betterment could produce moral improvement; that is, a transformed and planned urban environment could change the behavior and mold the character of urbanites.

These "positive environmentalists," as historian Paul Boyer has labeled them, embraced a series of reforms, including housing renewal, the creation of parks and playgrounds, sanitary improvements, smoke control, and city planning, as the means to regenerate urban society.[2]

The Pittsburgh Survey investigators fit the profile of Boyer's "positive environmentalists." They believed that environment could shape behavior for better or worse. They also believed that environment was malleable and could be reshaped by human intervention, often with technological assistance. Their improvement efforts targeted both the natural and the built environments, reflecting their belief that broad governmental programs could restructure the urban environment and influence the city's "moral destiny." The reformers argued that change would result from coalitions of "urban experts" (especially engineers and "social engineers" or social workers) and "enlightened" businessmen, who would join forces to promote the "new science and art of social up building," with the end of producing "a self-reliant, self-directing community."[3]

The Survey authors, however, as well as other progressive urban reformers, were overly sanguine (and perhaps naive) about the ability of urban experts and enlightened businessmen to overcome the environmental ills of industrial cities. Pittsburgh's reformers had actually begun to improve drinking water quality, reduce smoke, provide more parks and playgrounds, and upgrade sanitary conditions before the Survey was conducted, but had made only limited progress. The Survey highlighted these environmental issues and undoubtedly accelerated change in some areas, but permanent improvements often proved transitory or nearly impossible to achieve, as entrenched political forces and technological and economic limitations impeded change. By highlighting these environmental ills, however, the Survey provided a graphic record as ammunition for those who would renew the fight against them in the future.

This chapter explores several themes: the attitude of the Survey authors toward natural and manmade environments; their view of the "moral" importance of parks, gardens, and playgrounds to working class neighborhoods; and the ill effects of smoke and water pollution, inadequate sanitary facilities, and floods on the urban population. Finally, it considers why reformers found it so difficult to produce long-lasting and substantial improvements in these domains.

The Power and the Malleability of the Environment

The Survey investigators strongly associated human behavior with Pittsburgh's natural and built environments. They also believed that humans could alter these landscapes, for good or ill, with technology. While they perceived the city's rugged topography, massive industrialization, and haphazard growth patterns as formidable barriers to environmental improvement, they also believed that "human engineering" allied with modern technology could improve the quality of Pittsburgh life.

The compelling power and imagery of the physical features of the Pittsburgh site captured the imagination of the Survey writers. Boston settlement house director Robert A. Woods linked Pittsburgh's geography to destiny, arguing that "Physical environment, no less than racial stock and economic factors" was critical to community life. Woods admired the "involved panorama of the rivers, the . . . long ascents and steep bluffs, the visible signs everywhere of movement, of immense forces at work—the pillar of smoke by day, and at night the pillars of fire against the background of hillsides strewn with jets of light. . . ." City leaders were optimistic concerning their ability to conquer these landscape features, making Woods optimistic about Pittsburgh's future. The "sheer forces of physical setting and commercial need," he argued, drove Pittsburgh toward "urban coherence" and might lead the way toward a new "civic and human welfare movement" in American cities.[4]

Paul U. Kellogg, the Survey director, shared these beliefs, arguing that environment might be "destiny," but humans could still shape it. "Environment is inevitable as a selective agent," he wrote, "but the people once here, can by their willing, mold and perpetuate or destroy the holding power of the district."[5] Like other positive environmentalists, he believed that a socially responsible elite could control, improve, and shape the urban environment. Although Pittsburgh's three rivers were "not easy overlords," he wrote, the city's engineers had forced them to yield their "mastery," being "damed and sluiced and boiled and filtered to suit the demands of navigation and power and temperature and thirst." For Kellogg, a "matrix" of pipe lines, electric, telegraph and telephone wires, mile-long river barge tows, and extensive railroad tracks had almost completely obliterated the natural features of rivers, hills, and valleys. These technological sinews, he wrote, "bind here a district of vast natural resources into one organic whole," with the built environment replacing the natural. Thus, he observed, the city, not the wilderness, had become the "frontier."[6]

For the Survey writers, technology would enable Pittsburgh reformers to conquer their environment. The reprinting of Richard Realf's 1878 poem, "Hymn of Pittsburgh," embodied the strong belief that man could harness nature for his own ends. The poem was published in the January 2, 1909 issue of *Charities and The Commons* and also as a frontispiece for the volume of collected articles, *The Pittsburgh District: Civic Frontage.* The poem reads in part:

> My father was mighty Vulcan,
>   I am Smith of the land and sea . . . ,
> I am monarch of all the forges,
>   I have solved the riddle of fire,
> The Amen of Nature to need of Man,
>   Echoes at my desire . . . ;
> I quell and scepter the savage wastes,
>   And charm the curse from the soil;
> I fling the bridges across the gulfs,
>   That hold us from the To Be,
> And build the roads for the bannered march
>   Of Crowned Humanity.

Allen T. Burns, a Survey author and social worker, explicitly connected technology and domination of nature in his 1911 article, "Coalition of Pittsburgh's Civic Forces." Burns noted that, despite Pittsburgh's greater "physical obstacles" to communication than any other American city, bridges and tunnels were breaking down "the physical barriers" and bringing a "new spirit of co-operation . . ." between people and communities. By overcoming topographical obstacles and permitting flows of people and traffic, technology and physical infrastructure could unite the city's disparate groups.[7]

Putting City Dwellers in Touch with Nature: Parks, Gardens, and Playgrounds

Survey investigators followed the beliefs of Frederick Law Olmsted, the great nineteenth-century landscape architect, that the city would best meet human biological and social needs when it bridged the built and natural environments. Many early-twentieth-century urban intellectuals maintained that parks and other natural habitats could reduce urban stress. Exposure to nature, they argued, could promote the masses' moral growth; even small bits of nature in or around the home would be beneficial to family life.[8]

In his 1909 Survey essay, "Civic Improvement Possibilities of Pittsburgh," Charles Mulford Robinson, a self-educated architectural critic and an advocate of civic improvement, argued that the city should create parks in order to put Pittsburghers in touch with nature. Like Olmsted, Robinson believed in the redeeming character of natural environments and their ability to produce better citizens by awakening "high desires that had before been dormant. . . ."[9] Robinson, however, adapted Olmsted's ideas to include organized activities. He contended, as did other Progressive park designers, that mere exposure to nature was insufficient; parks needed to provide organized activity, such as directed sports, if they were to reform immigrant working-class behavior. As Galen Cranz observes, the reform park was a "moral defense against the potential for chaos" caused by unregulated spare time.[10]

Robinson noted that although "nature" had endowed Pittsburgh with one of the "most picturesque city sites in the world," the city had failed to develop fully its magnificent heritage. Thus, he added, a "wonderful natural picturesqueness is contrasted with the utmost industrial defilement, smoke and grime and refuse pervading one of the finest city sites in the world." Robinson suggested construction of a comprehensive network of parks and playgrounds along the river banks with a "connecting system of boulevards and parkways." These would open up the "beauty of nature" to the "mass of . . . workers" for whom the idea might be "a new thought."[11]

For most of the nineteenth century, Pittsburgh had no parks; by the time of the Survey it had acquired what Robinson called "four public reservations," as well as a few small parks or "ornamental spaces." These four were Schenley and Highland parks in Pittsburgh's East End and Riverview Park and "the reservation" or West Park. From Robinson's perspective, none of these parks met what Pittsburgh "ought to have." Robinson maintained that the four large parks, especially Highland and Schenley parks, served the city's elite rather than its masses. Highland Park, he wrote, with its monumental entrances, costly flower beds, and zoo, was "a pretty good park of its kind—a very costly luxurious kind." Although located five miles from city hall, many working people used it on holidays; thus the park did "some social work although far from the amount desired." Schenley Park, in contrast, benefited workers "very little": it was distant from working-class neighborhoods, difficult to reach by streetcar, and devoid of paths, walks, or playing fields. Schenley, noted Robinson, was the province of Pittsburgh's wealthy East Enders and was not intended to meet "the democratic needs of Greater Pittsburgh."[12]

Robinson observed that an alternative type of park, the "trolley park" built at the end of a streetcar line, held more appeal than parks devoted to nature. Robinson referred to Kennywood Amusement Park, located on the bluff behind the Homestead mill and across the river from the Edgar Thompson Works in Braddock. With its "swings and boats, slides and ponies . . . garish with a blaze of electric lights," it attracted far more users than did the large city parks. Robinson disliked the "trolley park," even though he realized that people enjoyed its "entertainment, vivacity, and brilliancy." (In a later essay, he would write that "lights and music" would sometimes lure tenement dwellers into "dangerous places.") To meet the challenge of the amusement park, Robinson advocated the construction of "two or three well distributed and readily accessible large parks that would be real municipal pleasure grounds." In these, "tired workers" could be entertained and "learn there the more tranquil pleasure of contemplating nature."[13] Ideally, they would gradually distinguish between the mere stimulation and excitement of amusement parks, appreciate nature's deeper values, and become better citizens.

How does one understand the amusement park's powerful attraction? Historian John F. Kasson explains the popularity of amusement parks, such as Coney Island, as providing a kind of "play" that appealed to people's imagination. "Instruments of production and efficiency were transformed into objects of amusement, and life around them lifted from dull routine to exhilarating pageantry."[14] Joseph Stella, who helped illustrate the Survey volumes, was supposedly "thrilled" with Coney Island and the "spectacle of a new world of steel and electricity, surging with drama and demanding to be translated into a new art." This perspective would certainly apply to Pittsburgh, where the city's trolley parks appealed to working-class people who lived and worked in neighborhoods dominated by the huge structures, noise, and rhythms of the mill.[15]

Amusement parks offended some Survey investigators as sites of immorality and drinking. In her article on the Skunk Hollow slum, located in the valley below Luna Park, another Pittsburgh amusement park, Florence Larrabee Lattimore concluded that "the fanciful towers of Luna Park peered *jeeringly* into this pest hole of neglect. . . ." According to Kasson, progressives were concerned that the demands of industry and the lack of creative leisure were driving the immigrant urban working class toward "strong and coarse entertainment"—such as the amusement park or the saloon or the dance hall—that promised "escape rather than renewal." Only nature could provide such renewal.[16]

Margaret Byington appreciated Kennywood Park more than Lattimore or Robinson. She found the trolley park to be "the scene of many school and church picnics and lodge gatherings." Here, she noted, the young people "find the skating rinks and dancing pavilions and the shrill music of the merry-go-rounds" all acceptable amusements. Much to her dismay, she also observed that workers flocked to the wilder rides, such as the roller coasters, that catered to their baser instincts and exposed them to dangerous thrills. Byington wanted workers to maintain a balance between acceptable amusements and those that threatened to destroy self-control.[17]

Byington's Homestead had but two small parks, and she may therefore have viewed Kennywood amusements more sympathetically than the other Survey authors because it offered a locale for community leisure-time activities. She perceived the town as primarily populated by agricultural people who had moved from "quiet villages . . . to this smoky town; from labor in the open fields to heavy work in the yards and thundering sheds of the mill." Immigrant rural backgrounds, Byington believed, explained Slav efforts to develop vestiges of nature under difficult conditions. She observed, for instance, that the Slavs cultivated gardens and trees or added "a little bed of lettuce with its note of delicate green or the vivid red of a geranium blossom, despite their smoky environment," as well as "bits of lawn and flowers" in front of small and "closely set" frame structures. Houses on the hills above the river flats where the steel mill was located had larger gardens, and some residents raised chickens to sell the eggs. Company houses in Munhall, owned by the Carnegie Land Company, had "squares of lawn and shade trees in front."[18]

Gardens interested Byington for several reasons. First, they put town dwellers in touch with nature and reminded them of their rural origins. They also broke the "monotony of street after street," served as "play places" for children, and offered "rest and refreshments" for grownups. Vegetable gardens were important as supplements to the family budget; and gardens also could enhance "neighborliness of spirit, since the women often discuss over the fences their horticultural ambitions." Byington complained, however, that few Homestead neighborhoods offered the possibility for gardens because rear houses, privies, or sheds filled most backyards. The houses were also so close to the street that "the tenant can scarcely have that bit of garden so dear to the heart of former country dwellers."[19]

Byington's concern for natural elements and gardens links her ideologically with the park reformers, since gardens could offer some of the same uplifting elements as a great park. Other Progressive reformers shared this belief; in 1902, for instance, New York park reformers established minuscule farms, each assigned to a child, in New York's DeWitt Clinton Park. Advocates of the positive effects of vegetable gardening spread the idea throughout the country. Many urban educators advocated school gardens, especially in city neighborhoods without space for home gardens. Progressive educators argued that gardens would teach children about the beauty of nature, the "miracle of seedtime and harvest," and the virtues of industry and future planning.[20]

In addition to advocating the creation of parks as a source of spiritual uplift and as a means of inculcating proper decorum, the Survey also discussed playground reform. Beulah Kennard, president of the Pittsburgh Playground Association and a major figure in the Playground Association of America, maintained in a 1909 article in *Charity and the Commons* that members of Pittsburgh's working class often lacked a play spirit because they had been recruited from "the oppressed and impoverished peasants of southeastern Europe . . . [who were] not rich in play traditions and customs." Such a spirit could not be found either in the "ugly and forlorn . . . city of iron whose monster machinery rested neither day nor night . . . [where] green things could not grow because of the pall of smoke . . . clouding the sunlight, and leaving a deposit of grime on everything, including the children." Because the city lacked recreational spaces, children played in the streets. Playground reformers considered streets dangerous because of traffic but also because their freedom might tempt unsupervised street children into forbidden activities.[21]

Kennard believed, however, that "schools of play" could teach children how to play properly. In 1896, as chairman of the Education Department of the Civic Club of Allegheny County, she led the move to open several schoolyard playgrounds for public use under authority of a new state enabling act. The Civic Club operated the playgrounds as the "agent" of the Central School Board. In 1901, the Civic Club and several Pittsburgh women's clubs opened several "recreation parks." These were small neighborhood parks containing recreational facilities such as field houses, gymnasiums, baths, swimming pools, game rooms, and libraries, as well as gardens and play areas. Private funding supplemented the city council's allocations for the recreation parks and Pittsburgh Central School Board allocations for school playgrounds.[22]

In 1906, the several groups interested in playgrounds formed the Pittsburgh Playground Association and opened five new recreation parks. In 1907, however, the city refused to provide a requested $2 million city allocation to open more parks. Kennard published her article on Pittsburgh playgrounds in May, 1909, and in November, 1910, voters approved a $1 million bond issue for playgrounds. Her article, as well as others in the press, undoubtedly influenced the vote. By 1911 the city had created twelve new recreation centers, making seventeen in total operated by the Pittsburgh Playground Association, with a separate program on the city's North Side.[23]

The playgrounds and recreation parks designed for organized play differed from Olmsted's conception of parks as passive. Playground reformers believed that play energies had to be guided in worthy directions and combined with educational activities for older children (over eight years of age). Thus, the Pittsburgh "vacation school" linked play with the study of carpentry, cooking, industrial work, music, nature study, and art. The larger purpose of playgrounds was to socialize children to modern urban society, to prepare them for citizenship, and to create a "bond of fellowship" among "the common interests of the poor, the rich . . . , [and] the wage earners."[24]

The city and the Playground Association continued to improve existing parks and establish new parks and playgrounds. In 1915, the city formed a Bureau of Recreation under W. F. Ashe, the respected former director of the Pittsburgh Playground Association, and park and playground progress seemed assured. Ashe resigned, however, within two years, and both parks and playgrounds began to suffer from bureaucratic inattention. By the 1920s, reformers were again criticizing Pittsburgh's poor recreational facilities compared to other cities'.[25]

The pertinent publications were the Citizen's Committee on City Plan of Pittsburgh's 1920 report on playgrounds and 1923 report on parks. These studies detailed the inadequacy and poor condition of Pittsburgh's parks and playgrounds. They found that, while the city council had appropriated adequate funds for operations and improvements, park management was riddled with inefficiency and political favoritism. Playground operation was fragmented among several different governmental units which competed for funds.[26]

Thus, while the Survey's articles on parks and playgrounds may have fostered some progress when they appeared, the Survey writers had clearly underestimated the obstacles that blocked permanent improvements and continued efficient operation. Like other Progressive Period reformers, they had trusted citizen awareness and involvement to protect improvements in the city environment and underestimated the power of politics and bureaucratic infighting to derail this purpose.

Pollution, Public Health, and Floods

Progressive Period social workers and reformers, including the Survey investigators, held an engineering conception of social work. They believed that diverse community institutions could be structured to work together for reform. They strongly urged that professional investigators use governmental power to reduce poverty and acculturate the immigrant.

Social engineering would establish standards for a decent quality of life, and social workers would assist in bringing them above this line. Central to this goal was the improvement of the physical environment through technology and planning, for without decent housing, clean water, adequate sanitary facilities, and recreation services, they believed, other reform efforts would fail. Pittsburgh's unique environmental and sanitary conditions made these needs all the more urgent.[27]

The difficulties of coping with Pittsburgh's rugged landscape and industry's domination and scarring of its river valleys and flood plains emphasized the need for environmental remediation. The interaction of industrialization, urbanization, and the natural environment produced, as Anne W. Spirn observes, an "ecosystem very different from the one that existed prior to the city."[28] This ecosystem was maintained by huge inputs of energy and materials, altering and distorting the flow of natural processes but seldom fully controlling them. Such distortion was obvious in Pittsburgh, where the growth of the city and industry triggered major pollution problems that negatively affected every city resident and, according to reformers, complicated their attempts to motivate immigrant workers to higher levels of citizenship.

During the late nineteenth and early twentieth centuries, Pittsburgh annexed nearby communities and extended its infrastructure into the new areas to facilitate growth.[29] But service provision was uneven, and rapid city population growth and industrial development caused water pollution, inadequate sanitation, and voluminous smoke, all of which lowered the quality of life. Severe floods increased the cost of business for firms with riverfront locations and posed hazards to the life and property of workers living in the flood plains. The Survey investigators found "a city struggling for the things which primitive men have ready to [sic] hand,—clear air, clean water, pure foods, shelter and a foothold of earth."[30] They acknowledged the city's gains, while highlighting its environmental deficits.

Smoke

Pittsburgh's most ubiquitous pollution problem was smoke. Pittsburgh was the "smoky city," with landscapes often obscured by a thick cloud of fog, dust, and grime. Robinson called smoke "Pittsburgh's most famous because most obvious drawback. . . ."[31] Three factors caused heavy smoke—the use of high-sulfur bituminous coal for energy by industries, railroads, river boats, businesses, and residences; inefficient combustion; and climatic inversions. As Survey investigators explored the various neighborhoods and trades, they found smoke ubiquitous, reducing the quality of life, increasing burdens and costs for all the city's inhabitants (but especially for women), and negatively affecting the landscape. A list of examples from the Survey follows:

THE LANDSCAPE[32]:

"the pillars of smoke by day, and at night the pillars of fire . . ." (Robert A. Woods)

"Then comes the city with its half-conquered smoke cloud, with its high, bare hills . . ." (Paul U. Kellogg)

"smoke and ore dust are very trying to vegetation under the most favorable conditions . . ." (Charles M. Robinson)

"On a little hill, barren as yet, after the wont of Pittsburgh hills . . ." (Elisabeth Butler)

"The trees are dwarfed and the foliage withered by the fumes; the air is gray, and only from the top of the hills above the smoke is the sky blue." (Margaret Byington)

THE HOUSEWIFE AND CHILDREN[33]:

"An enveloping cloud of smoke and dust through which light and air must filter made housekeeping a travesty in many neighborhoods." (F. Elisabeth Crowell)

"In this smoky town a double amount of washing and cleaning must be done." (Margaret Byington)

"In many places green things could not grow because of the pall of smoke which swept heavily down, clouding the sunlight, and leaving a deposit of grime on everything, including the children." (Beulah Kennard)

PEDESTRIANS AND WORKERS[34]:

"For the Pittsburgh fog is not the fog that a coast town knows; it is moisture permeated with coal dust and grime, perilous to the eyes and throat of the pedestrian, and of a fatal penetrating quality wherever open door or window gives it a chance to enter." (Elizabeth Butler)

"In Pittsburgh . . . the smoky atmosphere outside and the dirt from the operations conducted inside very quickly blacken the windows, the cost of keeping them clean may approximate that of burning electric light." (H. F. J. Porter)

"The absence of ventilation [in the workplace] may have been due to the amount of soot in the air of the Pittsburgh District." (H. F. J. Porter)

But while smoke (and soot) constituted a major problem, reformers failed to devise an effective or implementable control strategy. Although the city had enacted ordinances to curtail railroad and coke oven emissions as early as 1868 and 1869, these were ignored or ineffective. Atmospheric pollution intensified until the 1880s, when the discovery of local supplies of natural gas enabled many industries and residences to switch to the clean fuel, dramatically improving air quality. Even so, wasteful usage rapidly exhausted the supply. By the 1890s, many industries and households had returned to soft coal, resulting, as one resident put it, in Pittsburgh "going back to smoke." The decline in air quality also produced a serious campaign for smoke control. The Ladies' Health Association of Allegheny County, the Engineer's Society of Western Pennsylvania, and the Civic Club of Allegheny led the public protests. Beginning in the early 1890s, the city councils passed a series of smoke control ordinances, and in 1907 they established the Office of Smoke Inspector for enforcement purposes.[35]

The Pittsburgh Survey was conducted during the smoke control campaign, and Survey investigators expressed optimism about the city's "coalition of civic forces" challenging the smoke problem. In "Civic Improvement Possibilities," Charles M. Robinson noted that the smoke problem had been "tackled bravely by the Chamber of Commerce," resulting in the appointment of a chief smoke inspector and three deputies. These were attached, he added, to the Bureau of Health and would receive "large powers."[36]

Both the Survey authors and municipal reformers were overly sanguine about the possibilities of regulating smoke pollution. The problem was complex, involving a multitude of industrial, commercial, residential, and transportation sources, and uncertain methods of regulation and control. Reform strategies vacillated among educational campaigns, litigation, and technological fixes, but were never (individually or in combination) powerful enough to restrict or change industrial or domestic combustion practices. Complicating enforcement was that offenders could only be persecuted on the grounds that smoke was a nuisance rather than an identifiable health issue. The Survey investigators misperceived the scope and severity of the smoke problem and underestimated the effort that would be required to correct it. City Council finally approved a tough anti-smoke ordinance in 1941, over thirty years after the Survey. World War II interrupted the law's full application; but in the postwar years the city implemented the law, and a combination of legal sanctions and a shift from coal to clean natural gas eliminated heavy smoke from the Pittsburgh atmosphere.[37]

Water Supply, Sanitation, and Floods

At the turn of the century, Pittsburgh suffered from poor drinking water quality and inadequate sanitation, leading the nation in typhoid fever death rates. In addition, severe floods regularly plagued the city, causing both public health danger and property damage. In "The Civic Responsibilities of Democracy in an Industrial District," Paul U. Kellogg argued that these problems required increasing municipal administrative authority so that the efficiency of the city's social institutions could match that of its industrial enterprises. In this manner, the vitality of the region's work force (a "civic resource") could be conserved "to the utmost of its potential goods."[38]

The city's most severe public health problem resulted from its water supply. Pittsburgh had drawn its water supplies from the neighboring rivers since 1826, when it had constructed a municipal system. The city had expanded the network throughout the nineteenth century, and it had increased from 268 miles in 1895 to 743 miles in 1915. Most of the construction took place in middle-and upper-class neighborhoods, particularly in the parts of the city annexed in 1867, while working-class areas were poorly supplied. The unequal distribution resulted from an 1872 City Water Commission ruling that required that the size of the pipe laid on a particular street correspond to potential revenue. This so-called "segmented system," as Robin L. Einhorn calls it, was not unusual in American cities and was intended to "distribute costs and decision-making power among the propertied."[39] It also had the effect, however, of depriving poor neighborhoods, mainly occupied by renters, of sufficient water. Many working-class areas relied on pumps that drew from ground water; even when piped water was available, it was often accessed through a backyard spigot, frequently located near the privy vaults, and shared with other tenants. The absence of in-house water facilities meant that water for domestic uses such as cooking and washing had to be carried, and, as historian S. J. Kleinberg has shown, this burden fell with greatest intensity on women.[40]

Over the years, the rivers from which Pittsburgh drew its water supply became polluted as upstream communities, as well as Pittsburgh itself, constructed sewerage systems that discharged their untreated wastes into adjacent streams. Pittsburgh's water and sewer problem was, as Kellogg correctly perceived, a "water-shed problem," with 129 separate upstream municipalities and towns and a population of over 350,000 dumping their raw sewage into the rivers from which Pittsburgh drew its supply. Consequently, Pittsburgh had the highest typhoid fever death rate of the nation's large cities: 130 per 100,000 people from 1899 to 1907. By comparison, the average typhoid death rate for northern cities in 1905 was 35 per 100,000 persons. Later research would show that the typhoid fever frequency peaked in working-class immigrant neighborhoods, although the whole city suffered from substandard water.[41]

The Survey staff investigated Pittsburgh just after a long battle by reformers for a water filtration plant had been won. In the late 1890s, the nation's leading sanitary engineers had advised the city to construct a slow-sand water filtration plant, but political infighting and bureaucratic inertia blocked it for ten years. The plant finally came on line late in 1907, and typhoid fever rates sharply declined. Frank E. Wing (associate director of the Survey, secretary of the Pittsburgh Typhoid Fever Commission, and the superintendent of the Chicago Tuberculosis Institute) told the story of Pittsburgh's long fight for the filtration of publicly-supplied water in a Survey article that emphasized the interaction of "water, economics, and politics."[42]

Wing began his article by describing the frieze at the Pittsburgh Civic Exhibit, held in the Carnegie Institute in November, 1908, to publicize the Survey findings. The frieze contained 622 small silhouettes extending around the great hall, each representing a 1907 typhoid fever death. This dramatic mural, said Wing, symbolically represented the city's "most enduring disgrace," while the long-awaited water filtration system represented "one of Pittsburgh's greatest civic achievements." Paul Kellogg maintained that the case of typhoid fever in Pittsburgh lent itself to a form of "social bookkeeping" that would demonstrate "the larger waste of human life and private means; and will stand out not only for honesty and efficiency, but for the common well-being."[43]

The Survey staff, in collaboration with the Columbia Settlement, applied this type of social bookkeeping to help the public "grasp . . . the meaning of this typhoid scourge" in workers' lives. The Survey staff totaled the cost of lost wages, medical care, household expenses, and funeral expenses of typhoid victims. They estimated, as reported in the Wing article, that the 5,421 cases of typhoid fever in Pittsburgh and Allegheny City in 1907 drained the city of over $3 million. This method of social cost-accounting for public health had originated with Edwin Chadwick, the great British sanitary reformer, and was increasingly used in the United States during the late nineteenth and twentieth centuries. But, as both Wing and Kellogg noted, such statistical approaches failed to consider the "family readjustments and inconveniences, the distress of mind and unalloyed misery . . . [necessary] to form any adequate idea of what such sickness holds for a wage earning population."[44]

Wing attributed the unnecessary loss of life to "a lethargic public sentiment, selfish political purposes, and municipal shortsightedness." The filtration plant stood as an "object lesson of tardy justice and a monument to those hundreds of lives that paid the penalty . . . of an unaroused municipal conscience." For Kellogg, the typhoid fever problem was an example of a failure to meet the "civic responsibilities of democracy in an industrial district." The community could best meet these responsibilities by modeling itself after "a first-rate industrial concern." That is, "figure out the ground it can cover effectively and gear its social machinery so to cover it."[45] Thus, the model for social improvement should be patterned after the efficient business organization, using the best "technology" available in order to reach its ends.

All cases of typhoid did not originate with water supply. As Wing observed, "a favorable laboratory for the growth and dispersal of germs exists in the city's unsanitary dwellings. . . ."[46] Housing contamination resulted from insufficient water supplies for cleaning, shared water facilities, overcrowding, and ubiquitous privy vaults. The Survey staff often referred to overflowing privy vaults, inadequate and unclean toilet facilities, and open sewers. "Of these evils," wrote Emily Dinwiddie, "the vaults were without doubt the most noxious and omnipresent. Within a few blocks of the county court house . . . [were] antiquated and indescribably foul privies,—privies that were not only polluting the atmosphere but were contributing a large quota to the mortality and morbidity of the community by serving as breeding places of disease germs to be distributed by flies."[47]

Such sanitary conditions reflected, according to F. Elisabeth Crowell, "the glaring consequences resulting from the combination of private greed and public indifference." She argued that such conditions had to be eliminated because they complicated the maintenance of "standards . . . for . . . decency and morality among those who . . . dwell in the adverse environment. . . ."[48] Undoubtedly, for some, "decency and morality" ranked equally with public health hazards as a rationale for eliminating unsanitary conditions.

The prevalence of privy vaults and cesspools in working-class neighborhoods reflected the city's uneven distribution of sewer services. Construction of Pittsburgh's centralized sewerage system began in the late 1880s, and between 1889 and 1912 the city's new Bureau of Engineering constructed over 412 miles of sewers, almost all of the combined type. But because of the requirement that abutters pay for services, a common municipal practice in nineteenth-century cities, many working-class areas of the city were unsewered when the Survey was conducted. While the Board of Health tried to mandate household connections to the sewers, both homeowners and landlords of rental buildings resisted, and retained their old privy vaults and cesspools. Even working-class homeowners often avoided connections with the sewer system because of costs, preferring to invest in other forms of home improvements.[49]

In 1888, the councils barred the construction of cesspools where sewer service was available; in 1901 they outlawed water closets from draining into a privy vault and prohibited the connection of privy wells to a public sewer. The Bureau of Health ordered the cleaning or removal of thousands of privy vaults, although the effect of these orders was limited by a small staff of inspectors, as well as by collusion between inspectors and scavenger firms. While such problems were encountered in many cities, Pittsburgh's rugged topography increased the cost of infrastructure construction.[50]

Many years would pass before the city finally extended sewers to all working-class districts, and the sanitary problems presented by privy vaults persisted in Pittsburgh until after World War II. Sewage pollution of Pittsburgh rivers also remained for another half century. It would not be until 1959 that the Allegheny County Sanitary Authority finally went into operation, providing sewage treatment for the city as well as for other county municipalities. Here again, as with the issue of smoke, several factors combined to hinder environmental improvement. Among the most significant was the resistance of various economic and political elites to providing city-wide service amenities as opposed to outfitting middle- and upper-class neighborhoods alone. The same type of parochial attitude, although resting on a territorial rather than class or ethnic basis, was also responsible for delaying regional sewage treatment for more than half a century.[51] Although Survey writers had identified fragmentation as a regional problem, they had underestimated how persistent it would remain as a shaping factor in governmental change.

The Survey's other major environmental concern was floods. Other cities suffered from the problem, but the situation in Pittsburgh was extreme. Between 1832 and 1907, the waters of the city's three rivers rose eleven times above the twenty-five-foot flood stage. Floods caused millions of dollars of damage to factories and railroads and the destruction of hundreds of homes located on the river flood plains. The great flood of 1907 crested at 35.5 feet, submerging some 1,600 acres, including parts of the central business district and industrial plants along the rivers. Damages were valued at over $160 million, and over 100,000 workers were unemployed for more than a week. W. W. Ashe, a member of the U. S. Forest Service, wrote the key Survey article on floods. Ashe viewed the rivers as "at once the making and the menace of Pittsburgh," and floods as "the open expression of the rivers' authority." He argued that while natural factors, such as heavy rains and steep valleys, caused floods, human actions increased the risk. He recommended as preventative measures dense tree replanting and the construction of storage reservoirs that could also double as domestic water supply reservoirs and enhance water power development.

The Pittsburgh Flood Commission, created by the Pittsburgh Chamber of Commerce in 1908 to formulate strategies to solve the flood problem, echoed Ashe's recommendations in its 1912 report. The Flood Commission began with only private sector membership, but eventually expanded to include city and county officials. Members represented the city's leading reform organizations or were professionals who supported a policy of coordinated watershed development under a single federal agency. The Commission's report called for construction of seventeen storage reservoirs at the headwaters of the Allegheny and Monongahela rivers and reforestation of the Allegheny Mountain slopes. These recommendations complemented the progressive conservationist concept of the multi-purpose use of waterways.

Like smoke control and sanitary reform, however, Pittsburgh flood control proved far tougher than either Survey writers or Pittsburgh reformers had envisaged. While state foresters supported reforestation, the U. S. Corps of Engineers, the body responsible for maintaining river navigability, opposed the building of the reservoir system and argued that neither reforestation nor reservoirs could control floods. The essential reason for their opposition, however, was that the multi-purpose use of rivers threatened their autonomy. Thus, in spite of the Survey publicity about floods and the efforts of Pittsburgh reform and business elites to lobby the federal government for flood control, a well-entrenched federal bureaucracy refused to change its position. As with sewage treatment and smoke control, flood control would languish until the 1940s, when the federal government constructed the dams necessary to control flooding in Pittsburgh.[52]

Conclusions

The Pittsburgh Survey is rich with examples of the negative environmental consequences of urban industrialization, such as a polluted and scarred landscape and a dilapidated built environment—conditions especially noticeable in working-class neighborhoods. Pittsburgh's polluted rivers, rubbish-filled valleys, smoke-ridden atmosphere, and barren hills all testified to the extent to which the quality of the natural environment had been disregarded in the construction of a major industrial city. Working-class areas were often marked by wretched sanitary facilities, inadequate water supply, primitive methods of waste disposal, and limited facilities for play and recreation. Such environmental degradation increased the burdens of workers and housekeepers alike and fell with particular intensity on women and children. Other American industrial cities often had similar conditions, but Pittsburgh was perhaps unique in terms of its concentration of environmental problems exacerbated by the rugged character of its site.[53]

As "positive environmentalists," the Survey investigators firmly believed that environment and behavior were closely linked. J. T. Holdsworth, a University of Pittsburgh economist, concisely expressed this perspective in 1912 in his *Economic Survey of Pittsburgh,* a follow-up to the Survey: "It is coming to be recognized that environment is largely responsible for the character of . . . man, and that many of the evils of overcrowding, intemperance, and thriftlessness are properly chargeable to bad environment." Because the Survey writers believed that landscapes were malleable, especially with the use of technology, purposeful change could produce improved environments leading to higher levels of civic consciousness. Survey writers' advocacy of city parks, playgrounds, and gardens as providing uplifting and civilizing experience reflects their faith in the elevating effects of nature on human behavior.

While the Survey investigators believed that nature could enhance the quality of life, they also believed that solving urban industrial problems required political reform, large-scale governmental intervention, and planning. Because Pittsburgh was throbbing with reform during the period of the Survey investigations—one of the reasons why the Russell Sage Foundation chose it for examination—writers such as Allen T. Burns, Charles M. Robinson, and Robert A. Woods were sanguine about the future. Thus, they believed that a coalition of enlightened businessmen, politicians, social workers, engineers, and planners could position Pittsburgh in the forefront of "a great civic and human welfare movement."[54]

The Survey articles appeared at the time when Democratic mayor George W. Guthrie was leading a reform crusade. Guthrie was typical of the Pittsburgh elite reformers who were models for the Survey writers. Blaming corrupt politics and ruthless nineteenth century capitalism for the degradation of the industrial city, Guthrie pushed for a series of political and environmental changes. In 1909, inspired by the Survey's planning emphasis, Guthrie created the Pittsburgh Civic Commission to "plan and promote improvements in civic and industrial conditions which affect the health, convenience, education, and general welfare of the Pittsburgh industrial district. . . ." The commission's fourteen committees focused on a range of urban problem areas that replicated those considered by the Survey; on its advisory board sat several figures involved with the Survey, including Paul C. Kellogg, Allen T. Burns, and Robert A. Woods.[55]

The Pittsburgh Civic Commission produced several expert reports, including that of Frederick Law Olmsted, Jr. and engineers Bion J. Arnold and John R. Freeman, "Preliminary Report to Pittsburgh Civic Commission Upon Methods of Procedure in City Planning" (1909), and Olmsted's plan, *Pittsburgh, Main Thoroughfares and the Down Town District: Improvements Necessary to Meet the City's Present and Future Needs* (1910). Olmsted, the son of the great landscape planner Frederick Law Olmsted, Sr., had been active in the Pittsburgh region since 1901, landscaping and shaping the estates of the Pittsburgh elite. Edward K. Muller and John F. Bauman have argued that the "ethos" that the Olmsteds, father and son, "articulated in their private and public work flowed . . . from a concern for environmental beauty and order embedded in the late nineteenth and early twentieth century urban elite and upper middle-class mentality." Thus, the various urban and environmental reforms urged by the Pittsburgh reformers, both reflected and advanced in the Survey, illustrated the belief "that the antidote to urban violence and disorder existed in civic art and improvement."[56]

Survey writers and Progressive reformers in general realized that gaining control of the city's political structure and the regulation of land use were necessary for them to realize their goals, but they still underestimated the obstacles to change, exaggerated the strength of the reform thrust, and misperceived the attractiveness of their appeal. Their optimism about the possibilities of change and the attractiveness of the civic ideal, blinded them to the complexity of political and governmental institutions and the depth of city divisions. Their belief in the meliorative power of environmental improvement limited their ability to forecast future problems or to foresee that conditions might revert to where they had been before they were exposed.[57]

While some of Olmsted's specific planning suggestions for physical improvements were realized, environmental reforms were less successful. Gains made in areas such as improved parks and recreational playgrounds disappeared within a few years because of inadequate funding and a lack of professional attention. Major environmental problems such as air and water pollution and flooding proved almost intractable without public-private cooperation, major government regulatory intervention, and technological advances. It would not be until the Pittsburgh Renaissance of the post–World War II period that the city would finally control its smoke and floods, treat its sewage, and redevelop its downtown.[58]

The Pittsburgh Survey cannot be judged a document that led to the accomplishment of profound environmental or urban improvements. But, while the belief of its authors in environmental behaviorism was flawed, their efforts to highlight serious environmental, recreational and sanitary problems were ultimately beneficial. Their writings and photographs influenced both elite and professional attitudes and helped nudge the process of change. Rather than actual, concrete accomplishments, the Survey's environmental legacy may lie more in its influence on individuals who passed on the mission of environmental reform over generations. For the historian of the city and of the environment the legacy is somewhat different, for the Pittsburgh Survey provides a graphic record of the early–twentieth century industrial city's environmental conditions and a striking illustration of the belief of Progressive Period urban reformers in the link between environment and human behavior.

Notes

I would like to thank Karen Anderson Howes for her research assistance, Maurine Greenwald for her excellent editing of earlier drafts of this manuscript, Roy Lubove and Jeffrey Stine for their perceptive suggestions for reshaping the original paper, and David Hounshell for shaping the title.

1. See, for instance, David Brody, *Steelworkers in America: The Nonunion Era* (NY: Harper Torchbooks, 1969); Francis G. Couvares, *The Remaking of Pittsburgh: Class and Culture in an Industrializing City 1877–1919* (Albany: State University of New York Press, 1984). Roy Lubove, *Twentieth Century Pittsburgh: Government, Business and Environmental Change* (New York: John Wiley & Sons, 1969), is an exception to the generalization.

2. Paul Boyer, *Urban Masses and Moral Order in America, 1820–1930* (Cambridge: Harvard University Press, 1978), 220–251. See also Roy Lubove, "Pittsburgh and the Uses of Social Welfare History," in Samuel P. Hays (ed.), *City at the Point: Essays on the Social History of Pittsburgh* (Pittsburgh: University of Pittsburgh Press, 1989), especially pp. 296–300, and Stanley K. Schultz, *Constructing Urban Culture: American Cities and City Planning 1800–1920* (Philadelphia: Temple University Press, 1989), 109–150.

3. Edward T. Devine, "Pittsburgh: The Year of the Survey," in Paul U. Kellogg (ed.), *The Pitts-*

burgh District: Civic Frontage (New York: Survey Associates, 1914), 6; Robert A. Woods, "Pittsburgh: An Interpretation of Its Growth," in ibid., 21–42; and Allen T. Burns, "Coalition of Pittsburgh's Civic Forces," in ibid., 60. See also Boyer, *Urban Masses and Moral Order in America*, 190.

4. Robert A. Woods, "Pittsburgh: An Interpretation of Its Growth," *Charity and The Commons* (henceforward cited as C&C) (Jan. 2, 1909). See also Woods, "Pittsburgh: An Interpretation of Its Growth," in Kellogg (ed.), *The Pittsburgh District*, 19–20, 41–43, which combines the former article with "A City Coming to Itself," *C&C* (Feb. 6, 1909).

5. Paul U. Kellogg, "The Civic Responsibilities of Democracy In An Industrial District," C&C (Jan. 2, 1909), 638; Clarke Chambers, *Paul Underwood Kellogg and the Survey* (Minnesota: University of Minnesota Press, 1973). Robert A. Woods used the same type of terminology in describing Pittsburgh's population—he found it to be "a sort of natural selection of enterprising spirits from out of every European nation and tribe. . . ." See "Pittsburgh: An Interpretation of Its Growth," *C&C* (Feb. 6, 1909), 528, 532. In his essay on the Pittsburgh Survey, John F. McClymer calls this a "benign Darwinism." See "The Pittsburgh Survey, 1907–1914: Forging An Ideology in The Steel District," *Pennsylvania History* XLI (April 1974): 176.

6. Paul U. Kellogg, "The Pittsburgh Survey," *C&C* (Jan. 2, 1909), 520, 526.

7. Allen T. Burns, "Coalition of Pittsburgh's Civic Forces," in Kellogg (ed.), *The Pittsburgh District*, 44–45; John R. Commons, "Wage Earners of Pittsburgh," *C&C* (Mar. 6, 1909), 1051–1052; and F. Elisabeth Crowell, "What Bad Housing Means to Pittsburgh," *C&C* (Mar. 8, 1908), 1684. Peter Roberts maintained that "The geographical contour of the region . . . had its influence in keeping the foreign population within certain limited districts" and near the mills where they worked. See "The New Pittsburghers," *C&C* (Jan. 2, 1909), 540.

8. Boyer, *Urban Masses and Moral Reform*, 237–240; Roy Lubove (ed.), *Landscape Architecture as Applied to the Wants of the West*, by H.W.S. Cleveland (Pittsburgh: University of Pittsburgh Press, 1965), x–xi; Peter J. Schmitt, *Back To Nature: The Arcadian Myth in Urban America* (New York: Oxford University Press, 1969), xvi–xix; and Galen Cranz, *The Politics of Park Design: A History of Urban Parks in America* (Cambridge: MIT Press, 1989), 62. Nature was also considered essential to realization of the suburban ideal, and designers and architects believed that contact with nature on a regular basis was a necessity. See Mary Corbin Sies, "The City Transformed: Nature, Technology, and the Suburban Ideal, 1877–1917," *Journal of Urban History* 14 (November 1987): 101–104.

9. Charles Mulford Robinson, "Civic Improvement Possibilities of Pittsburgh," C&C (Feb. 6, 1909), 801–826. The author of several influential studies including *Modern Civic Art or The City Made Beautiful* (1903) and *The Improvement of Towns and Cities; or, the Practical Basis of Civic Aesthetics* (1910), as well as other books on municipal planning, Robinson influenced the redefinition of the City Beautiful movement, the major urban planning effort at the beginning of the twentieth century. For Robinson and the City Beautiful Movement, see William H. Wilson, *The City Beautiful Movement* (Baltimore: Johns Hopkins University Press, 1989), 45–48, and M. Christine Boyer, *Dreaming the Rational City: The Myth of American City Planning* (Cambridge: MIT Press, 1983), 132–134. Quoted in Boyer, *Urban Masses and Moral Order*, 265.

10. Robinson, "Civic Improvement Possibilities," 824; Boyer, *Urban Masses and Moral Order*, 240–242; David Schuyler, *The New Urban Landscape: The Redefinition of City Form in Nineteenth-Century America* (Baltimore: Johns Hopkins University Press, 1986), 184–185; and Cranz, *The Politics of Park Design*, 62.

11. See the Hine photograph, "Natural Beauty vs. Industrial Odds," in Robinson, "Civic Improvement Possibilities of Pittsburgh," 801–803, 824.

12. Ibid., 823. The Survey published a short article on a small park, "Little Jim Park," created by unemployed workers on Pittsburgh's South Side. See Leroy Scott, "Little Jim Park," C&C (Feb. 1909), 911–912, and Kleinberg, *The Shadow of the Mills*, 123–124. For other discussions of Schenley Park, see Barbara Judd, "Edward M. Bigelow: Creator of Pittsburgh's Arcadian Parks," *The Western Pennsylvania Historical Magazine* 58 (January 1975):53–67; Couvares, *The Remaking of Pittsburgh*, 107–110.

13. Robinson, "Civic Improvement Possibilities," 824–825. See John D. Fairfield, *The Mysteries*

of the Great City: The Politics of Urban Design, 1877–1937 (Columbus: Ohio State University Press, 1993), 103–104, for the Robinson quotation. In the 1914 essay, Robinson called for the construction of civic centers that would "visibly dominate" the city. Robinson, "Civic Improvement Possibilities," 824–825.

14. John F. Kasson, *Amusing The Millions: Coney Island at the Turn of the Century* (New York: Hill & Wang, 1978), 73–74.

15. See ibid., 88–90.

16. Emphasis added. See Florence Larrabee Lattimore, "Skunk Hollow: A Pocket of Civic Neglect in Pittsburgh," C&C (Feb. 6, 1909), 896. In addition, see the photograph after 890, and its caption, "Looking Down on Skunk Hollow. Luna Park is seen on the sky-line at the right." Kasson, *Amusing the Millions,* 99–101; and Cranz, *The Politics of Park Design,* 13–19, 28.

17. Margaret Byington, "Homestead: A Steel Town and Its People," C&C (Jan. 2, 1909), 618. The Hine photograph illustrating Homestead amusements features a night picture of Kennywood with its "garish" lights. See also Addams, *The Spirit of Youth and the City Streets,* 15–16.

18. Byington, *Homestead,* 47–48, 132, 136. In his essay on the "The New Pittsburghers," Peter Roberts also emphasized the rural origins of the immigrants. See *C&C* (Jan. 2, 1909), 539, 543. In contrast, Alis B. Koukol, in "The Slav's A Man For A' That," *C&C* (Jan. 2, 1909), 590, noted that it was "rarely true" that Slavs went "straight from their villages to Pittsburgh." Polish immigrants at this time often had some urban and industrial experience before migrating to the city. See John Bodnar, Roger Simon, and Michael P. Weber, *Lives Of Their Own: Blacks, Italians, and Poles in Pittsburgh, 1900–1960* (Urbana: University of Illinois Press, 1982), 37–41. See the Hine photographs opposite p. 48 on "Back Yard Possibilities"; for a discussion of gardening and urban social change, see Sam Bass Warner, Jr., *To Dwell Is To Garden: a History of Boston's Community Gardens* (Boston: Northeastern University Press, 1987).

19. Byington, *Homestead,* 46–48, 132–136; Schmitt, *Back to Nature,* 90–92; and, Jon A. Peterson, "The Impact of Sanitary Reform Upon American Urban Planning," *Journal of Social History* 13 (Fall 1979): 83–103.

20. Cranz, *The Politics of Park Design,* 77.

21. Beulah Kennard, "The Playgrounds of Pittsburgh," in Kellogg (ed.), *The Pittsburgh District,* 306, 313; Cranz, *The Politics of Park Design,* 80–81.

22. Kennard, "The Playgrounds of Pittsburgh," 312, 316–319.

23. Ibid., 321–323.

24. Ibid., 306, 312–313; Addams, *The Spirit of Youth and the City Streets,* 97–100; Lubove, *Twentieth Century Pittsburgh,* 51; and, Boyer, *Urban Masses and Moral Reform,* 242–251.

25. "Charles F. Ball Appointed Head of City Playgrounds," *Pittsburgh Gazette Times,* Sept. 13, 1918; "City Backward in Playground Work, Charged," *Pittsburgh Sun,* Dec. 20, 1920.

26. Citizens Committee on City Plan of Pittsburgh (CCCP), Pittsburgh Playgrounds: A Part of The Pittsburgh Plan (Pittsburgh: June, 1920), and Parks: A Part of The Pittsburgh Plan (Pittsburgh: Sept., 1923). The philosophy of the CCCP reformers in regard to the uses of parks and playgrounds, however, had not changed from that expressed earlier by the Survey authors. According to the CCCP, public recreation was required so that "the spare hours from childhood to maturity may be properly and profitably occupied. . . . under proper administration, playground activities furnish opportunity for children and youth to secure invaluable training, co-operative competition taking the place of gang antagonism. . . . The justification for so great a public undertaking is its ultimate economy in the upbuilding of a citizenship which shall be sound physically and morally." See CCCP, Playgrounds, 9.

27. Roy Lubove, *The Progressives and the Slums: The Tenement House Reform in New York City 1890–1917* (Pittsburgh: University of Pittsburgh Press, 1962), 189–215; Stephen Turner, "The Pittsburgh Survey and its Intellectual Content," this volume. For a discussion of attempts to achieve these goals in other cities, see Martin V. Melosi (ed.), *Pollution and Reform in American Cities, 1870–1930* (Austin: University of Texas Press, 1980), and Schultz, *Constructing Urban Culture,* 151–209.

28. Anne W. Spirn, *The Granite Garden: Urban Nature and Human Design* (New York: Basic Books, 1984), 12–14.

29. Joel A. Tarr, "Infrastructure and City-Building in the Nineteenth and Twentieth Centuries," in Hays (ed.), *City at the Point*, 235–238.

30. "Pittsburgh Survey: Introductory To This Issue," *C&C* (Feb. 6, 1909), 784.

31. Robinson, "Civic Improvement Possibilities of Pittsburgh," 816. See also Kleinberg, *In the Shadow of the Mills*, 65–66.

32. Woods, "Pittsburgh: An Interpretation of Its Growth," 19; Kellogg, "The Pittsburgh Survey," 520; Robinson, "Civic Improvement Possibilities of Pittsburgh," 816; Elizabeth Beardsley Butler, *Women and the Trades: Pittsburgh, 1907–1909* (New York: Charities Publication Committee, 1909), 329; and, Byington, *Homestead*, 3.

33. F. Elisabeth Crowell, "The Housing Situation in Pittsburgh," *C&C* (Feb. 6, 1909), 871; Byington, *Homestead*, 137; and Kennard, "The Playgrounds of Pittsburgh," 306.

34. Butler, *Women and the Trades*, 165; H.F.J. Porter, "Industrial Hygiene of the District," in *Wage-Earning Pittsburgh* (New York: Survey Associates, Russell Sage Foundation, 1911), 223–226.

35. Joel A. Tarr and Bill C. Lamperes, "Changing Fuel Use Behavior and Energy Transitions: The Pittsburgh Smoke Control Movement, 1940–1950—A Case Study In Historical Analogy," *Journal of Social History* 14 (Summer 1981): 562; John O'Connor, Jr., "The History of the Smoke Nuisance and of Smoke Abatement in Pittsburgh," *Industrial World* (March 24, 1913). The Ladies' Health Association was absorbed into the Civic Club.

36. Robinson, "Civic Improvement Possibilities of Pittsburgh," 816–817.

37. Tarr and Lamperes, "Changing Fuel Use Behavior," 561–587; Lubove, *Twentieth Century Pittsburgh*, 48–49. The substitution of natural gas for coal after 1947 also played a definitive role.

38. Kellogg, "Civic Responsibilities of Democracy," 630–631.

39. See Robin L. Einhorn, *Property Rules: Political Economy in Chicago, 1833–1872* (Chicago: University of Chicago Press, 1991), 14–22.

40. Kleinberg, *In the Shadow of the Mills*, 88–90.

41. Tarr, "Infrastructure and City Building," 235–236; and Clayton R. Koppes and William P. Norris, "Ethnicity, Class, and Mortality in The Industrial City: A Case Study of Typhoid Fever in Pittsburgh, 1890–1910," *Journal of Urban History* 11 (May 1985): 259–279.

42. Frank E. Wing, "Thirty-Five Years of Typhoid: The Fever's Economic Cost to Pittsburgh and The Long Fight for Pure Water," *C&C* (Feb. 6, 1909), 923–939.

43. Ibid., 923; Kellogg, "Civic Responsibilities of Pittsburgh," 633.

44. Ibid., 630–634; Wing, "Thirty Five Years of Typhoid," 930–932.

45. Ibid., 939; Kellogg, "Civic Responsibilities of Pittsburgh," 638.

46. Wing, "Thirty-Five Years of Typhoid," 931–935. In the spring, 1908, the Russell Sage Foundation, funder of the Pittsburgh Survey, provided a grant to the city for investigation of typhoid fever sources. Mayor George W. Guthrie appointed a Pittsburgh Typhoid Fever Commission in April, 1908, chaired by Dr. James F. Edwards, former superintendent of the Bureau of Health, that included some of the nation's leading public health experts. The Commission concluded that the polluted water supply was the prime cause of typhoid, and that other possible causes, such as inferior sanitary conditions or the milk supply, were relatively unimportant.

47. Dinwiddie and Crowell, "The Housing of Pittsburgh's Workers Discussed From The Standpoint of Sanitary Regulation and Control," Kellogg (ed.), *The Pittsburgh District*, 92–97.

48. Crowell, "What Bad Housing Means to Pittsburgh," 1683–1684.

49. Terry F. Yosie, "Retrospective Analysis of Water Supply and Wastewater Policies in Pittsburgh, 1800–1959," (unpublished Doctor of Arts Dissertation, Carnegie Mellon University, 1981), 52–55, 68–70, 106; Tarr, "Infrastructure and City Building," 236–238; Roger D. Simon, *The City Building Process: Housing and Services in New Milwaukee Neighborhoods, 1880–1910* (Philadelphia: American Philosophical Society, 1978), 40; and Einhorn, *Property Rules*, 14–22.

50. Yosie, "Retrospective Assessment," 104–119.

51. Joel A. Tarr, "Disputes Over Water Quality Policy: Professional Cultures in Conflict,

1900–1917," *American Journal of Public Health* 70 (April 1980): 427–435; Tarr, "Infrastructure and City Building," 248–61; and Shelby Stewman and Joel A. Tarr, "Four Decades of Public-Private Partnerships in Pittsburgh," in R. Scott Fosler and Renee A. Berger (eds.), *Public-Private Partnership in American Cities* (Lexington, MA: Lexington Books, 1982), 59–128.

52. Smith, "Pittsburgh Flood Control," 9–24, and "The Politics of Pittsburgh Flood Control, 1936–1960," *Pennsylvania History* 44 (Jan., 1977): 3–24. See Samuel P. Hays, *Conservation and the Gospel of Efficiency* (Cambridge: Harvard University Press, 1959), 109, 204–211.

53. See, for instance, the essays in Melosi (ed.), *Pollution and Reform in American Cities.*

54. Paul Boyer notes that progressives of all distinctions believed that society had the right to intervene when the well-being of its members was threatened. Boyer, *Urban Masses and Moral Order,* 197–198. Woods, "Pittsburgh: An Interpretation of Its Growth," 43; Burns, "Coalition of Pittsburgh's Civic Forces," 59–60; and Robinson, "Civic Improvement Possibilities in Pittsburgh." See also Fairfield, *The Mysteries of the Great City,* 104–105.

55. Lubove, *Twentieth Century Pittsburgh,* 24–40; John F. Bauman and Edward K. Muller, "The Olmsteds in Pittsburgh: (Part II) Shaping the Progressive City," *Pittsburgh History* 76 (Winter 1993/94):192.

56. The Olmsted, Arnold, and Freeman "Report to Pittsburgh Civic Commission" is reprinted as Appendix D, in Kellogg (ed.), *The Pittsburgh District,* 489. See also Bion J. Arnold, *Report on the Pittsburgh Transportation Problem* (Pittsburgh: City of Pittsburgh, 1910). Edward K. Muller and John F. Bauman, "The Olmsteds in Pittsburgh: (Part I) Landscaping the Private City," *Pittsburgh History* 76 (Fall 1993): 123, 138; Boyer, *Urban Masses and Moral Order,* 280.

57. See Wilson, *The City Beautiful Movement,* 74–86; Boyer, *Urban Masses and Moral Order in America,* 280–283.

58. Wilson, *The City Beautiful Movement,* 301–02; Joel A. Tarr, *Transportation Innovation and Changing Spatial Patterns in Pittsburgh, 1850–1934* (Chicago: Public Works Historical Society, 1978), 26–27; and Bauman and Muller, "The Olmsteds in Pittsburgh (Part II)," 194–195. Lubove argues that the "quest for environmental regeneration" failed because it was "identified with a business and professional elite whose ideal of bureaucratic rationalization was compromised by a reluctance to encroach upon the prerogatives of voluntary interests." Pittsburgh accomplished these improvements during the so-called Pittsburgh Renaissance, beginning in 1945. See Stewman and Tarr, "Four Decades," and Lubove, *Twentieth Century Pittsburgh,* 106–176.

PART II Water Pollution

Water pollution has the longest and most extensive history of any pollution problem in the United States. Water is a basic element of existence—essential for drinking and cleanliness, cooking, fire fighting, and manufacturing—but humans often take its presence and availability for granted. Visible contamination of drinking water supplies in cities such as New York and Philadelphia appeared as early as the middle of the eighteenth century, and pollution increased throughout the nineteenth and twentieth centuries. By the 1920s, many rivers, lakes, and estuaries had become almost unusable for fishing or swimming, while drinking-water supplies had to be extensively treated to avoid adverse health effects.

Three major societal processes—natural resource exploitation, industrialization, and urbanization—were largely responsible for the pollution. But while activities such as lumbering and mining caused the despoliation of many lakes and rivers, urbanization and industrialization were the primary offenders. By the late nineteenth century, municipal and industrial effluents had ruined many drinking-water sources, destroyed fish populations and habitats, and raised costs for many water-using industries. Sanitarians in the public health movement, scientists, and engineers increasingly believed that water pollution and epidemic diseases such as cholera, typhoid, and yellow fever were related. In the late nineteenth century, they launched campaigns for improved water supplies and construction of municipal sewerage systems as means to protect the public health.

In the last decades of the nineteenth century, under the guidance of a group of new professionals called sanitary engineers, hundreds of towns and cities constructed sewerage systems. Because it was the cheapest means of disposal, and because it was widely believed that running water purified itself, cities discharged millions of gallons per day of untreated sewage into rivers and lakes. Many of these waterways were used as water-supply sources by downstream cities, and in the late nineteenth-century typhoid death rates rose sharply in a number of these communities. It is one of the great ironies in the history of technology and its relationship to the environment that a technology designed to improve local health conditions and eliminate nuisances—water-carriage technology or sewerage—had extremely devastating effects on both the environment and human health.

Many cities and states passed regulations to attempt to deal with these problems, and eventually water filtration and chlorination sharply reduced the most severe public health effects. The focus on drinking-water quality, however, did little to improve the degraded environmental quality of waterways, and the more thorough cleaning had to await the passage of far-reaching federal legislation, beginning in 1948 with the first federal water pollution control act and culminating in the clean water acts of the 1970s. In the process of constructing sewerage systems and in dealing with their effects, major technological and policy choices, large-scale institutional developments, and social changes have occurred. These critical decisions, as well as their broader social effects, are explored in the following chapters.

The first chapter, "Historical Decisions about Wastewater Technology, 1800–1932," presents what I believe were the major municipal decisions made in regard to technology and its pollution effects in the late nineteenth and early twentieth centuries: to adopt the water-carriage technology of waste removal; to build largely combined rather than separate sewer systems; and, in most cities, to treat and purify drinking water rather than treat sewage. The remaining three chapters in this section explore these decisions in varying amounts of detail, looking not only at the health and environmental consequences but also at the sweeping social and institutional changes caused by the technology's adoption.

Photo II.1. "Pumping Station, Toledo." Source: J. T. Fanning, *A Practical Treatise on Hydraulic and Water-Supply Engineering* (New York: D. Van Nostrand Co., 1895).

Photo II.2. "The 'Town Pump,'" supplying drinking water for the 568 people living in "Painter's row," Pittsburgh, 1908. Source: Paul Underwood Kellogg (ed.), *The Pittsburgh District: Civic Frontage, The Pittsburgh Survey* (New York: Survey Associates, 1914).

Photo II.3. "Extreme Civic Neglect." Privy vaults and pump in Pittsburgh, 1909. Source: Paul Underwood Kellogg (ed.), *The Pittsburgh District: Civic Frontage, The Pittsburgh Survey* (New York: Survey Associates, 1914).

Photo II.4. A Baltimore scavenger, 1911. Source: *Survey*, 1911.

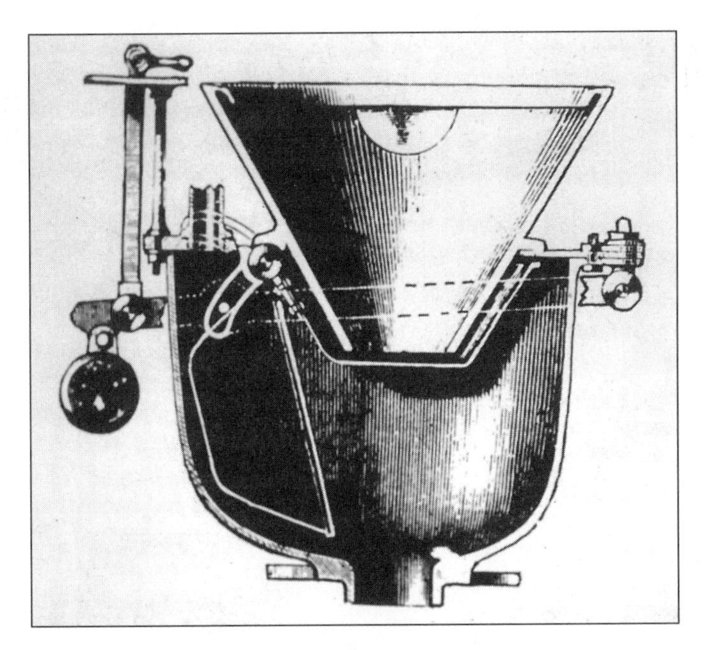

Photo II.5. "Pan Closet: Pan, Plunger, and Plug." A common nineteenth-century type of water closet. Source: S. S. Hellyer, *The Plumber and Sanitary Houses* (London, 1877).

Photo II.6. "Made in Baltimore." An odorless excavator, used in cities to pump out vaults and cesspools. The odorless excavator was actually a Paris innovation. Source: *Survey*, 1911.

Photo II.7. An open sewer in Pittsburgh. Source: Paul Underwood Kellogg (ed.), *The Pittsburgh District: Civic Frontage, The Pittsburgh Survey* (New York: Survey Associates, 1914).

Photo II.8. Sewer construction, 1910. Source: Leonard Metcalf and Harrison P. Eddy, *American Sewerage Practice: Volume II Construction of Sewers* (New York: McGraw-Hill, 1915).

CHAPTER IV

Decisions about Wastewater Technology: 1850–1932

Introduction

The decade of the 1970s is identified with a major involvement of the federal government in environmental regulation. The U.S. Environmental Protection Agency was created by presidential order in December 1970 to mount a coordinated attack on environmental problems. The broad legislative authority for the EPA programs comes primarily from nine separate acts: the Clean Air Act; the Federal Water Pollution Control Act; the Safe Drinking Water Act; the Solid Waste Disposal Act; the Federal Insecticide, Fungicide, and Rodenticide Act; the Public Health Service Act; the Noise Control Act; the Marine Protection Research and Sanctuaries Act; and the National Environmental Policy Act. The titles of the acts suggest the many avenues of response found necessary to address existing environmental problems.

This paper is focused on a specific collection of environmental problems that were first addressed by the United States over a century ago. It will examine, with an historical perspective teased by our present concerns, the turning points in the selection of certain technologies for the collection and treatment of domestic wastewaters in the United States from 1850 to the early 1930s. This is the time period in which the federal government did *not* play the major role that it now does. Then, as well as now, the problems were nevertheless complex.

The duration of this study is conveniently divided into three relatively well-defined periods: 1850–80, 1880–1900, and 1900–32. Three technological system controversies will be examined, one for each time period. The resolution of each of these selected controversies led to the emergence of new problems.

The period 1850–80 may be characterized by the choice to build collection systems for domestic wastes that depend on water for waste transport, so-called water-carriage systems. This was the choice and adoption of a wholly new technology. Between 1880 and 1900, the technical alternatives of developing sewers that transported sanitary wastes separately or in combination with stormwater drainage highlighted a second turning point. This choice was conditioned by existing pre–germ theories on the etiology of disease. Lastly, the period from 1900–1932 was marked by the conflict over whether to treat collected domestic sewage before discharge to natural waterways as well as to treat (filter) raw water supplies, or do the latter alone. Economic and institutional arguments dominated the controversy.

During each of these periods, the different technological, institutional, economic, and medical factors had varying weights and interacted to produce a particular orientation and policy regarding waste disposal and water and wastewater problems. Each of the periods was marked by controversy among competing systems and approaches, as adherents of one system or technology strove to dominate the field. By the end of the period, however, a particular approach was normally dominant. In each case, however, it was the application of this system or approach that led to the major policy controversy of the next period, as the dominant system developed problems and anomalies, some of which were unanticipated.

Selection of Sewerage Based on Use of Water for Transporting Domestic Wastes

The conventional explanation for the adoption of water as the transport system for waste removal in American cities during the nineteenth century is related to public health. Historians and others have argued that cities adopted this new technology in order to escape from the hazards of the cholera and yellow fever epidemics that they experienced at periodic intervals throughout the century. This explanation, however, is based almost entirely on the writings of public health authorities, as well as special cases like Memphis, Tennessee. While this explanation has some validity, it is so general that it does little to help us fully understand the causes of the breakdown of the previous technological system, or the costs of this breakdown to society or individuals.

This was a major decision, and it initiated a very capital-intensive system of public works. Sewers placed below the ground surface must be sufficiently deep to drain the lowest levels of the connecting buildings. The sewer system is a gravity-flow system, and the slope, as well as the size of the pipe, affects the transport capacity of the system. Each increase in the required depth of burial of the pipe adds to the system cost.

Presewer Cesspool and Privy Vault System

The period of 1800–80, what we might call the presewer period, was marked by a relatively primitive system in regard to the technology of human-waste removal and disposal. Human wastes were primarily deposited in cesspools and privy vaults. In many cities, when the cesspool or privy vaults became full, they were cleaned by municipal or by private scavengers, usually at the owner's cost. The wastes were then utilized as fertilizer on nearby farms or dumped in adjacent water courses and on vacant land. For most of the period, there were no public sewers for human-waste removal; the only sewers that existed were for the removal of storm or surface water.[1]

Government involvement in waste disposal was minimal. Although there were many municipal regulations concerning the building and cleaning of cesspools and privy vaults, they were seldom enforced. City health departments, when they existed, had little money and limited powers. Usually only the threat of an epidemic could stimulate them to action. Medical theory in regard to the cause of epidemics was divided, with some physicians arguing that disease was the result of miasmas produced by decaying organic matter (anticontagionists), while others maintained that specific contagia, probably animate and usually imported, were responsible (contagionists). Reflecting this divided opinion, the response of city officials to epidemics was usually twofold: (1) clean the city and (2) institute a quarantine.[2]

Introduction of Running Water at Household Tap

A major technological change was introduced during the 1800–1865 period—a continuous supply of water, freely running at the household tap. Until the beginning of the nineteenth century, all American cities obtained their water from local sources such as wells, cisterns, and springs. Population growth, the pollution of local sources, and a rising demand for more water for drinking, cleaning, and fire fighting resulted, however, in the construction of water systems that obtained their supplies from outside the city. Philadelphia was the first city to install such a system, followed by other large cities such as New York, Boston, and Baltimore. By 1860, the nation's sixteen largest cities, as well as many smaller ones, had waterworks of some kind. In total, there were 136 American waterworks in 1860, 58 percent of which were owned privately and the remainder public, although the privately owned waterworks were more characteristic of the small rather than the large communities.[3]

The introduction of running water into cities resulted in a huge increase in water consumption: per-capita water consumption increased from about 2–3 gallons per day (gal/day) to between 50–100 gal/day. Boston, for instance, introduced water from the Cochituate Aqueduct into the city in 1848, and by 1855 the system had nearly 18,000 household connections, over 1,200 hydrants, and over 110 miles (180 km) of distribution pipes. The demand was so great that the construction of a second reservoir was required. Similar expansion of demand occurred in other cities.[4] But, although many cities introduced running water during this period, no city simultaneously made provision for means to remove the water. It was expected that the previous means of water disposal—street gutters or cesspools—would deal with the problem.

Installation of Water Closets

Lack of provision to handle the increase in the daily load of wastewater from kitchen and household use could be expected to add to cesspool or privy-vault overflows. Difficulties such as these were amplified with the introduction of another new technology—the water closet. This device was an old idea, but it had to await the availability of running-water supplies to enable its widespread adoption. (Before the complete acceptance of the water closet, some sanitarians attempted to persuade the public to adopt the earth closet, which was used in both Great Britain and in Europe. They argued that with the earth closet the valuable fertilizing materials in human waste could be saved. However, the earth closet never secured wide acceptance in this country since it could not compete with the convenience of the water closet.) No water closet patents were issued in the United States until 1833. Once running water was available, however, many affluent urbanites installed the "modern" convenience of a bathroom with a water closet. In Boston, for instance, in 1864 (population approximately 180,000), there were over 14,000 water closets; while in Buffalo in 1874 (population approximately 125,000), there were over 3,000 water closets in use. By 1880, roughly one-third of urban households had water closets.[5]

The installation of water closets in the absence of sewerage systems resulted in the adaptation of elements of the old waste-removal system to the new technology. The wastes from the water closets were now run into the cesspools and the privy vaults that retained their use as collectors. Even though some cities had stormwater sewers, the law usually forbade the placing of human excrement in them, although householders occasionally made surreptitious connections. The result of directing the flow from water closets into cesspools and privy vaults was to overwhelm the capacity for infiltration of wastewater into the soil around the cesspools or simply to overflow the privy vaults. This happened in city after city in the nineteenth century. As late as 1894, Benjamin Lee, the secretary of the State Board of Health of Pennsylvania, complained that, while water supply companies besieged Pennsylvania municipalities to adopt their systems and contractors encouraged home builders to install "modern" bathroom appliances, neither warned householders that the old "leaching cesspool" was inadequate to handle the resulting flood. "Copious water supplies," warned Lee, "constitute a means of distributing fecal pollution over immense areas . . . and no water closet should ever be allowed to be constructed until provision has been made for the disposition of its effluent in such a manner that it shall not constitute a nuisance prejudicial to the public health."[6]

Incentives for Sewers Based on Water Transport

Reports on the necessity for urban sewerage systems in the nineteenth century confirm that this typical sequence of events normally led to the demand for water-carriage removal of human wastes or the conversion of existing storm sewers to a combined purpose: cities introduced running water; householders installed water closets as well as other water fixtures and connected them to cesspools; the cesspools overflowed. The overflow from the cesspools, however, resulted not only in an unsanitary nuisance but also in a cost to the householder. Overflowing cesspools now had to be emptied far more frequently than before, sometimes at the insistence of city inspectors, and with a much higher annual cost. In the 1880 report on a sewage plan for Newport, Rhode Island, for instance, the following was noted: "There are in our town residents who having introduced water into their dwellings find no way of ridding themselves of it. Their cesspools . . . that formerly required cleaning out twice or three times in a season must be emptied as often as every ten days. The expense of removing the contents of the cesspool at one house last summer was $300. The amount in dollars we can appreciate, but the danger to health cannot be realized."[7] This $300 charge for cesspool cleaning is an extremely high figure and not typical of such charges, although statements in a number of other urban sewerage reports confirm the higher cost of cleaning once water was introduced in a city.

The previous analysis illustrates how the introduction of a new technology, running water, resulted in the adoption of another new technology—the water closet. The unanticipated result of the adoption of these technologies was to upset the existing system of waste disposal and to create problems of both sanitary nuisance and higher out-of-pocket costs for cesspool cleaning. There were attempts to preserve the existing system by using "odorless evacuators" (vacuum pumps) to empty cesspools, but this technology was both costly and undependable.[8] Increasingly, in the period after the Civil War, as more cities installed running-water supplies and householders adopted water closets, urban policymakers became convinced that the only method to deal with the wastewater problem was to adopt another new technology—the installation of sewers that transported wastes by water carriage.

Variations in Sewer Technologies: Separate or Combined Sewer Debate, 1880–1900

The decision to abandon cesspools and privy vaults as collectors of human waste and wastewater and to substitute a sewerage system utilizing water-carriage removal resulted in the emergence of a new system for removal and disposal. Within the general outlines of this system, however, there were important debates about sewer technology, sewage disposal, the etiology of disease, and the role of government. Utilization of water-carriage removal, for instance, caused a shift in concern about the locus of pollution and media that were being polluted. With the cesspool–privy vault system, pollution was localized—the earth around cesspools was saturated, and wells and groundwater supplies were endangered with contamination. With the water-carriage system, the locus of pollution was shifted to streams and rivers often distant from the vicinity of the household. In addition, the disposal of one city's wastes often created problems for other downstream cities. Whereas in the period before the Civil War boards of health had only been local, it now became evident that extralocal authorities—either metropolitan or state boards of health—were needed to deal with pollution questions.

Origins of Separate Versus Combined Sewer Debate

The dispute included questions of disease etiology, methods of disposal, and cost differentials for dual-pipe systems. Municipal and engineering decisions made during this period favoring the installation of combined rather than separate systems represented a choice based on perceptions of local needs and fiscal constraints. These decisions, however, produced pollution problems of a magnitude that was unanticipated and led to a new system for dealing with wastewater problems in the next period.

The debate over separate and combined sewers was an old dispute that went back to the 1840s in Great Britain and essentially involved the question of whether human wastes and surface runoff from storms should be carried in the same pipe channels. The original urban sewers were for stormwater only. In Great Britain and Europe, storm sewers became combined sewers when households connected up to the existing sewers after the installation of running water and water closets.[9]

The improper working of these systems, however, due to poor design, led to the idea of a separate pipe system for household and human wastes. In addition, some sanitarians argued that the separate system would make possible the retrieval of human wastes for fertilizing purposes—an opportunity for resource recovery in addition to simple waste removal. A few cities in Great Britain and Europe did construct sanitary sewers and allowed their combined sewers to revert back to their original stormwater purpose. But the great majority of cities, faced with the problem of providing for both stormwater and household waste removal at the cheapest cost, built combined sewerage systems.[10]

To a great extent, American sanitary engineers looked to Great Britain and Europe for models on which to base their designs. From 1857, the date at which the first sewerage system designed to handle sanitary wastes as well as stormwater was constructed, in Brooklyn, to 1880, combined systems were typically constructed in American cities, although a separate system was built in the small resort town of Lenox, Massachusetts, in 1876.[11] While American engineers were aware of separate systems and the debate over their use in Great Britain, they argued that the need to provide for stormwater removal as well as household wastes made the combined system the most economical.

Waring Separate System and Its Impact

The first major application of a separate system came in 1880 in Memphis, Tennessee. Conditions there were unusual in that the system was built as a direct result of a severe yellow fever epidemic that occurred in 1878–79. Another unusual circumstance was the involvement of a federal agency, the short-lived National Board of Health, which entered the situation as the result of an invitation from local officials. After surveying the city and considering a number of sewerage plans, a commission appointed by the National Board of Health recommended the plan for a separate sewerage system conceived by one of its own members, Colonel George E.

Waring, Jr., a well-known sanitarian who had constructed the Lenox separate sewer system.

Waring's small-pipe system, in contrast to other sewerage systems, was based on a theory of disease etiology—a version of the filth theory—rather than engineering design principles. Waring maintained that unless human fecal wastes were quickly transported out of the city in "fresh condition" they would "undergo putrefaction and give off objectionable gases." These sewer gases would be lethal to the extent that fecal matter had decomposed. In order to speed the removal of human wastes, the Waring system utilized small vitrified pipes (4-inch [100-mm] house branches, 6-inch [150-mm] laterals) and automatic flush tanks. Waring criticized combined sewers as too large and as having insufficient velocity to prevent sewer gas-producing deposits from forming. Stormwater removal, he insisted, was only a secondary function of sewers, and stormwater could safely be permitted to run off along surface channels. A further feature of his system, said Waring, was that, in contrast to the combined system, it permitted land treatment of sewage where neighboring streams provided insufficient dilution for disposal.[12] This was an opportunity to select an alternative method of disposal at the end of the pipe.

The Waring system attracted a great deal of attention, not the least of which was generated by Waring himself in a series of articles in popular and scientific journals. By 1892, approximately twenty-two towns and cities had constructed Waring systems. Important in its adoption was Waring's success in cleaning up Memphis as well as concerns of towns or cities over the evil effects of sewer gas. One civil engineer maintained, for instance, that if cities adopted the combined system rather than the Waring separate system, it would "largely increase . . . 'zymotic' diseases, such as typhoid, diphtheria, scarlet fever, etc." In addition, the relatively inexpensive cost of the separate ($7,000/mile [$4,300/km]) sanitary sewer only, compared to the combined system ($28,000/mile [$17,000/km]), recommended it to many city officials. Most cities that installed a Waring system made no provision for the underground removal of stormwater. Some, however, did provide for treatment of their sewage by broad irrigation or sewage farming.[13]

Typical Late-Nineteenth-Century Sewerage Systems

Despite the alleged sanitary virtues of the Waring system, no large city installed a separate system of any design in the years up to 1900. Most engineering opinion followed the course laid down by Rudolph Hering in his famous report of 1881 to the National Board of Health. Hering took the position that the combined and separate systems had equal sanitary value and that the construction of one system or another depended on local conditions. Generally, he maintained, the combined system was best suited and cheapest in large, densely built-up cities where surface drainage of rainwater was unacceptable. The separate system, on the other hand, was suitable for cities and towns that did not require underground removal of stormwater and where sewage required treatment before discharge to a receiving waterway.[14]

By the last decade of the nineteenth century, therefore, an accepted model had emerged in regard to the building of sewer systems. Large cities built combined systems and disposed of household wastes and stormwater in nearby water courses to undergo treatment by dilution. Most small cities and towns (population less than 30,000) built separate systems, but usually with sanitary sewers only, allowing stormwater to run off on the surface. The most important factors in the choice of one system over another were local conditions of population and traffic density, the costs of the alternative systems, and availability of water courses for disposal.

Impacts of Combined Sewer Technology

While combined sewerage systems simplified the problems of transporting household wastes and stormwater, they complicated the question of disposal. As sanitary engineer Moses N. Baker commented in the 1905 *Social Statistics of Cities,* "The general rule observed by American cities of all sizes is to discharge their sewage into the nearest available water until the nuisance becomes intolerable to themselves, and then to divert it from their own shores, resting content with inflicting their wastes on neighbors below, until public protest or lawsuits make necessary adoption of remedial measures."[15]

Cities disposed of their sewage through dilution in water courses because it was convenient and cheap and because the technology of sewage treatment was expensive and often uncertain. They operated on the rationale of the theory of the self-purification of streams, that running water purified itself within a given distance. What distances were involved and at what stream velocity in regard to a given amount of sewage, however, was unclear. What was known was that it was far cheaper to treat sewage from separate sanitary sewers than from combined sewers. In 1902, for instance, of the ninety-five cities with over 3,000 population that treated a portion of their sewage, only two—Worcester, Massachusetts, and Providence, Rhode Island—had combined sewers.[16] Although exact information is lacking, the sources suggest that the great majority of the smaller cities that treated their sewage had separate sanitary systems.

The fact that sewage disposal in water courses could, if populations rose, cause pollution problems cannot be called an unanticipated result. There were early cases of stream pollution in Great Britain in the 1850s and 1860s, and Massachusetts actually took action against such pollution in the 1870s. A number of well-known sanitary engineers such as Rudolph Hering, Moses N. Baker, and Colonel Waring warned of the threat of dangerously polluting inland rivers through sewage pollution.[17] Waring's solution to the problem was to build separate sanitary sewers and treat the collected sewage through land disposal. During the 1890s, a number of sanitary engineers took the same position. But while Waring's motivation in urging the separate system was his concern over sewer gas, as well as the need for sewage treatment, engineers such as Baker and George Rafter who took the same position in the 1890s were primarily concerned with bacterial contamination.

While sanitary engineers were aware that rivers could become overloaded with sewage and present health hazards as well as nuisances, they did not know how much pollution constituted a hazard to downstream users nor how far downstream the hazard persisted. Until approximately 1890, chemical analysis was the only means to test water quality. In the late 1880s, however, bacteriologic research played a more prominent role in defining water quality and gradually resulted in the substitution of the germ theory of disease for the filth theory. In the early 1890s, the work of William T. Sedgwick at the Massachusetts Board of Health laboratories clarified the etiology of typhoid fever. The clear identification of the waterborne nature of communicable disease produced an increased concern with methods of sewage "purification."[18] And, given the nature of existing technology and the high volume of wastewater produced by combined sewers, purification required the installation of separate sanitary sewers.

Needs to Modify the Nineteenth-Century System

By the end of the century, sanitary engineers were confronted with a dilemma. They had enthusiastically urged the replacement of the cesspool–privy vault system of waste removal by the water-carriage system. Faced with a choice of system design, they had widely recommended the combined sewer over the separate sewer system, because the combined system handled storm runoff in addition to household wastewaters at less cost than a dual system. The reliance on running water in surface streams to purify itself was not understood quantitatively and could not be included in any reliable systematic engineering design.

The 1880s and 1890s witnessed rising typhoid death rates in downstream cities that drew their water supplies from rivers in which upstream cities discharged their sewage. Bacterial analysis confirmed the link between sewage pollution of rivers and typhoid fever. The system of waste disposal that had replaced the cesspool—privy vault system—a system that primarily used combined sewers to remove both household wastes and stormwater and that relied on dilution as a disposal method—had broken down and recreated the health hazards of the old system.[19] Increasingly, the health hazards caused by the sewage pollution of rivers initiated the development of new institutions—state boards of health with enforcement powers—that stood ready to force change in the system.

Debate over Wastewater Treatment and Water Filtration: 1900–1932

Evolution of State Regulations

The sewage pollution of streams and the resulting typhoid death rates focused the attention of public health officials and sanitary engineers on the power of the state to prevent disposal in streams. Massachusetts established the first state board of health in 1869, and this example was followed by a number of other states in the late nineteenth century. More significantly, during the 1890s and the beginning of the twentieth century, state legislatures in states like New Jersey, Ohio, and Pennsylvania equipped their state boards of health with power to forbid sewage pollution of streams by municipalities.[20]

In Pennsylvania, for instance, the State Board of Health was established in 1885 as the result of a severe typhoid fever epidemic that swept through the mining community of Plymouth. The epidemic had clearly demonstrated that state authority in public health matters was required because the state had sole power to enforce regulations beyond municipal boundaries. The board's role, however, was limited to the investigation and abatement of nuisances and the recommendation of new legislation, and it received no specific powers in regard to stream pollution.[21]

In 1903, however, a severe typhoid epidemic caused nearly 1,400 cases of illness and 111 deaths in the town of Butler. The Pennsylvania legislature responded to the Butler epidemic by passing a stream pollution law. This law forbade the discharge of sewage into the state's waterways without treatment, although cities that had sewerage systems prior to the passage of the law were exempted from its effects. The law, however, did apply to extensions of existing sewerage systems. In addition to the stream pollution legislation, the legislature also replaced the State Board of Health with the Department of Health with the authority to enforce the legislation.[22]

By 1905 there were several other states, such as Minnesota, New Jersey, and Ohio, that had laws comparable in strength to the Pennsylvania law.[23] High typhoid death rates, although not necessarily epidemics, were the motivating cause. In these states, as in Pennsylvania, legislation did not apply to systems constructed before the enactment of the law but did apply to extensions.

Progress in Treatment Technologies: Sewage and Water

The passage of laws forbidding sewage pollution of streams assumed the existence of a technology available at a reasonable cost capable of preventing such pollution. At the end of the 1890s and the beginning of the twentieth century, both sanitary engineers and public health officials were quite optimistic in this regard. In 1898, for instance, the Pennsylvania Board of Health stated that the "purification of the sewage of large towns, is a matter of not the slightest difficulty" and recommended that all cities possessing public water-supply systems install sewage-treatment facilities. Two years later, *The Engineering Record* confirmed this judgment in an editorial: "the resources of the sanitary engineer are sufficient to bring about the purification of sewage to any reasonable degree. This costs money . . . , but not so much as is often believed."[24] This optimism, however, was clearly unwarranted.

As mentioned previously, by 1902 ninety-five cities treated some portion of their sewage, but only eleven of these had populations of more than 30,000, and all but a small fraction had separate sewerage systems. The major sources of river pollution, however, were the large cities and, as of 1905, New Orleans was the only city with a population above 30,000 to have a separate system. The remainder of the large cities, all with combined systems, disposed of their sewage by dilution. They followed this practice because it was the cheapest form of disposal and because sewage pollution of water supplies was considered primarily a problem for downstream cities.[25]

Even if these cities had wanted to treat their sewage and reduce the danger of typhoid epidemics, their combined sewerage systems and the uncertain and changing state of sewage-treatment technology would have made this immensely costly. The first methods utilized in this country were broad irrigation and sewage farming. In the early 1890s, however, experiments at the Massachusetts Board of Health's Lawrence Experiment Station had proven the feasibility of intermittent filtration, and this method became the dominant technology utilized in the 1890s. Intermittent filtration, however, required both the availability of soil with a high infiltration rate and large tracts of land. Both factors restricted its feasibility for many cities. In the beginning of the twentieth century, other sewage-treatment technologies such as household septic tanks and central treatment plant trickling filters were introduced.[26]

Simultaneous with the changes in sewage-treatment technology, important developments were taking place in the area of water filtration. Here, as with sewage treatment, the Lawrence Experiment Station of Massachusetts led in the development of technology. In 1893, a slow sand filter designed by Hiram F. Mills of the Lawrence Station was installed in the city of Lawrence, and in five years the typhoid death rate was observed to decrease by 79 percent. The development of other types of filters, such as the mechanical filter and the rapid sand filter, followed and were installed for water-supply treatment in a number of cities. By 1900, 28 percent of the urban population was drinking filtered water, and typhoid death rates in cities were dropping quickly.[27]

The striking success of water-filtration processes in lowering typhoid fever rates, as compared with the slow progress in sewage-treatment development, caused a crucial shift in engineering opinion. As *The Engineering Record* editorialized in 1903, "'Many engineers believe today that in some cases it is often more equitable to all concerned for an upper riparian city to discharge its sewage into a stream and a lower riparian city to filter the water of the same stream for a domestic supply, than for the former city to be forced to put in sewage-treatment works."[28] Not only was water filtration more certain in terms of lowering typhoid death rates, but it also could be provided at a far cheaper cost than sewage treatment.

Conflict over Enforcement: The Pittsburgh Case

While engineering opinion was shifting toward the choice of water filtration over sewage treatment in order to protect drinking-water supplies, state legislatures and the courts were moving in a different direction. By 1909 a number of states, such as California, Indiana, Minnesota, New Jersey, and Ohio, had passed laws forbidding the discharge of raw sewage into streams from new municipal systems. These laws were designed, in most cases, to be administered by state boards of health staffed, to a large extent, by physicians. The medical position, in contrast to that of most sanitary engineers, was that the public health would best be protected if no municipality was permitted to discharge raw sewage into streams. And they stood ready to use the authority of the state to enforce their directives.[29]

The critical clash between the state boards of health and the sanitary engineers came in the city of Pittsburgh. In 1910, Dr. Samuel G. Dixon, the Pennsylvania commissioner of health, required Pittsburgh to submit within one year a comprehensive plan providing for sewage treatment. The preparation of this plan was a condition for receiving a temporary discharge permit for extensions of the city's combined sewer system. The city hired two well-known sanitary engineers, Allen Hazen and George Whipple, to make recommendations. In January 1912, after one year of investigation, Hazen and Whipple made what *The Engineering Record* called "The Most Important Sewerage and Sewage Disposal Report Made in the United States."[30]

The Hazen and Whipple report concentrated on the economic feasibility of constructing separate sanitary sewers and a sewage-treatment plant in Pittsburgh. They estimated that replacing Pittsburgh's combined sewers with a separate system and building a treatment plant would cost Pittsburgh taxpayers a minimum of $46 million with no provision for the costs of disruption caused by construction of the system. They also noted that Pittsburgh's water-filtration plant had cost $7 million and calculated that the twenty-six towns downstream from Pittsburgh on the Ohio River could provide filtered water for their residents for far less than the $46 million it would cost Pittsburgh to build its treatment plant. No precedent existed, they argued, "for a city's replacing the combined system by a separate system for the purpose of protecting water supplies of other cities." Given this situation, Hazen and Whipple concluded that "no radical change in the method of sewerage or of sewage disposal as now practiced by the City of Pittsburgh is now necessary or desirable."[31]

Engineering opinion overwhelmingly supported the Hazen and Whipple report and viewed the controversy as an issue of "how far engineers are at liberty to exercise their own judgment as to what is best for their clients and how far they must give way to their medical colleagues." Faced by this opposition and uncertain about his ability to secure enforcement of his order compelling Pittsburgh to build a separate system and treat its sewage, Dixon retreated and issued the city a temporary discharge permit. The state commissioner of health continued to issue such permits to the city until 1939.[32]

In the years after the Pittsburgh case, engineering opinion coalesced around the view, as expressed by sanitary engineer Allen Hazen, that "a dollar spent in water purification goes much father toward protecting a community from the dangers of sewage pollution in its potable water supply than a dollar laid out in sewage-treatment works." From 1900 into the 1920s, the adoption of filtered water supplies proceeded at a much faster rate than the selection of technology for sewage treatment, especially for large cities with combined sewerage systems. Many smaller communities, however, installing sewerage systems for the first time, followed the dictum of state boards of health and built separate sanitary sewers with at least primary treatment. But, in many of these cities, no provision was made for stormwater.[33]

Heritage of Decision to Treat Water but Not Sewage

In the period from approximately 1900 into the 1930s, engineers were again confronted with major policy choices in regard to wastewater treatment and the protection of public water supplies. The major choice made in the previous period, 1880–1900, to build combined rather than separate sewer systems in large cities, had resulted in extensive stream pollution by the turn of the century. Engineers initially responded to this situation by promoting various forms of sewage-treatment technology, but limitations of the technology, plus the expense of treating outflow from combined systems, restricted this option. Increasingly, engineers turned toward water filtration, a less expensive and more cost-effective means of purifying public water supplies.

The pollution problems produced by combined sewers, however, spawned not only sewage- and water-treatment technologies but also water pollution laws enforced by activist state boards of health. These boards of health, often staffed by public health doctors representing the so-called "New Public Health," concerned themselves with controlling the causes of disease at the source. For them, this meant keeping all human wastes from contact with public water supplies. In a number of states these boards of health compelled small cities to stop discharging raw sewage into streams and to build treatment plants. The crucial question, however, involved the major cities that had combined sewer systems. Could the state force them to convert their combined sewer systems to separate systems and build treatment plants, often at a cost that exceeded their bonded debt limits? Engineering opinion took the position that such demands were "radical if not . . . quixotic." Dilution, they insisted, was "sound in principle and safe in practice if carried on with proper restrictions," especially since it was "physically impossible to maintain waterways in their original and natural condition of purity."[34] In addition, economics militated against so-called "purification" of sewage that could only take place at enormous expense.

The focus on bacteriological pollution associated with sewage wastewaters and the cost-effective alternative of treating water at the point of use was expedient and proper for a limited set of water-quality concerns. It neglected, however, chemical pollutants associated with industrial sources and initiated an overdependence on dilution in streams.[35] In addition, it did not anticipate the need for waste-treatment technologies for waste constituents that proved refractory to water-filtration plants. Tastes and odors were the most common nuisance, although they did not serve as specific evidence or correlate with health hazards.

Conclusion

The adoption of a technology and its development in each historical period strongly influenced choices in subsequent periods. Decisions had impacts, many unanticipated, in a variety of technological, environmental, public health, and governmental areas.

The introduction of running water at the tap led to the installation of water closets. Increased domestic water use overloaded existing wastewater disposal systems and led to the building of water-carriage systems for domestic waste removal. Periodic batch cleaning of individual private privy vaults and cesspools was replaced by flushable water closets with continual removal via direct connections to sewers. Water supply and waste disposal changed from a private responsibility to a public and governmental function.

Waste disposal and resource recovery by land application of domestic wastes as fertilizers gave way to disposal in waterways, leading to high typhoid death rates. Pollution of rivers used for water supply divided environmentalists into two groups. Some, mainly public health physicians, favored sewage treatment as well as water treatment, while others, mainly sanitary engineers, emphasized water treatment at a higher priority than waste treatment. The engineers, influenced by municipal fiscal constraints, urged water treatment first, as the available technology was more cost-effective.

The examination of the record shows the complexity of past decision making in the area of wastewater collection and disposal. Choices made were often based on a combination of local needs and current scientific ideas. Although the adoption of a new collection technology caused problems to become regional rather than local, financing remained available solely on a municipal basis. Only the intervention of either regional authorities or the federal government, with its higher funding basis, could begin to cope with the pollution problems that were the heritage of the original adoption of a water-carriage system for domestic waste removal.

Notes

1. J. A. Tarr, "From City to Farm: Urban Wastes and the American Farmer," *Agricultural History* 49 (October 1975): 599–611.

2. J. Duffy, *A History of Public Health in New York City: 1625–1866* (New York: Russell Sage Foundation, 1968); C. E. Rosenberg, *The Cholera Years* (Chicago: University of Chicago Press, 1962).

3. N. M. Blake, *Water for the Cities* (Syracuse: Syracuse University Press, 1956).

4. Ibid.

5. "Annual Report of the Cochituate Water Board to the City Council of Boston for the Year 1863" (Boston, Mass., 1864).

6. W. H. Bent, et al., "Report of Special Committee on Sewerage for City of Taunton, with Report and Proposed Plan by J. H. Shedd" (Taunton, Mass., 1878); H. A. Carson, "Separate vs. Combined Sewer," *Sanitary Engineer* 4 (February 15, 1881): 143; "Report of the Selectmen of Sewerage" (Medford, Mass., 1873); J. Duffy, *History of Public Health in New York;* and B. Lee, "The Cart Before the Horse," *Papers and Reports of the American Public Health Association* 20 (American Public Health Association, 1895): 34–36;

7. E. S. Chesbrough, "Report . . . on Plan of Sewerage for the City of Newport" (Newport, R.I., 1880).

8. A. Ames, "The Removal and Utilization of Domestic Excreta," *Papers and Reports of the American Public Health Association* 4 (1877): 65–80.

9. R. Hering, "Report of the Results of an Examination Made in 1880 of Several Sewerage Works in Europe," *Annual Report of the National Board of Health 1881* (Washington, DC: National Board of Health, 1882), 99–137.

10. Ibid.; J. A. Tarr, "The Separate vs. Combined Sewer Problem: A Case Study in Urban Technology Design Choice," *Journal of Urban History* 5 (May 1979): 308–339.

11. "Sewage Purification in America," *Engineering News* 28 (July 14, 1892).

12. F. S. Odell, "The Sewerage of Memphis," *Transactions, ASCE* 10, no. 216 (February 1881): 23–31.

13. R. Hering, "Six Years Experience with the Memphis Sewers," *The Engineering and Building Record and The Sanitary Engineer* 16 (November 26, 1887): 738–739; G. E. Waring, Jr., "The Memphis System of Sewerage at Memphis and Elsewhere," *American Public Health Association* 18 (1893): 153–168; O. Chanute, "The Sewerage of Kansas City," *American Contract Journal* 11 (February 16, 1884): 81; G. W. Rafter and M. N. Baker, *Sewage Disposal in the United States* (New York: D. Van Nostrand and Company, 1894); and B. Williams, "Separate vs. Combined Systems of Sewerage," *Sanitary Engineer* 11 (April 16, 1885): 413–414.

14. R. Hering, "Report of the Results of an Examination," 99–137.

15. U. S. Department of Commerce and Labor, *Statistics of Cities Having a Population of over 30,000: 1905* (Washington, DC: U. S. Government Printing Office, 1907).

16. J. Slater, "The Self-Purification of Rivers and Streams: A Study of the Transformation of Water Safety Standards in the Late Nineteenth Century," *Synthesis* 2 (Autumn 1974): 41–53; "Sewage Purification and Water Pollution in the United States," *Engineering News* 47 (October 3, 1902): 275–276; and Department of Commerce and Labor, *Statistics of Cities, 1905*.

17. R. Hering, "Notes of the Pollution of Streams," *Papers and Reports of the American Public Health Association*, vol. 13 (American Public Health Association, 1888): 272–279.

18. B. G. Rosenkrantz, *Public Health and the State* (Cambridge, MA: Harvard University Press, 1972).

19. W. T. Sedgwick, *Principles of Sanitary Science and the Public Health* (New York: The Macmillan Company, 1918).

20. "The Pollution of Streams," *Engineering Record* 60 (August 7, 1909): 157–159.

21. F. H. Snow, "Administration of Pennsylvania Laws Respecting Stream Pollution," *Proceedings of the Engineers' Society of Western Pennsylvania* 23 (July 1907): 267–270.

22. Ibid.

23. S. D. Montgomery and E. B. Phelps, "Stream Pollution: A Digest of Judicial Decisions and a Compilation of Legislation Relating to the Subject," *Public Health Bulletin* (Washington, DC: U.S. GPO, 1918).

24. *Annual Report of the Pennsylvania State Board of Health, 1898* (1898), 1:5; "The Water Supply of Large Cities," *Engineering Record* 41 (January 27, 1900): 73.

25. U.S. Department of Commerce and Labor, *Statistics of Cities Having a Population of over 30,000: 1909* (Washington, DC: U.S. GPO, 1913).

26. G. W. Fuller, *Sewage Disposal* (New York: McGraw-Hill Book Company, Inc., 1912); Tarr, "From City to Farm," 599–611.

27. A. Hazen, *Clean Water and How to Get It* (New York: John Wiley and Sons, Inc., 1907).

28. "Sewage Pollution of Water Supplies," *Engineering Record* 28 (August 1, 1903): 117–118.

29. H. W. Hill, "The Relative Values of Different Public Health Procedures," *Engineering News* 66 (October 12, 1911): 436; "The Need for a More Rational View of Sewage Disposal," *Engineering News* 64 (October 13, 1910): 394–395.

30. "The Most Important Sewerage and Sewage Disposal Report Made in the United States," *Engineering Record* 65 (January 24, 1912): 209–212.

31. "Pittsburgh Sewage Disposal Reports," *Engineering News* 67 (February 29, 1912): 398–402; G. P. Gregory, "A Study in Local Decision Making: Pittsburgh and Sewage Treatment," *Western Pennsylvania Historical Magazine* 57 (January 1974): 25–42.

32. "Compulsory Sewage Purification," *Engineering Record* 65 (February 24, 1912): 198; Gregory, "A Study in Local Decision Making," 25–42.

33. Hazen, *Clean Water and How to Get It;* H. P. Eddy, "Use and Abuse of Systems of Separate Sewers and Storm Drains. Can Their Failure be Prevented?" *Proceedings of the American Society for Municipal Improvements* 29 (October 1922): 128–131.

34. "A Plea for Common Sense in State Control of Sewage Disposal," *Engineering News* 67 (February 29, 1912): 412–413; G. W. Fuller, "Is It Practicable to Discontinue the Emptying of Sewage into Streams?" *The American City* 7 (March 1912); and G. C. Whipple, "Sewage Treatment vs. Sewage Purification," *Engineering News* 68 (October 3, 1912): 388–389.

35. A. Wolman, "Domestic and Industrial Wastes in Relation to Public Water Supply: A Symposium," *American Journal of Public Health* 16 (August 1926): 777–804.

CHAPTER V

The Separate vs. Combined Sewer Problem

A Case Study in Urban Technology Design Choice

Introduction

Public capital-spending projects involving technologies such as water supply, street paving, and sewers were characteristic of American cities during the late nineteenth and early twentieth centuries. All involved questions of technology adoption and design considerations. The decision by a municipality to construct a sewerage or water-carriage system of waste removal involved a critical design choice between two forms of sewer technology—the combined sewer and the separate one. In normal use the combined system carries both household wastes and stormwater in one large pipe. The separate system theoretically provided two sets of pipe: a small-diameter pipe for household wastes, called a sanitary sewer, and a larger-diameter pipe for stormwater from streets, roofs, and yards. In many cases, however, cities with separate systems only constructed sanitary sewers and made no provision for underground removal of stormwater.

This paper focuses on the elements involved in decisions about sewerage system design choice in the later nineteenth and early twentieth centuries. In order to clarify the dimensions of these choices, I will first consider the factors leading to the decision to replace the existing privy vault–cesspool system of waste removal with the sewerage or water-carriage system.

The Decision to Abandon the Privy Vault–Cesspool System and Adopt Water-Carriage Removal

The method of human-waste collection replaced by sewerage technology in American cities in the late nineteenth and early twentieth centuries was the privy vault–cesspool system. The privy vault and cesspool were essentially holes in the ground, often lined with stone, located close by residences or even in cellars. Although much of the contents of a well-constructed cesspool or privy was often absorbed by surrounding soil, the receptacles still needed periodic emptying. In some cities, the contents of the vaults or cesspools were removed by scavengers or farmers, often under contract to the municipality. In many locales, however, householders merely covered the full vaults with dirt and dug new receptacles. The privy vault–cesspool system, therefore, had the following characteristics: it was locally based, removal was inefficient and labor intensive, and the system was largely privately maintained. Whatever sewers existed in cities were for stormwater removal. Most municipalities utilized surface gutters for this purpose, while some larger cities had underground channels. Regulations, however, usually prohibited householders from placing excremental matter in them.[1]

As cities developed and grew in the nineteenth century, a combination of demographic and technological factors caused the privy vault–cesspool system to become increasingly inadequate to deal with waste-disposal problems. Urban population and density growth enlarged the pressure on existing facilities, increased the frequency of cleaning, and necessitated the digging of new privies in urban alleys, backyards, and cellars. Located close together and serving larger populations, these receptacles often overflowed, causing nuisances and aesthetic problems. The night soil carts of the scavengers, utilized to contain and remove the wastes from the privies, created similar difficulties. Soil saturated with fecal wastes contaminated groundwater supplies and wells and ponds, often badly polluting sources of water supply.[2]

But while demographic factors were an important element in increasing the nuisances and problems caused by the privy vault–cesspool system, the adoption by American cities of the new technology of piped-in water was equally critical. Cities built water systems to obtain larger supplies of potable water for drinking and household purposes, to fight fires, and to use for street flushing at times of concern over epidemics. Philadelphia constructed the first waterworks in 1802, and by 1860 the sixteen largest cities in the nation had followed its example. In 1860, there were a total of 136 systems in the country, a number that had increased to 598 by 1880. The availability of a constant supply of water in the household caused a rapid expansion in per-capita use, as demand interacted with supply. Chicago water usage, for instance, went from 33 gallons per capita per day in 1856 to 144 in 1882; Cleveland increased from 8 gallons per capita per day in 1857 to 55 in 1872; and Detroit from 55 gallons per capita per day in 1856 to 149 in 1882.[3] These figures include industrial and other nonhousehold uses, but they are still indicative of greatly increased household water consumption over a relatively short span of time.

But while hundreds of cities and towns installed waterworks in the first three-quarters of the nineteenth century, no city simultaneously constructed a sewer system to remove the water (some short, private sewers were built). In most cities with waterworks, wastewater was initially diverted into cesspools or existing stormwater sewers or street gutters, often causing serious problems of flooding and disposal. The adoption by urban households of the water closet and the discharge of large quantities of fecally polluted water into cesspools and privy vaults designed for much smaller volumes greatly exacerbated the problems. Soil became saturated, cellars were "flooded with stagnant and offensive fluids," and vaults needed much more frequent emptying. Urban public health officials, especially those imbued with a belief in the anticontagionist "filth theory" of disease, were especially concerned. They warned that overflowing cesspools and privies with water closet connections posed an especially dangerous threat to a healthy environment.[4]

The adoption of the two new technologies of piped-in water and the water closet, therefore, combined with higher urban densities to cause the breakdown of the privy vault–cesspool system of waste removal and to increase its productivity of both nuisance and of real and perceived health hazards. City councils, sanitary committees, and health and engineering groups throughout the nation engaged in discussions concerning the various alternatives to the privy vault–cesspool system.[5] Most groups ultimately agreed that, compared to other options, the so-called water-carriage or sewerage system of waste disposal provided the most benefits and the lowest costs.

In contrast to the privy vault–cesspool system, sewerage systems were capital rather than labor intensive and required the construction of large public works. They operated in automatic fashion, almost eliminating the need for human decisions and actions to remove the wastes and offered greatly improved conditions of convenience, cleanliness, and the elimination of nuisance. And sewerage systems solved both the collection and transportation problems, moving the wastes in a wastewater stream from the immediate to a distant locality. Proponents of the system maintained that municipalities should adopt them primarily for three reasons: the capital and maintenance costs of sewerage systems would represent a saving over the annual cost of collection and cleaning with the privy vault–cesspool system; sewerage systems would create greatly improved sanitary conditions and result in lowered morbidity and mortality from infectious disease; and because of improved sanitary conditions, cities that constructed sewerage systems would attract population and industry and grow at a faster rate than those that did not.[6]

Starting in the 1850s with Brooklyn and Chicago and increasingly in the late nineteenth and early twentieth centuries, city after city made the decision to abandon the privy vault–cesspool system of waste collection and to construct water-carriage systems. An integral and inseparable part of the discussions and decisions over system change was the design choice over whether to build combined or separate sewers.

The decision between types of sewers would appear to be one of simple engineering design, based on an evaluation of the needs of a city from the perspective of some model of engineering choice. Such a model, based mainly on rudimentary cost-benefit calculations, was primarily utilized in making the choice for sewerage. No such model, however, existed for choice of design. Its absence is partially explained by the relative newness of sewerage systems and the lack of a community of practitioners who agreed on basic criteria for implementation. As Leonard Metcalf and Harrison P. Eddy argued in the 1914 edition of their classic work on sewerage, "American Sewerage Practice is noteworthy among the branches of engineering for the preponderating influence of experience, rather than experiment, upon the development of many of its features."[7]

The debate over design criteria, however, concerned more than the newness of the field. Also involved were basic differences over theories of disease etiology, the removal of stormwater, and the disposal and treatment of sewage. Complicating the efforts to arrive at a solution to these questions were the important roles played by a charismatic individual, Colonel George E. Waring, Jr., and a crisis event, the yellow fever epidemic that ravaged the Mississippi Valley in 1878 and 1879. Waring's success in publicizing his own ideas about sewerage retarded the development of a "rational" engineering design model for implementation of the alternative sewerage systems. (By "rational" model, I mean one based on empirical considerations, such as costs and rainfall and traffic conditions necessitating stormwater removal, rather than considerations of disease etiology not empirically verifiable.) The final development of this "rational" model, however, and the adoption of one system or another according to its formula, were to have unanticipated consequences of a magnitude requiring modification. In addition, these unexpected impacts necessitated the evolution of a public policy regarding water-quality issues.

Early Experience with the Separate and Combined Systems

The design of sewerage systems in the United States during the first period of their implementation in the latter half of the nineteenth century was largely based on previous European and British experience. While technology transfers were modified by local conditions, there was no development in American sewerage that was not previously anticipated in Great Britain. The first combined sewerage systems originated in both Europe and Great Britain when households, newly supplied with running water, tapped into existing drainage sewers to dispose of household and water closet wastes. In some cases, this required alteration of statutes forbidding the placing of excremental wastes in the sewers. The American record in regard to the original combined sewers replicated the European experience.[8]

Small-pipe separate sewers seem to have been advocated first by the British sanitarian Sir Edwin Chadwick in 1842. Chadwick's motivation was twofold: to remove excremental matters from the household speedily and with a minimum effort and to remove them in a form that permitted their agricultural use as fertilizer. The first motivation stemmed from a belief that human wastes, if left to putrefy, produced disease-propagating miasmas; the second from the conviction that the valuable ingredients in human wastes should be preserved and that their utilization would make a system of arterial sewers self-financing.[9]

When the city of London finally constructed a sewerage system in 1858, however, it was based on the plans of engineer Joseph Bazalgette providing for a combined system constructed of brick that would accommodate both household wastes and stormwater. The concept of the separate system was rejected because of the need to remove stormwater from the surface of crowded London streets and because of the argument that, primarily because of animal excrement, "rain falling upon the streets is as much polluted as sewage, and ought to be treated as such."[10]

The debate in London between advocates of the separate and combined systems had a profound impact on sewerage building in the United States. The two major sewerage systems constructed in the United States before the Civil War—that in Chicago, constructed by E. S. Chesbrough, and that in Brooklyn, built by Julius W. Adams, were of the combined type. In his 1857 report to the Brooklyn Commissioners of Drainage, Adams referred to the London debate and rejected the separate system on the grounds that only the intended agricultural use of sewage justified its construction. He argued that if the combined system were built, the sewage could be disposed of in the ocean and stormwater would also be eliminated from the streets. After examining European sewerage systems in 1859, Chesbrough also decided against the separate system. He told the Chicago Board of Sewerage Commissioners that the separate system entailed great expense and that "it would not result in freeing the sewers intended for surface water from the introduction of substances that render them offensive."[11]

In the 1860s and the early 1870s, a number of cities in the United States constructed sewerage systems, but always of the combined type. The choice of this technology was dictated by several factors: there was no European or American precedent of a successful separate system, and engineers were reluctant to experiment with large capital works; from a cost-benefit perspective, the combined system was cheaper than a separate system that provided separate pipes for household wastes and rainwater and more practical for densely built-up cities that required street drainage; and engineers rejected the concept of the agricultural use of sewerage that required a separate system because they maintained that the costs of the system outweighed the benefits to be derived from resource recycling. They believed that sewage could be safely deposited in nearby waterways, a belief based on the theory that running water purifies itself. The methods of chemical analysis of water quality used in the 1860s and 1870s gave this theory scientific validity.[12]

Colonel George E. Waring, Jr., the "Sewer Gas" Theory, and the Memphis Separate Sewer System

George E. Waring, Jr., was the most noted sanitarian of the late nineteenth century. Trained primarily as a scientific agriculturist with some work in engineering, he had lectured to farmers on agricultural methods, managed several large farms, and published books on husbandry before the Civil War. In 1857, Waring was appointed drainage engineer of New York's Central Park, and during the Civil War he served as a cavalry officer. In the period after the Civil War, Waring became actively involved in the sanitary movement, constructing sewerage systems and writing and speaking extensively about his theories.

Waring combined a belief in the agricultural utility of human wastes with a devotion to the theory that "sewer gas" produced by putrefying fecal wastes was the cause of "zymotic" or infectious disease. In the latter half of the nineteenth century, medical theory was in a state of confusion revolving around contagionist and noncontagionist theories of disease etiology. From the 1860s to the 1870s, the dominant belief was the so-called "filth" or "pythogenic" theory, which held that disease evolved de novo from putrefying organic matter. This theory was critical in stimulating the sanitary movement and in arousing the public to the dangers of poor sanitary facilities.[13] Waring wrote prolifically in popular and professional journals about the importance of sanitary reforms, reflecting the contemporary state of the art in terms of disease etiology.

Waring's early writings on sanitation focused on the importance of drainage and the agricultural use of wastes. In 1867 he published *Draining for Profit and Draining for Health,* and in 1868 he began promoting the use of the earth closet as an alternative to the "sewer gas"–producing water closet. The earth closet, however, secured little public acceptance due to the greater efficiency and self-working nature of the competing water closet, and in the 1870s Waring became a promoter of the separate system of sewers in conjunction with improved water closets. Waring's rationale for the adoption of separate as opposed to combined sewers was that they provided for swifter removal of wastes from the household and the city. He argued that unless human feces was transported out of the household in a "fresh condition," it would "undergo putrefaction and give off objectionable gases." These sewer gases would be lethal to the extent that the fecal matter had decomposed. Waring maintained that the separate system was more sanitary and efficient than the combined system and was also cheaper to build and to maintain.[14]

During the mid-1870s, Waring published a number of influential articles on sanitation in such popular journals as the *Atlantic Monthly, Harper's Magazine,* and *Scribners,* as well as in more specialized journals such as *The American Architect and Building News.* In these articles, he described the dangers of sewer gas and espoused the adoption of separate sewer systems with flush tanks as the surest means of guaranteeing a healthy sanitary environment. In addition, he argued for agricultural use of sewage as a means of economy. In 1875–76, he constructed a separate system in the small Massachusetts resort town of Lenox. The Lenox system was not only the nation's first separate system, but it also had the first disposal plant, a subsurface irrigation system.[15] Waring's greatest fame, however, was to come from his work in Memphis, Tennessee.

The city of Memphis is located on the east bank of the Mississippi River, halfway between St. Louis and New Orleans. In 1880 it had about forty thousand inhabitants and covered approximately four square miles. A private water company supplied the city with piped-in water, but most people obtained their water from cisterns or wells. Prior to 1880, there were only four miles of private sewers in the city; all surface water was delivered to the river or nearby bayou by street gutters. Human and household wastes were deposited in about seven thousand privy vaults and cesspools, many located in the basements of houses and filled to overflowing. Rotten wooden streets in the densely settled areas and frequent flooding in the low-lying sections worsened the city's sanitary condition.[16]

In both 1878 and 1879, yellow fever ravaged Memphis, as well as other cities in the Mississippi Valley, causing over five thousand deaths in Memphis alone. At the request of the city officials, the National Board of Health, a short-lived body founded in 1879, appointed a special commission, including Waring, to make recommendations for sanitary improvements. The commission considered a number of plans for the sewerage of Memphis, including one by Waring for a separate system at a cost of approximately $200,000 and another for a combined system at a cost of $1 million, drawn up by E. S. Chesbrough, the engineer of the Chicago sewerage system, and George Hermany, a Louisville civil engineer.[17]

Over the opposition of a number of civil engineers, the commission accepted Waring's plan, and it was constructed within a matter of months. The Waring system had the following characteristics: it provided for house sewage only, excluded all rainwater, and made no provisions for stormwater; the lateral sewers were six-inch vitrified pipe and the house branches were four inches in diameter; Field's automatic flush tanks of 112-gallon capacity were placed at the head of each branch; manholes were omitted and inspection provided through occasional hand-holes; sewers were ventilated through an open soil pipe at each house and fresh air inlets; subsoil drainage was provided by agricultural tiles laid in the same ditch with the sewer; and the sewage was discharged through an outlet into the Wolf River. The Memphis system differed from other separate systems in that it excluded all rainwater, even that from roof gutters, used automatic flushing tanks for cleaning the sewer, and had no manholes for inspection or to remove obstructions.

While a great deal of controversy developed over these design features of the Waring separate system, the construction of sewerage as well as other sanitary improvements caused a marked improvement in the health of the town. By transforming a "fever-plagued spot into a healthy city," and by doing it at a cost affordable to a nearly bankrupt city, Waring greatly publicized the benefits of sewerage technology to the nation at large and centered attention on himself.[18]

The Hering Report and the Attempt to Promulgate a Rational Engineering Model

The construction of the Memphis separate sewer system and the considerable improvements it created in sanitary conditions within the city stimulated a great debate in the engineering press and in engineering circles about the virtues of the separate and combined systems and of the special features incorporated by Waring. This debate was complicated by questions of personality, professional ethics, and medical theory as well as matters of engineering design, although few questioned the benefits of water-carriage removal. The important part played by nonengineering factors in the decision to implement sewerage technology and the confusion over design are characteristic of the early stages of new technology implementation.

In 1881, the National Board of Health, undoubtedly concerned with its role in the separate vs. combined system controversy, sent sanitary engineer Rudolph Hering to Europe to report on sewerage experience there. Unlike Waring, Hering was a trained engineer and had graduated from the German Royal Polytechnical School in 1867. He had worked from 1876 to 1880 as assistant city engineer in Philadelphia in charge of bridges and sewers and had written a number of well-regarded articles on sewerage problems.[19]

In his exhaustive report, Hering carefully explored the variations in European sewerage and suggested a model for the choice between combined and separate systems. Neither system, he said, had a greater sanitary value. Implementation, therefore, should depend on local conditions and financial considerations, not on a supposed sanitary advantage. Hering maintained that the combined system was best suited and most economical for large, densely built-up cities that had to be concerned with stormwater runoff as well as household wastewater. In these cities, the combined system would permit underground removal of stormwater, thereby eliminating a traffic hazard. In smaller cities, on the other hand, where traffic was not a problem and where household wastes were the chief concern, the separate system without underground removal of stormwater could be constructed. "The factor," he concluded, "which will mainly govern a preference is less the sanitary value . . . than the cost of construction and maintenance."[20]

Hering had attempted to put the question of sewer design and implementation on a rational engineering basis within a cost-benefit framework. The personalities involved, questions of professional ethics, and confusion over theories of disease etiology, however, prevented acceptance of the model until nearly the end of the century. In this next section, the issues that were exogenous to the rational engineering design model but that intruded on its formulation will be considered. The subsequent section will examine factors that were endogenous to the engineering model, such as design features, provision for stormwater, and disposal.

Exogeneous Factors Affecting the Rational Model: Personality, Professional Ethics, and Disease Etiology

Personality

George Waring, Jr., was a great publicist as well as a great sanitarian. His achievement in Memphis received attention throughout this country and in Europe, and Waring did not hesitate to capitalize on it. He spoke before many medical, sanitary, and other professional groups, and he published numerous popular and professional articles describing his system and defending it against its detractors. He developed a loyal following, including engineers and public health authorities, who defended his principles and pushed adoption of his system. Even though many engineers resented his fame and believed that he made unwarranted claims for his system, his ability to convince decision makers of the correctness of his position hampered the adoption of the rational model.[21]

Professional Ethics

Just as controversial as Waring's design was his patenting of the features of his separate system and his formation of a company, the Drainage Construction Company, to sell it to the cities of the nation, even though the general features of the system had been known for some years. At a time when the engineering profession was attempting to establish its professional status, many engineers considered Waring's patents as reprehensible behavior. In 1881, one civil engineer wrote to the *Engineering News* that Waring's actions were "of serious importance to the profession and the public generally, because if this instance is made a precedent and the idea should become fixed, that no city or town can be properly sewered without having recourse to Mr. Waring or his patent rights, the profession is unfairly and heavily handicapped, and the ignorant public imposed upon." In addition, in his report to the National Board of Health, Hering observed that there were "direct personal interests concerned" in the espousal of the separate system and that those who advocated it exclusively, were "generally not concerned in directing, managing, nor are always familiar with the engineering question in a large and populous city."[22] As in the case of Waring's personality, his patents and financial involvement in his system amplified difficulties of engineers and policymakers in reaching a rational engineering approach to the problems of sewerage implementation.

Disease Etiology

The third exogenous factor involved was that of disease etiology. Historians and sociologists of history and technology have argued about whether or not complex technological change is based on previous scientific "discoveries."[23] The public health and sanitation movement of the late nineteenth century rested to a large extent on a belief in the filth theory of disease and stimulated the construction of sewerage systems of all types in order to achieve local health benefits. The case of the Waring separate system, however, furnishes a clear example of a direct rather than a general relationship between a theory of disease etiology—the sewer gas theory—and a specific form of technology design—the Waring separate system.

The link between sewer gas theory and the Waring separate system had a direct impact on public policy in regard to the construction of sewerage systems in the 1870s and the 1880s. In 1882, for instance, the New York State Board of Health recommended that towns adopt the separate system of sewerage with flush tanks (the Waring system) because it avoided the "evils of sewer gas." In Stamford, Connecticut, a town that suffered from a high death rate from infectious disease in the 1870s and 1880s, industrialist and engineer Henry R. Towne, a local influential, became a devout follower of Waring. In 1885 he persuaded the town council to hire Waring to construct a separate system for the city because of its superior health benefits and elimination of sewer gas. In other cities where Waring constructed systems, such as Norfolk, Virginia, and Newport, Rhode Island, the policy issue was phrased in the same terms.[24]

For most engineers, however, the varying theories of disease etiology debated by physicians and scientists were irrelevant to the question of good engineering design. They regarded matters of cost and the need for stormwater drainage and for sewage disposal as the important considerations in their choice. They agreed, as Hering observed in his report, that both systems were equally sanitary if they were well designed and adequately maintained. Sir Robert Rawlinson, a pioneer in British sanitary engineering, cogently expressed this lack of concern with disease theory in an 1882 letter to the *Sanitary Engineer.* After noting that he owed "little of my knowledge to tables or to books," Rawlinson observed that

> As to the so-called causes of disease, by the growth of low forms of organic life, "bacteria," etc., in sewers and drains, I am contented to leave their study to medical men, chemists, and microscopists. My aim is to construct sewers and drains of absolute truth in line, grade, and sectional form, having smooth and vitreous surfaces, and to so proportion them to the flow of sewage and flushing that in work they shall remain absolutely clean. I then leave rats, bacteria, germs, and other organisms to take care of themselves, as I know by experience that such cannot find any abiding place in sewers and drains so constructed and so managed.

A well-known American civil engineer who expressed similar sentiments in 1883 went on to note that "nine of ten of the best engineers of the United States" believed that neither the separate nor combined system was superior from a sanitary perspective.[25]

But although the great majority of engineers believed that the two systems had equal sanitary capabilities, elements of the public health profession and the public continued to be concerned about the health dangers of combined sewers and to advocate the building of separate sewers to avoid sewer gas hazards. By 1892, twenty-two towns and small cities had constructed Waring separate systems, while public health authorities had advocated these for large cities, such as Pittsburgh and Kansas City. While it is not clear that those municipalities who built separate rather than combined sewers did so because of concern over sewer gas, it was obviously an issue in some cases. The concern over the health dangers of combined sewers therefore impacted public policy on both a local and a state level in regard to system implementation and retarded full acceptance of the rational model.[26]

Endogeneous Factors Affecting the Rational Model: Waring System Design Features, Costs, Stormwater Disposal, and Sewage Treatment

This section will examine those factors that might be labeled endogeneous to those seeking to make decisions concerning the implementation of one form or another of sewerage technology. The elements that will be considered are the special design features of the Waring system, comparative costs, stormwater disposal, and sewage treatment.

Waring System Design Features

Much of the criticism of the Waring separate system revolved around the special design features that distinguished it from the state-of-the-art characteristics of the sewerage systems that had been formerly constructed. Initial comment focused on the small size of the sewer laterals and mains, the lack of manholes to remove obstructions, the inadequacy of the flush tanks to scour the sewers, and the absence of provision for stormwater. Several engineers predicted that the mains would become clogged and that without manholes, the streets would have to be dug up to remove the obstructions. The extent to which Waring had deviated from the state-of-the-art of engineering practice was reflected by a letter written in 1881 to the *Sanitary Engineer* by noted Boston civil engineer, Edward S. Philbrick: "Colonel Waring launched out upon new and unbeaten paths, pursuing methods which had been discarded as bad ones by nearly all engineers who had had much experience in the construction and maintenance of city sewers during the past generation, both in this country and in England."[27]

In 1887, seven years after installation of the Memphis system, sanitary engineer Rudolph Hering subjected it to a thorough evaluation. Hering found that all of the original criticisms made by engineers of the special design features of the system had proven valid and that none had worked as Waring had predicted. The small-pipe sewers did not flow smoothly and were subject to frequent stoppages. Since there were no manholes, the streets had to be torn up to clear blockages, and the city installed forty-three manholes in 1883. Hering found that the flush tanks, justified by Waring as necessary to speed the sewage to its outlet before putrefaction, were inefficient and unnecessary because of the slope of the sewers. And, while Waring had intended to keep all rainfall from his sewers, this had proven impossible in practice because of surreptitious connections of drains by householders. The city, therefore, was forced to construct overflows and an intercepting sewer to accommodate excess water volume.

Hering concluded his study by making a strong statement defending the state-of-the-art of sewer design and implicitly justifying the rational model he had presented in his 1881 report to the National Board of Health: "Every one of the deviations made by Colonel Waring from the principles previously adopted," said Hering, "have . . . proved to be either of minor importance or wholly objectionable." Whatever success the system had, was due to "those principles of sewer construction which had been known and practiced elsewhere for many years," as well as to a competent maintenance force.[28] The implications of Hering's criticisms were clear—the "novelties" popularized by Waring had retarded progress toward the development of a rational model to help engineers and municipal policymakers determine which particular sewerage system design suited their city's needs.

Costs

In comparison with the alternative system of combined sewers, the separate system with only sanitary sewers offered the advantage of eliminating privy vaults and cesspools for a comparatively low cost. This was one of the principal selling points of the Waring system, rivaling its boasted health benefits. There is evidence, for instance, that the Memphis decision makers who chose the Waring plan found its cheapness as attractive as its sanitary qualities.[29] A separate system that only provided small diameter sanitary sewers and made no provision for underground removal of stormwater would obviously be cheaper than a combined system that required pipes large enough for both sanitary wastes and stormwater. At the time of the Memphis crisis, however, because of lack of precedent, no other engineer besides Waring had the daring to suggest such a system.

Waring made the most of his system's low costs and glossed over the lack of provision for stormwater. He argued, for instance, that there was "no instance recorded of the greater cost of the sewerage of a city by the separate system than by the combined system, and it is doubtful whether one-half of the cost has ever been reached." The comparative examples he cited—$7,000 per mile in Memphis and in Keene, New Hampshire, for separate sanitary sewers and $25,600 per mile in Brooklyn and $34,550 per mile in Providence for combined sewers—neglected to note that the towns with separate systems had no stormwater sewers. Sanitary engineer Benezette Williams, who had himself installed a separate system at Pullman, Illinois, but with storm sewers for underground removal of stormwater, pointed out the fallacy of Waring's position in an 1885 article: "Neither Col. Waring nor any other advocate of the separate system as cheaper at all times and in all places, has ever advanced an argument tending to alter the conclusion, that a given quantity of sewerage and a given quantity of stormwater can be carried at less expense in one conduit than in two, and that all arguments used for the all-time and all-place cheapness of this system start with the fallacy of . . . ignoring stormwater drainage entirely." The confusion over relative costs and benefits, however, persisted into the 1890s and complicated the decision-making process for municipalities trying to make policy choices regarding what type of system to install. In 1886, for instance, two engineering followers of Waring published a text on the *Separate System of Sewerage* that went into a second edition in 1891. In the text, the authors argued that taxpayers should find the separate system attractive because of its low cost compared with the combined system. They maintained that the only reason for cities to continue to build combined systems was "that engineering precedent carries great weight with it among engineers, and a venerable error, even, is hard to put down."[30] The appearance of such arguments in reputable engineering texts made more difficult the acceptance of a rational cost model by both practitioners and policymakers.

Stormwater Disposal

The problem of stormwater had essentially two dimensions. The first concerned the actual physical nuisances created during heavy rains, while the second involved the chemical composition of stormwater. Among the physical difficulties created by stormwater were street and basement flooding and the disruption of traffic. These problems were amplified when cities paved their streets, a development usually viewed as an improvement. Paved streets, as compared with unpaved streets, resulted in less ground absorption, rapid runoff, and the flooding of low-lying neighborhoods.[31] Municipal policymakers facing the stormwater problem usually considered building combined sewers or leaving rain to run off in surface channels, only constructing underground pipes in heavy traffic areas.

Waring maintained that stormwater should be kept on the surface as long as possible in order to restrict the production of sewer gas, although he admitted that in a few cities heavy traffic necessitated underground removal. Even in these cases, however, he argued that "very shallow sewers would be as effective as deep ones." Hering, on the other hand, maintained in his report that in densely built-up districts and rapidly expanding cities, stormwater had to be removed and that the combined system did this most expeditiously. Most engineers agreed with Hering's position. In small cities, however, and especially those with unpaved streets where mud and silt would be washed into the sewers, engineers maintained that stormwater should be removed on the surface.[32]

The question of the chemical composition of stormwater posed a different set of problems. For some time, sanitary engineers had observed that stormwater flowing through the streets accumulated all kinds of waste and debris. These wastes included organic materials, such as animal excrement and food particles, as well as mud, leaves, and grit. The first sewers constructed were intended for the drainage of surface waters, and they naturally became the receptacles for the wastes collected by street runoff. Occasionally, households surreptitiously connected their water closets to these storm sewers, increasing their organic content. Because of their large size, poor design, and slow flow, the accumulated wastes decayed, producing, according to the medical theory of the day, disease-laden "sewer gas." "A sewer was synonymous," noted Hering, "with all that is repulsive, filthy, and disgusting,"[33] and they were often referred to as "elongated cesspools."

While the construction of smaller-size and better designed sewers, as well as an increase in water flow, reduced the accumulation of organic wastes, the polluted character of the street runoff still posed problems. Chemical analysis of street runoff in London showed that the first flush from the streets during a storm was approximately as polluted, due mainly to animal excrement, as ordinary sewage. The use of catch basins, intended to prevent debris from entering and clogging the sewer, increased street runoff pollution. The catch basins retained organic materials that became septic and that heavy rains washed into the sewers. The close comparison in chemical nature between household sewage and street runoff was often used as an argument for building a combined sewerage system. Waring argued, however, that the solution to the polluted character of street runoff was to clean the streets effectively rather than leaving that function to stormwater. Stormwater could then drain on the surface into adjacent water courses without any adverse effect.[34]

The overflow problem was a further argument for surface removal of stormwater. Rainfall figures were imprecise, and to build combined sewers large enough to cope with any possible amount of storm runoff would have been prohibitively expensive. Combined sewers were, therefore, provided with storm overflows that acted automatically to remove sewage from the pipe above a certain volume. These overflows were usually located on sewer mains, on the intercepting sewer, or before the treatment plant, if one existed, and the overflow led directly into nearby water courses. Since this overflow consisted of stormwater mixed with household sewage, it created both water pollution and nuisances in parts of rivers that were normally free of a high content of wastes. Since overflow drains were often close to water intake pipes or heavily populated areas, a serious hazard could be created.[35]

To a large extent, however, the rational model disregarded the potential dangers of the overflow condition because of the belief that the sewage would be very diluted. While Hering described the construction of overflows in his 1881 report, he did not regard them as a pollution danger and observed that it was "generally permissible to let a slight amount of sewage flow into the river together with a large amount of stormwater."[36] Waring took a more skeptical view, and in 1886 he warned of the dangers of the "relief overflow." Most engineers, however, viewed overflows as a means of dealing with the probability of very heavy rains and restricting the size and costs of combined sewers. From the purely technical perspective, they were a mechanism to prevent the type of backup flooding from storms that had marked the early combined sewers.

Sewage Treatment

Waring and other proponents of the separate system often maintained that it simplified and lowered the costs of sewage pumping and treatment. Waring believed that his system facilitated sewage purification by land treatment, and his Lenox plan of 1876 had this feature. Other advocates of the separate system maintained that it made possible sewage farms, from which revenue could be produced to pay for maintenance. In 1881, for instance, engineer C. S. Latrobe, in a report to the Baltimore city council, took this position. By 1892, twenty-seven cities and towns treated their sewage, twenty-one by some method of land treatment and five by chemical precipitation; of these twenty-seven, twenty-six had separate sewers.[37]

Up until approximately 1890, however, engineers and urban policymakers viewed sewage treatment primarily as a means to avoid nuisance. They recognized that locating a sewer outlet near a water intake often posed a health danger, but they also believed that separating the two by a reasonable distance could resolve this problem. What distance was safe, however, was unclear. The justification for disposing of raw sewage into streams was the theory that running water purified itself. This theory had been challenged in both Great Britain and the United States in the 1870s, but it was still accepted by most American engineers. In 1877, for instance, in their report on a sewerage system for Quincy, Illinois, John Nichol and Charles Macritchie noted that "with its rapid current, [the Mississippi River] will quickly carry away the discharge from the sewers, and from the quantity and nature of the river water in constant and rapid motion, will quickly dilute and deodorize the sewage, so that all traces of it will disappear in a short distance below the port of discharge." Even Massachusetts, a pioneer in water pollution control, permitted waste disposal in streams twenty miles above any public use and exempted large rivers from the regulation.[38]

Thus, even though the separate system did facilitate sewage treatment, the rational engineering model held that treatment was only necessary in a few special cases of inland cities not located on large streams. There was no universal need for treatment because of the theory of the self-purification of streams, and city policymakers could feel free in adopting combined sewers if their outlets were into comparatively large bodies of water.

The Rational Model and the Challenge of Water Pollution

By 1890, ten years after the construction of the Memphis system, most engineers, with the support of the major civil engineering publications, had accepted the rational model of design choice articulated by Hering in his 1881 report. They believed that since neither the separate nor combined system had unique sanitary qualities, implementation decisions should be based on the needs of local areas within the parameters set by costs. The various exogenous factors that had been created by the Memphis experience and Colonel Waring's philosophy and style had ceased to play a role in decisions involving sewerage system choice. The so-called "novelties" utilized by Waring at Memphis and in the other systems he constructed had been largely discredited. As for Waring himself, while some engineers regarded him as the exemplar of bad engineering design, others gave him credit for removing "old prejudices" against small-pipe sewers and emphasizing their "manifest advantages."[39] To the public at large, he retained the image of the great sanitarian of his time.

The concept of the rational model of design choice, however, rested on the assumption that large cities that constructed combined sewers could safely dispose of their sewage in adjacent waterways. At approximately the same time that the rational model was winning wide engineering acceptance, new evidence was emerging that challenged its validity. During the late 1880s, bacteriologic research supplemented chemical analysis in the determination of water quality and gradually resulted in the substitution of the germ theory of disease for the filth and sewer gas theories. In the early 1890s, the work of William T. Sedgwick at the Massachusetts Board of Health laboratories, as well as other bacterial researchers, clarified the etiology of typhoid fever and its relationship to sewage-polluted waterways. This clear identification of the waterborne nature of communicable disease raised a serious challenge to the rational model and to the safety of discharging sewage into streams utilized for water supply by downstream cities. It also caused increased interest in methods of sewage "purification."[40]

Outputs from combined sewers were the cause of the pollution of most streams and, by 1893, leading engineering opinion was calling for reexamination of the separate system because of its utility in facilitating purification. As sanitary engineers Moses N. Baker and George Rafter noted in their 1893 path-breaking work, *Sewage Disposal in America,* purification was necessitated by "the recent extensions of knowledge of the causation of typhoid fever, and the other water-born communicable diseases."[41] Engineering supporters of the rational model had previously rejected the concept that one system provided more sanitary benefits than another, but, in ways unforeseen, this assumption was being challenged by the concern over treatment. The new bacterial findings posed a direct challenge to the further utility of the model as a means by which engineers and municipal policymakers could decide between sewerage alternatives.

Although the vulnerability of the rational model was apparent in the 1890s, engineers and municipal policymakers continued to make sewerage decisions by its precepts. (By the 1890s, mainly due to the work of Rudolph Hering, the hypothesis of the self-purification of streams had been replaced by the concept of dilution. Dilution rested on the empirical evidence that, under certain conditions, streams would actually be self-purifying through natural processes. However, in the late nineteenth century, estimates as to the exact conditions under which dilution took place altered. In addition, increases in miles of sewerage systems, in population connected to sewers, and in municipalities discharging into particular waterways overwhelmed the dilution capacity of many streams.) From 1890 to 1905, the total mileage of sewers of all types increased from 6,005 (cities over 25,000) to 19,460 (cities over 30,000). Of the 1905 total, there were 14,856 miles of combined, 3,756 miles of sanitary, and 847 miles of stormwater sewers (see table 1). Thus, under the precepts of the rational model, cities continued to build many more miles of combined than separate sewers.[42]

The rational model had always been sensitive to cost considerations in its choice between sewerage systems, but it had assumed the safety of stream disposal. If, however, stream disposal was limited for health or other reasons, and if treatment became a necessity, the installation by large cities of higher-cost separate systems became a possible public policy alternative. (Waring had proposed the installation of separate systems for New York, Philadelphia, and Washington in the 1880s.) But since large cities already had combined systems, and the costs of replacement were ex-

Table 5.1. Sewer Mileage by Type and Population Class—1905, 1907, 1909

	1905			
Population Group	*Sanitary Sewers*	*Storm Sewers*	*Combined Sewers*	*Total Mileage by Population Group*
1. >300,00	335.2	157.0	8,229.9	9,422.6
2. 100,000–300,000	809.0	120.5	2,961.0	4,101.5
3. 50,000–100,000	965.8	242.8	2,491.1	3,709.7
4. 30,000–50,000	1,313.4	326.6	1,507.4	3,147.7
Total by Type	3,756.2	846.9	14,856.5	20,381.5
	1907			
Population Group	*Sanitary Sewers*	*Storm Sewers*	*Combined Sewers*	*Total Mileage by Population Group*
1. >300,00	554.8	352.0	9,242.3	10,149.1
2. 100,000–300,000	1,300.0	262.9	3,690.5	5,253.4
3. 50,000–100,000	1,097.1	181.6	2,627.1	3,905.8
4. 30,000–50,000	1,611.3	383.9	1,562.9	3,558.1
Total by Type	4,563.2	1,180.4	17,122.8	22,866.4
	1909			
Population Group	*Sanitary Sewers*	*Storm Sewers*	*Combined Sewers*	*Total Mileage by Population Group*
1. >300,00	789.5	349.9	9,834.3	10,973.7
2. 100,000–300,000	1,404.4	284.2	4,405.8	6,094.4
3. 50,000–100,000	1,831.5	384.2	2,615.5	4,831.2
4. 30,000–50,000	1,232.9	333.8	1,505.9	3,072.6
Total by Type	5,258.3	1,352.1	18,361.5	24,971.6

SOURCE: Bureau of the Census, U.S. Dept. of Commerce and Labor, *Statistics of Cities Having a Population of over 30,000: 1905;* Moses N. Baker, "Appendix A: Sewerage & Sewage Disposal," (Washington, D.C., 1907), 342–1247; Bureau of the Census, *Statistics of Cities Having a Population over 30,000: 1907,* "Special Reports" (Washington, D.C.: U.S. Dept. of Commerce and Labor, 1910), 458–63; idem, *General Statistics of Cities with Populations Greater Than 30,000: 1909,* "Special Reports" (Washington, D.C.: U.S. Dept. of Commerce, 1913), 88–93.

tremely high, cities continued well into the twentieth century to build combined sewers and to discharge their untreated wastes into adjacent waterways. The incentive to construct sewage-treatment facilities, except in cases of obvious nuisance, was actually diminished by the development of methods of water filtration and chlorination, which reduced the hazard from waterborne infectious disease. As sanitary engineer Allen Hazen commented in his 1907 book, *Clean Water and How to Get It,* "the discharge of crude sewage from the great majority of cities is not locally objectionable in any way to justify the cost of sewage purification." Rather, said Hazen, downstream cities should filter their water to protect the public health, and sewage purification should only be utilized in order to prevent nuisance.[43]

The Rational Model and Separate System Problems

Cost considerations also constrained the model in regard to smaller cities and towns. In these municipalities, engineers created a priority of goals that caused them to ignore underground removal of stormwater and to construct sanitary sewers only. These limited systems were especially characteristic of inland towns that had to treat sewage because of lack of access to large waterways. But these partial solutions came to haunt the policymakers who had chosen the least costly option. As sanitary engineer Harrison P. Eddy maintained in 1922, "neglect or postponement of provision for prompt removal of storm and surface water while offering a happy solution of immediate financial problems, may have been a policy of doubtful merit."[44]

One of the chief problems resulting from the construction of separate sewers without provision for stormwater, Eddy noted, derived from the unintended discharge of storm and groundwater into sanitary sewers. This discharge occurred when ground water infiltrated through inferior sewer joints or when roof and cellar drains and street inlets were connected with the sanitary sewer. In some cities, storm and groundwater so surcharged sanitary sewers that the sewage flowed back into house cellars or through manhole covers into the streets. In others, the volume of water overwhelmed the capacity of treatment plants, resulting in the diversion of raw sewage into streams. "Great care," concluded Eddy, "should be exercised in the selection of the type of sewer system to be installed. It should not be assumed that separate sewers are more advantageous, but conclusions should be reached only after careful study of all the conditions."[45]

Conclusions

The case of combined vs. separate sewers provides useful insights into the complicated interplay between changing scientific theories, engineering practice and choice of technological design, and impacts on cities. During the last quarter of the nineteenth century, sanitary engineers struggled to define a rational engineering model for sewerage system design choice within the parameters of municipal budgetary constraints and conflicting scientific theories about disease etiology. George E. Waring, Jr.'s advocacy of the separate system with special design features, based on the sewer gas theory of disease etiology, complicated attempts to define questions of design choice. Also important in the debate were probability problems, such as the expected amount of storm runoff, predicted levels of traffic, and anticipated volumes of water usage.

By the last decade of the century, however, the rational engineering model, based on empirical considerations of urban need, had triumphed, and a community of engineering practitioners had evolved who followed its precepts in design implementation. Essentially, this meant that large cities having a need for subsurface stormwater removal as well as the removal of household wastes constructed combined sewers, while smaller cities built separate systems with sanitary sewers only and surface removal of stormwater. By 1909, cities over 30,000 population had 18,361 miles of combined sewers, but only 5,258 miles of sanitary sewers and 1,352 miles of storm; in cities over 100,000, miles of combined sewers outnumbered those of sanitary sewers by about a 7-to-1 margin (see table 1).

But while the rational model provided a means to choose the sewerage design best suited for the local needs of a particular city, it created problems for downstream communities. Again the question of disease etiology was crucial. Engineers had originally maintained that sewage disposal into waterways was safe because of the theory that running water purified itself. Bacterial research during the 1880s and 1890s, however, refuted this theory and generated concern over the role of sewage-polluted water in generating typhoid epidemics. In addition, this research stimulated a reappraisal of the separate system because of its facilitation of sewage treatment. Large cities, however, having built thousands of miles of combined sewers, resisted the pressure of public health authorities to convert to separate sewers because of the costs involved. Engineers and municipal authorities maintained that the dilution capacity of streams should be utilized to its fullest and that downstream cities should adopt retrofits, such as filtration and chlorination, to protect water supplies against infectious disease.

While the "pendulum" of engineering opinion (pressed by directives from state boards of health) swung somewhat in the direction of the construction of separate sewers in the second and third decades of the twentieth century, thousands of miles of combined sewers remained in place.[46] They constituted a heritage to future policymakers seeking to deal with problems of water pollution and as testimony to the earlier triumph of the rational model of sewerage construction.

Notes

This article is based on research supported by the National Science Foundation under Grant No. ERP-75-08870. Any opinions, findings, and conclusions or recommendations expressed in this publication are those of the author and do not necessarily reflect the views of the National Science Foundation. The author would like to thank James McCurley, Frances Clay McMichael, Clay McShane, Terry Yosie, and Edward W. Constant II for their comments and criticisms of earlier drafts of this essay.

1. B. A. Segur, "Privy-Vaults and Cesspools," *Papers and Reports of the American Public Health Association* (hereafter cited as *APHA*) 3 (1876): 185–87; Henry I. Bowditch, *Public Hygiene in America* (Boston, 1877), 103–9; Mansfield Merriman, *Elements of Sanitary Engineering* (New York, 1906), 139–42; and Leonard Metcalf and Harrison P. Eddy, *American Sewerage Practice,* 2 vols. (New York, 1928), I:15–19.

2. All of the references cited above make comments about fecal pollution. See also George E. Waring, Jr., "The Sanitary Drainage of Houses and Towns, II," *Atlantic Monthly* 36 (October 1876): 434.

3. Nelson M. Blake, *Water for the Cities* (Syracuse: Syracuse University Press, 1956); "Ownership of American Water Works," *Engineering News* 27 (January 23, 1892): 83–86; J. T. Fanning, *A Practical Treatise on Hydraulic and Water-Supply Engineering* (New York, 1886), 625.

4. See, for example, E. S. Chesbrough, "The Drainage and Sewerage of Chicago," *APHA 4* (1878): 18–19; Town of Pawtucket, Committee on Sewers, *Report, 1885* (Pawtucket, 1885), 15; and Town Improvement Society of East Orange, *The Sewerage of East Orange* (East Orange, N.J., 1884). In Boston, for instance, in 1863 (population ca. 178,000) there were over 14,000 water closets out of a total of approximately 87,000 water fixtures; in Buffalo, in 1874 (population ca. 118,000), there were 5,191 dwellings supplied with water and 3,310 with water closets. By 1880, although the data are imprecise, it can be estimated that approximately one-quarter of urban households had water closets while the remainder depended on privy vaults and cesspools. See Cochituate Water Board, *Annual Report, 1863* (Boston, 1864), 43; City of Buffalo, *Sixth Annual Report of the City Water Works 1874* (Buffalo, N.Y., 1875), 47. The estimate for 1880 is based on information in U.S. Department of the Interior, Census Office, *Tenth Census of the United States, 1880, Report of the Social Statistics of Cities,* 2 vols., comp. by George E. Waring, Jr., (Washington D.C., 1887); Charles V. Chapin, "The End of the Filth Theory of Disease," *Popular Science Monthly* 60 (January 1902): 234–39; George E. Waring, Jr., "The Sanitary Drainage of Houses and Towns, III," *Atlantic Monthly* 36 (November 1875): 535–51; and Benjamin Lee, "The Cart Before the Horse," *APHA* 20 (1895): 34–36.

5. Azel Ames, "The Removal and Utilization of Domestic Excreta," *APHA* 4 (1877): 65–80.

6. These characteristics of water carriage are a summary of information from many sources. See, for example, Julius W. Adams, *Report of the Engineers to the [Boston] Commissioners of Drainage, September 10, 1857* (Brooklyn, 1857); J. Herbert Shedd, *Report on Sewerage in the City of Providence* (Providence, R.I., 1874); Ames, "The Removal and Utilization of Domestic Excreta," 69–70; and Rudolph Hering, *Report on a System of Sewerage for the City of Wilmington, Delaware* (Wilmington, 1884). These benefits and costs are described in more detail in Joel A. Tarr and Francis Clay McMichael, "The Evolution of Wastewater Technology and the Development of

State Regulation: A Retrospective Analysis," in Joel A. Tarr, ed., *Retrospective Technology Assessment—1976* (San Francisco, 1977), 174–78.

7. Leonard Metcalf and Harrison P. Eddy, *American Sewerage Practice, Vol. I—Design of Sewers* (New York, 1914), 1.

8. Rudolph Hering, "Report of the Results of an Examination Made in 1880 of Several Sewerage Works in Europe," *Annual Report of the National Board of Health 1881* (Washington, D.C., 1882), 101–4 (hereafter cited as *Hering Report, 1881*).

9. Edwin Chadwick, *Report on the Sanitary Condition of the Labouring Population of Great Britain*, ed. M. W. Flinn (Edinburgh, 1965), 58–73.

10. Francis Sheppard, *London 1808–1870: The Infernal Wen* (Berkeley: University of California Press, 1971), 253–58.

11. Adams, *Report of the Engineers to the Commissioners of Drainage*, 4–7; E. S. Chesbrough, *Chicago Sewerage: Report of the Results of Examinations Made in Relation to Sewerage in Several European Cities 1856–7* (Chicago, 1858), 66.

12. "Errors in Engineering," *The Plumber and Sanitary Engineer* 2 (June 1, 1879): 181; "Alfred S. Jones to Editor," ibid. (August 1, 1879): 27; and "Separate Drainage and Sewerage," ibid. (October 1, 1879): 350. For self-purification of streams, see William T. Sedgwick, *Principles of Sanitary Science and the Public Health* (New York: The Macmillian Company, 1918), 128–29, 233–46.

13. For an appreciation of Waring as an environmentalist, see Martin V. Melosi, "Pragmatic Environmentalist: Sanitary Engineer, Col. George E. Waring, Jr.," *Essays in Public Works History* 4 (Washington, D.C.: Public Works Historical Society, 1977). For a more critical appraisal of Waring, see James H. Cassedy, "The Flamboyant Colonel Waring: An Anti-Contagionist Holds the American Stage in the Age of Pasteur and Koch," *Bulletin of the History of Medicine* 36 (March/April 1962): 163–76.

14. See Waring, "The Sewering and Draining of Cities," 34–40.

15. "Sewage Purification in America," *Engineering News* 28 (July 14, 1892): 33.

16. For descriptions of Memphis, see "The Sanitary Condition of Memphis," *Bulletin of the National Board of Health* (Washington, D.C., 1879) 1:187–89; "Memphis," in Waring, comp., *Social Statistics of Cities* 2:144–47; idem, *The Sewerage of Memphis, U.S.A.* (London, 1881).

17. "Memphis," *Social Statistics of Cities* 2:144–47.

18. William Paul Gerhard, "A Half-Century of Sanitation—II," *American Architect and Building News* 43 (March 4, 1899): 67.

19. Frank A. Taylor, "Rudolph Hering," in Dumas Malone, ed., *Dictionary of American Biography*, (New York: Scribner's, 1933), 8:576–77; "Rudolph Hering," Committee on History and Heritage of American Civil Engineering, American Society of Civil Engineers, *A Biographical Dictionary of American Civil Engineers* (New York, 1972), 58–59.

20. *Hering Report, 1881*, 115. See, also, Rudolph Hering, "Sewerage Systems," *Transactions of the American Society of Civil Engineers* (1881), 10:361–86; and "Report on European Sewerage Systems, with Special Reference to the Needs of the City of Philadelphia," *Journal of the Franklin Institute* 114 (July, December 1882): 186–201. Another statement of the model is Howard A. Carson, "Separate vs. Combined Sewers," *Sanitary Engineer* 4 (February 15, 1881): 142–43.

21. See the discussion, for instance, following Frederick S. Odell, "The Sewerage of Memphis," *Transactions of the American Society of Civil Engineers* (1881), 32–52.

22. "Col. Waring's Sewerage Patents," *Engineering News* 8 (October 22, 1881): 425–27; *Hering Report, 1881*, 112. Hering made reference to European systems, such as the Shore or Liernur Compressed Air Systems.

23. See, for instance, "The Interaction of Science and Technology in the Industrial Age," *Technology and Culture* 17 (October 1976).

24. James T. Gardiner, *Report to the State Board of Health on the Methods of Sewerage for Cities and Large Villages, in the State of New York*, no. 42 (1882); "New York State Board of Health on Sewerage," *Sanitary Engineer* 5 (March 23, 1882): 337–38. For Stamford, see Estelle F. Feinstein, *Stamford in the Gilded Age* (Stamford, Conn.: Stamford Historical Society, 1973), 166–85.

25. Robert Rawlinson, "Sewerage and Drainage," *Sanitary Engineer* 5 (April 27, 1882): 450;

Charles E. Chandler, *Plan of Sewerage for Greenville, Conn.* (Norwich, 1883), 5. Chandler also noted, "The germ theory of disease, the generation of bacteria, and the comparative safety of a large sewer with an intermittent flow, and a small sewer with a nearly uniform flow, are subjects which I freely confess my inability to discuss to advantage. It is my opinion, however, that either of the three systems properly carried out, can be relied on to safely and satisfactorily perform the offices for which they are designed, provided the plumbing and house drainage is properly designed and honestly constructed."

26. George E. Waring, Jr., "The Memphis System of Sewerage at Memphis and Elsewhere," *APHA* 18 (1893): 153–68.

27. See discussion following Odell, "The Sewerage of Memphis," 32–52. See also letters to the editor, *Sanitary Engineer* 4 (April 1, 15, May 15, and June 15, 1881), and 5 (April 27, 1882).

28. Rudolf Hering, "Six Years' Experience with the Memphis Sewers," *Engineering & Building Record and the Sanitary Engineer* 16 (November 26, 1887): 738–39.

29. Odell, "The Sewerage of Memphis," 26. See also, "Memphis," *Social Statistics of Cities* 2:145–46.

30. See the previous citations on Memphis by Waring as well as, "The Memphis System of Sewerage at Memphis and Elsewhere," 153–68. Quoted in Benezette Williams, "The Separate vs. Combined System of Sewerage," reprint from *Journal of the Association of Engineering Societies* 4 (1885): 17–18; see also "Separate Versus Combined Systems of Sewerage," *Sanitary Engineer* 11 (April 16, 1885): 413–14; and Cody Staley and G. S. Pierson, *The Separate System of Sewerage* (New York, 1891), 50.

31. Henry N. Ogden, *Sewer Design* (New York, 1899), 70–71.

32. Odel, "The Sewerage of Memphis," 45–46; *Hering Report, 1881,* 108.

33. *Hering Report, 1881,* 103; Waring, "The Sanitary Drainage of Houses and Towns, II," 437–42.

34. *Hering Report, 1881,* 105; George W. Rafter and M. N. Baker, *Sewage Disposal in America,* 150–51; Waring, "Storm Water in Town Sewerage," 326–27; and Staley and Pierson, *The Separate System of Sewerage,* 39–42.

35. George E. Waring, Jr., "Mechanical Appliances in Town Sewerage," *Journal of the Franklin Institute* 121 (1886): 276–77.

36. *Hering Report, 1881,* 107, 124–26. *The Sanitary Engineer* claimed that overflow from combined sewers was not a serious problem since "the amount of sewage actually coming in fresh to be mixed with the rain fall, is so small in comparison with the latter, and the dilution so enormous, that no appreciable pollution of the open channels actually occurs." See, "Errors in Engineering," 181.

37. George Waring, Jr., "Out of Sight, Out of Mind," *Century* 27 (1893–94): 943–46; Rafter and Baker, *Sewage Disposal in the U.S.,* 361–559. For Baltimore, see "The Sewerage of American Cities and Towns," *Engineering News* 8 (December 3, 10, 1881): 486–87, 496–97. See also John H. Gregory, "Separate and Combined Sewers in Relation to Sewage Disposal," ibid. 70 (October 1913): 875–76, for a later statement of this position.

38. Charles Macritchie and John Nichol, *Report on the Sewerage of the City of Quincy, Illinois* (Milwaukee, 1877) 4; Jay Slater, "The Self-Purification of Rivers and Streams: A Study of the Transformation of Water Safety Standards in the Late Nineteenth Century," *Synthesis* 2 (Autumn 1974): 41–45; and Barbara Rosenkrantz, *Public Health and the State* (Cambridge, Mass.: Harvard University Press, 1972), 81.

39. See, for instance, Gerhard, "A Half-Century of Sanitation," 67.

40. Rosenkrantz, *Public Health and the State,* 103–6.

41. "Sewage Purification and Storm and Ground Water," *Engineering News* 28 (August 25, 1892): 180–81; Gregory, "Separate and Combined Sewers in Relation to Sewage Disposal," 875–76. Hering held that combined sewerage systems were as suitable for sewage treatment as were separate systems. He advocated treating the dry weather flow plus the first flush from the streets in case of a storm, allowing the remainder of the storm water to be diverted to nearby rivers by an

overflow. He warned, however, that local conditions would ultimately determine the best system. See *Hering Report, 1881,* 107. Rafter and Baker, *Sewage Disposal in America,* 150.

42. Hering, "Notes on the Pollution of Streams," *APHA* 13 (1888): 272–79. For a useful brief description of dilution and problems with changing standards, see Milton J. Rosenau, *Preventive Medicine and Hygiene* (New York: Appleton-Century, 1927), 1009, 1099–1103. For data on cities with separate or combined systems, see Table 41, "Length and Classes of Sewers," in *Statistics of Cities Having a Population of Over 30,000: 1905,* 342–47; Gregory, "Separate and Combined Sewers in Relation to Sewage Disposal," 875–76.

43. Moses N. Baker, "Sewerage and Sewage Disposal," in *Statistics of Cities Having a Population of Over 30,000: 1905,* 103–6; "Sewers and Sewer Service," in *General Statistics of Cities: 1909,* 31. In 1914 engineers from the Department of Public Works in Philadelphia prepared a report on collection and disposal of sewage for that city. In this report they argued that even though the separate system had advantages in terms of treatment and protection of water courses from pollution, a number of factors militated against its installation. These factors were (1) the city already had 1,300 miles of combined sewers; (2) sewage in the combined system was not as concentrated as the separate; (3) the combined system occupied less space in the streets; (4) citizens would have to rearrange their plumbing if the separate system was installed; and (5) the streets would have to be torn up and repaved at a great cost if the separate system was installed. See Bureau of Surveys, City of Philadelphia, *To Protect The Public Health* (Philadelphia: Department of Public Works, 1914), 32–35; Allen Hazen, *Clean Water and How to Get It* (New York, 1907), 34–37. In 1903, the *Engineering Record* editorialized that "it is often more equitable to all concerned for an upper riparian city to discharge its sewage into a stream and a lower riparian city to filter the water of the same stream for a domestic supply, than for the former city to be forced to put in sewage treatment works." See "Sewage Pollution of Water Supplies," *Engineering Record 28* (August 1, 1903): 117.

44. Harrison P. Eddy, "Use and Abuse of Systems of Separate Sewers and Storm Drains: Can Their Failure be Prevented?" *Proceedings of the American Society for Municipal Improvements* 29 (October 1922): 128–29.

45. Ibid., 137; W. H. Dittoe, "Prevention of Misuse of Sewers," *Proceeding of the American Society of Civil Engineers* 47 (December 1921): 642–44; and "Poor Sewers Worse Than None," *Engineering News* 177 (March 22, 1917): 193.

46. In 1923 Harrison P. Eddy wrote, "It is an interesting fact that human thought moves to extremes, like a pendulum. At first sentiment was strongly in favor of combined sewers; later it swung to the other extreme, favoring separate sewers and tending to discourage the construction of combined sewers. But improper use of separate sewers and a restudy of the economics of the two systems indicate that the time may be approaching when the pendulum will swing in the other direction." See Eddy, "Present Status of Sanitary Engineering," reprint no. 831, *Public Health Reports,* U.S. Public Health Service, (April 20, 1923): 826–37.

CHAPTER VI

Disputes over Water-Quality Policy

Professional Cultures in Conflict, 1900–1917

Introduction

The existence of a body of legislation on the statute books concerned with water-quality, such as the Safe Water Act and Federal Water Pollution Act, is striking evidence of the extent to which the determination and maintenance of standards in this vital area has become the province of the state. Such a role for government on the local, state, and then federal levels evolved gradually in the late nineteenth and twentieth centuries as a response to deteriorating water-quality, epidemics of infectious waterborne disease such as typhoid, and the nuisances generated by the utilization of waterways for the disposal of untreated human and industrial wastes. Public policy has therefore assumed the responsibility for questions relating to water-quality.

Within the area of water quality, once the principle of state intervention had been established, the question of the direction and timing of policy often produced sharp conflict among involved professional groups. These clashes stemmed from different professional perspectives on the effectiveness of various strategies and the relative costs and benefits of actions taken to protect the public health. In regard to wastewater disposal and water-quality policy, a critical division occurred at the beginning of the century between physicians involved in the public health field and members of the newly emerging profession of sanitary engineering. While they shared important areas of agreement within the public health area, prominent members of these professions often differed on matters of technology application, priority and standard setting, and professional competence. Such differences were revealed early in the 1880s in the dispute over the health-efficacy of separate or combined sewers.[1] They reached new heights in a controversy over water-quality policy that occurred in the years from about 1903 to 1917. The essential question dividing the two groups was which profession was most qualified to determine standards and policy to protect the public health from waterborne disease within the framework of municipal cost constraints.

Approaches to Water Pollution Control, 1878–1905

Late-nineteenth-century municipalities constructed sewerage systems in order to eliminate nuisances created by overflowing cesspools and privy vaults and to protect the public health. Except for a few small cities and towns, they disposed of their untreated wastes in adjacent rivers, lakes, harbors, or tidal estuaries under the assumption of the self-purifying nature of running water. While local health conditions improved for upstream cities that constructed sewerage, they deteriorated greatly in downstream or lake cities that derived their water supplies from water courses into which other cities discharged their sewage. Some municipalities, such as Newark and Jersey City, New Jersey, which had drawn their water from sewage-polluted rivers (in this case, the Passaic), established protected sources of supply in distant watersheds, but this option was either not available or too costly for many municipalities.[2]

During the years from the late 1870s to the 1890s, the increasing pollution of inland streams and lakes that served as sources of municipal water supplies and a consequent increase in typhoid fever morbidity and mortality led to attempts in a number of states to secure legislation to protect water-quality. Several acts appeared in the 1878–93 period. Massachusetts led the way with an 1878 law, followed by an act to protect the purity of inland water in 1886. Also in 1886, New York passed a statute requiring submission of sewerage and drainage plans to the State Department of Health; and Ohio passed similar legislation in 1893. Aside from occasional action by Massachusetts, however, restrictions on enforcement procedures and a lack of funds prevented any substantive moves by these states to halt water pollution.[3]

Another wave of legislation followed in the late 1890s and early twentieth century, and this second group of laws often contained stricter enforcement provisions than the earlier statutes. In 1905, the U.S. Geological Survey published a *Review of the Laws Forbidding Pollution of Inland Waters in the United States,* which listed thirty-six states having some legislation protecting drinking water and eight states with "unusual and stringent" laws. The new laws were badly needed: as of 1904, inland municipalities with a population of 20,400,000 discharged raw sewage into neighboring lakes or streams; seacoast municipalities with a population of 6,500,000 discharged raw sewage into harbors or tidal estuaries; while municipalities with a population of only 1,100,000 subjected their sewage to treatment.[4]

A trend toward a sterner attitude by the state courts in water pollution cases accompanied the passage of the above laws. In a number of cases decided between 1800 and 1905, the courts generally held that discharge of sewage into a river by an upper riparian user, if it constituted a nuisance and thereby violated the rights of a lower riparian user, entitled the latter to recover damages. In one 1899 case involving sewage pollution of the York, Pennsylvania, water supply, for instance, the Pennsylvania Supreme Court held that "No prescription or usage can justify the pollution of a stream by the discharge of sewage in such a manner as to be injurious to the public health . . . to deposit [sewage] in a natural water course, in close proximity to a source of supply from which the water is used for domestic purposes . . . is a public nuisance." In a Connecticut case of 1900, the State Supreme Court noted that the "right to pour into rivers surface drainage does not include the right to mix with that drainage noxious substances in such quantities that the river cannot dilute them, nor safely carry them off without injury to the property of another is an invasion of his right of property." In some cases, the awarding by courts of damages against municipalities because of the creation of downstream nuisances or property damage resulted in the construction of municipal sewage-treatment plants.[5]

On the other hand, given the difficulty in identifying the exact sources of waterborne disease, the courts refused to hold that upper riparian users were responsible for damages to the health of lower users in regard to waterborne disease. The courts were clearly not administratively capable of deciding cases involving complex scientific and medical information. They were, therefore, most important in cases involving damage to property or the creation of nuisances and relatively unimportant in cases involving hazards to health through waterborne disease.[6]

Supervision of water quality, whether through merely advisory powers or through stricter enforcement provisions, was usually entrusted to state boards of health. Massachusetts created the first of these bodies in 1869, and all states had them by 1909. State boards of health had a wide range of duties, including the collection of vital statistics, concern with the sanitation of dwellings, and the supervision of food marketing, as well as concern for water quality. Reflecting the health-related nature of many of these responsibilities, thirty-one states in 1912 required that the head of the state board of health be a physician; thirty states also required that a majority of the board be physicians. Only seven states in 1912, on the other hand, required the presence of a civil or sanitary engineer on the board. The trend, however, on the most activist boards, was to include a sanitary engineer as a member and to utilize engineering advice, even though physicians held the ultimate responsibility for decisions. Engineering periodicals advocated the formation of state boards of health and generally supported the extension of their powers in regard to mitigating water pollution hazards. They also vigorously demanded the inclusion of sanitary engineers on the boards.[7]

During the 1890s, several critical scientific and technological developments took place in the area of water pollution control. One was the clarification, through the use of bacteriological research, of the etiology of typhoid fever and its relationship to sewage-polluted waterways. This work took place primarily at the Lawrence Experiment Station of the Massachusetts Board of Health under the direction of epidemiologist William T. Sedgwick. Also performed at the Lawrence Laboratories was research into methods of sewage "purification" and the further development of intermittent filtration as an effective means of treatment. Intermittent filtration and sewage farming ("broad irrigation")—both land methods and both more feasible with sewage output from separate rather than combined sewers—were the most widely utilized techniques of sewage treatment in the 1890s. The success of these methods in treating sewage, especially intermittent filtration, suggested that their use offered a sound approach to protecting water supplies from sewage pollution. Sanitary engineers Moses N. Baker and George Rafter noted in their 1893 pathbreaking work, *Sewage Disposal in America,* that "the recent extensions of knowledge of the causation of typhoid fever, and the other water-born communicable diseases," required that cities seriously consider sewage purification. In 1900, the *Engineering Record* boasted that "the resources of the sanitary engineer are sufficient to bring about the purification of sewage to any reasonable degree. This costs money . . . , but not so much as is often believed."[8]

In the late 1890s, however, another technological option appeared for municipalities seeking to protect their water supplies. This alternative was water filtration. Although some small cities, as well as paper mills, had success with water filtration methods as early as the 1880s, no large city had been able adequately to filter its water before the 1890s. Experiments at the Lawrence Station, however, resulted in the development of a successful slow sand filter and its application at the city of Lawrence, while other experiments at Louisville, Kentucky, in 1895–97 led to the effective use of mechanical filters. The success of these methods in treating sewage-polluted water led many inland cities to install sand and mechanical filters in the years after 1897, resulting in an impressive decline in morbidity and mortality rates from typhoid fever as well as other diseases.[9]

By 1900, therefore, sanitary engineers and public health officials found themselves faced by new alternatives in regard to the problem of sewage pollution and drinking-water supplies. The options essentially concerned the means available to municipalities to obtain a potable water supply—to both filter water and treat their sewage so as to endeavor to protect both their own and the water supply of downstream cities or to filter the water alone, leaving it to downstream users to discover a means to guarantee the safety of their water supplies. Debate over which of these options was preferable from the perspective of protecting both the public health and the financial structure of municipalities caused a major division within the public health movement between physicians on the one hand and sanitary engineers on the other.

Water Filtration vs. Sewage Treatment Plus Water Filtration as a Means to Protect Water-quality

As experience with filtration methods conclusively demonstrated that sewage-polluted waters could be treated and utilized without risk from infectious disease, engineering opinion shifted on the preferred means to achieve safe water supplies. During much of the 1890s, the effectiveness of intermittent filtration in producing a relatively "pure" effluent had convinced many engineers and public health officials that sewage "purification" was feasible at a reasonable cost. By approximately 1900, however, a number of sanitary engineering spokesmen, impressed by the success of water filtration in reducing waterborne disease, were questioning the necessity for all cities to assume the cost of treating their sewage in cases where downstream communities drew their water supplies from the stream used for disposal. *Engineering Record* summarized this position in 1903 when it noted that "it is often more equitable to all concerned for an upper riparian city to discharge its sewage into a stream and a lower riparian city to filter the water of the same stream for a domestic supply, than for the former city to be forced to put in sewage treatment works."[10]

The dominant sanitary engineering position on water filtration and sewage treatment was precisely stated in Allen Hazen's 1907 book *Clean Water and How to Get It.* Hazen noted that sewage and manufacturing wastes could be purified before discharging them into rivers, and this made them "more desirable as sources of public water supply." But, he added, because of the expense and difficulty of sewage treatment, large cities only purified their sewage where it created a nuisance. In most cases, they discharged their sewage in adjacent waterways to be disposed of through dilution. To protect water supplies through sewage purification, Hazen said, would require treating the wastes of thousands of cities and towns where sewage did not constitute a nuisance. Therefore, "the discharge of crude sewage from the great majority of cities is not locally objectionable in any way to justify the cost of sewage purification." Hazen claimed that if one examined the question as a "great engineering problem," it was apparent that "it is clearly and unmistakably better to purify the water supplies taken from the rivers than to purify the sewage before it is discharged into them." The cheapness of this approach, its effectiveness in disease control, and its guarantee that all water intended for consumption would be treated made this method preferable. Since, he calculated, one dollar spent in purifying water "would do as much as ten dollars spent in sewage purification," keeping all sewage out of rivers was "not a practical proposition and it is not necessary. It is not even desirable, when the greater good to be secured by a given expenditure in other directions is taken into account."[11]

However, as sanitary engineers adopted the position on water-quality policy enunciated by Hazen, considerable public and professional opinion was moving in a different direction. The years from approximately 1900 to 1914 are known as the Progressive Era in American history, a period when a number of reform-oriented and structural changes took place in American society and government. A key component of the movement was the thrust toward conservation of the nation's natural resources, public lands, and waterways, often through the utilization of scientific knowledge and expertise. Essentially the conservation movement involved two dimensions: a preservationist orientation and the more effective and efficient use of the nation's resources. Waterway improvement, particularly in regard to navigation, was a major part of the movement, and concern over the growing sewage pollution of waterways as both a hazard to health and to shipping was a natural corollary. Conservation and environmental protection received the support of important politicians such as President Theodore Roosevelt, citizens groups, business associations, influential journals of opinion, and newspapers.[12]

In a number of instances between 1906 and 1914, business groups such as the Pittsburgh Chamber of Commerce and the Merchants Association of New York, physicians groups such as the Section on Public Health of the New York Academy of Medicine, and important representatives of the media such as the *New York Times* and the *Survey* protested against sewage disposal by dilution and called for the construction of sewage-treatment plants by municipalities. This issue arose in New York State in two cases—the further pollution of New York Harbor by the joint outlet sewer from the Passaic Valley Sewerage Commission and the pollution of Lakes Erie and Ontario by Rochester and Buffalo sewage. In these instances, the press and physicians and business groups called for a greater degree of sewage treatment or "purification" than engineering opinion thought warranted.

They maintained that treatment was necessary to protect the public health from the threat of sewage pollution of water supplies. In addition, they raised the new but potentially powerful argument that the sewage pollution of waterways diminished their recreational value as well as their usefulness as suppliers of drinking-water supplies. The role of waterways in serving multipurpose uses assumed increasing importance in the coming years.[13]

Progressive-minded politicians were quick to perceive the importance of these issues. In 1909, for instance, New York's governor, Charles Evans Hughes, maintained that the state could "no longer afford to permit the sewage of our cities and our industrial wastes to be poured into our watercourses." The following year, speaking in Buffalo, former president Roosevelt urged the protection of the purity of the Great Lakes and declared that "civilized people should be able to dispose of sewage in a better way than by putting it into drinking water." Roosevelt called for both state and national legislation to end water pollution and linked the need to control the pollution at its source with the obligation of the individual citizen to maintain high standards of public honesty and decency.[14]

At the federal level, attempts were made to increase the powers of the Public Health and Marine-Hospital Service to investigate the pollution of interstate waterways and even to create a Department of Public Health. The partial success of these efforts came in 1912 when the Public Health Service Act initiated official federal concern with the consequences of water pollution on a national basis. The U.S. Public Health Service proceeded to establish research facilities for scientific analyses of water pollution and conducted a number of important studies on the capacity of interstate streams for self-purification. But, while the many technical studies of the Public Health Service greatly advanced basic knowledge in the area of water pollution after 1912, attempts at control remained at the state level.[15]

In the years after the turn of the century, Connecticut, Minnesota, New Hampshire, New Jersey, New York, Ohio, Pennsylvania, and Vermont joined Massachusetts in giving state boards of health increased power to control sewage disposal in streams. As in many public policy situations, a crisis precipitated legislative action. In this case, the crisis was a series of unusually severe typhoid epidemics stimulated by adverse weather conditions in the winter of 1904. These outbreaks, plus the normal endemic typhoid situation at cities such as Allegheny, Pittsburgh, and Philadelphia, increased pressure from both the public and public health authorities on state legislatures to take substantive action. Typical of these laws was the 1905 Pennsylvania act "to preserve the purity of the waters of the State for the protection of the public health," which the legislature passed in response to the severe Butler typhoid epidemic of 1903. It forbade the discharge of any untreated sewage into state waterways by new municipal systems, and, while it permitted cities already discharging to continue the practice, it required them to secure a permit from the State Commission of Health if they extended their systems.[16]

The important policy question involved in statutes of this type was how state commissioners and boards of health would utilize the discretionary powers given them in the newly approved legislation. The report of the Committee on the Pollution of Streams at the 1908 Conference of State and Provincial Boards of Health indicated some directions. After aligning themselves with President Roosevelt's work for the conservation of natural resources, the committee argued that in no case should rivers be used for sewage disposal of untreated wastes if they also served as sources of water supply. The committee advocated that a "double safeguard" be used to protect the public health: sewage should be purified and water filtered in order to create two barriers to infectious disease. Keeping waterways free of pollution, added the committee, was important not only for health reasons but also for recreational purposes.[17]

The controversy between the sanitary engineering position on the protection of drinking-water supplies and that of the more aggressive state boards of health came to a head over a case involving the city of Pittsburgh in the years from 1910–13. Due to rapid industrial development, the city had reached a population of 533,905 by 1910, with approximately another half million in the metropolitan area outside its boundaries. In the late 1880s, Pittsburgh began constructing a combined sewerage system, and by 1909 it possessed 538 miles of combined sewers. These sewers discharged their untreated contents into the conjoined Allegheny, Monongahela, and Ohio rivers.

Due to the sewage disposal practices of both Pittsburgh and neighboring municipalities, typhoid fever was rampant in the region. The city and its suburbs had drawn their water supply from the Allegheny and Monongahela rivers since 1828. In 1900, 350,000 inhabitants in seventy-five upriver municipalities discharged their untreated wastes into these waterways, and Pittsburgh had an average death rate from typhoid fever of approximately 100 per 100,000 from 1880–1900. In order to deal with this health hazard, in 1907 the city erected a slow sand filtration plant, and by 1910 deaths from typhoid had been reduced to approximately 22 per 100,000.[18] Pittsburgh itself, however, continued to dump its untreated sewage into the rivers, endangering the water supply of downstream communities where typhoid rates remained high.

On January 31, 1910, the Pennsylvania Department of Health, headed by Dr. Samuel G. Dixon, responded to a request from Pittsburgh to extend its sewer lines with a requirement for a "comprehensive sewerage plan for the collection and disposal of all of the sewage of the municipality" by December 1, 1911. F. Herbert Snow, chief engineer of the department, maintained that this plan was needed to protect the public health of communities that drew their water supplies from the rivers downstream from Pittsburgh. "The baneful effect of Pittsburgh's sewage on the health of the brightest citizens at her door," wrote Snow, "admonishes city and state authorities alike of the futility of defying nature's sanitary laws." In addition, the Department of Health argued that in order to attain efficiency of treatment, the city should consider changing its sewerage from the combined to the separate system.[19]

The city responded to Dixon's order by hiring the engineering firm of Allen Hazen and George C. Whipple to act as consultants for the required study. Hazen and Whipple were among the most distinguished sanitary engineers in the nation and were already known for their espousal of water filtration as an alternative to sewage treatment to protect drinking-water quality. They based their study primarily on an evaluation of the costs of building a treatment system and of converting Pittsburgh sewers to the separate system. They compared these costs with those generated by maintaining the city's present system of combined sewers and disposal by dilution, with several improvements intended to facilitate the prevention of nuisance.

The engineers issued their report to the city on January 30, 1912. Hazen and Whipple argued that Pittsburgh's construction of a sewage-treatment plant would not free the downstream towns from threats to their water supplies nor from the need to filter them, since other communities would continue to discharge raw sewage into the rivers. The method of disposal by dilution, they maintained, sufficed to prevent nuisances, particularly if storage reservoirs were constructed upstream from Pittsburgh to augment flow during periods of low stream volume. In two key statements, Hazen and Whipple expressed the rationale for their position. First, they argued that there was no case "where a great city has purified its sewage to protect public water supplies from the stream below." Neither, they stated, was there any "precedent for a city's replacing the original combined system by the separate system, for the purpose of protecting water supplies of other cities taken from the water course below."[20]

Hazen and Whipple's most powerful argument concerned the lack of economic feasibility of converting Pittsburgh's sewerage system to separate sewers and building a sewage-treatment plant. They calculated that the costs of these measures would be approximately $37 million in capital charges for a population of 800,000, plus annual maintenance of $435,000; financing such a project would have caused the city to exceed its municipal indebtedness level and violate state law. The engineers maintained that, because the sewage-treatment plant was intended for the protection of the downstream communities, Pittsburgh would not receive any direct benefits from it. The downstream cities, however, would still have to filter their water, making the treatment plant unnecessary and an inefficient use of the city's limited resources. Considering these factors, Hazen and Whipple concluded that "no radical change in the method of sewerage or of sewage disposal as now practiced by the city of Pittsburgh is necessary or desirable."[21]

The engineering press received the Hazen-Whipple report with enthusiasm. *Engineering Record* called it "the most important sewerage and sewage disposal report made in the United States," and *Engineering News* entitled it "A Plea for Common Sense in State Control of Sewage Disposal." The latter journal accused the Pennsylvania Department of Health of having "joined blindly in . . . the doctors' or physicians' campaign against the discharge of untreated sewage into streams, with little or no regard to the local physical and financial conditions." In order to prevent such problems in the future, concluded the editorial, questions of water supply and sewage disposal should be decided by "engineers, not by physicians."[22]

Pennsylvania health commissioner Dixon, however, called the Hazen and Whipple report an insufficient response to his original instructions requesting Pittsburgh to develop a comprehensive sewerage plan based on long-range planning. He maintained that he had envisioned a report that would take a regional rather than a local approach to western Pennsylvania water pollution problems. He argued that the problem of water pollution had to be viewed from the perspective of health rather than from that of nuisance abatement and that the immediate costs of sewage treatment would be outweighed by the long-range health benefits. The time had come, Dixon stated, "to start a campaign in order that the streams shall not become stinking sewers and culture beds for pathogenic organisms."[23]

Given the political context, however, and the financial limitations on the city, Dixon had no realistic means by which to enforce his order. In 1913 he capitulated and issued Pittsburgh a temporary discharge permit. The city continued to receive such permits until 1939, and it was not until 1959 that Pittsburgh and seventy-one other Allegheny County municipalities ceased discharging raw sewage into the abutting rivers and began treating their wastes.[24] Thus, nearly half a century was to pass before Dixon's vision of sewage-free rivers would even begin to be realized.

Professional Cultures in Conflict

An analysis of the Pittsburgh case and of the relative positions taken in this period on the question of how best to deal with the health and nuisance problems posed by sewage-polluted waterways reveals primarily three major issues. These issues pertained to the relative authority of competing professional groups within the larger public health field; the question of which profession could best provide direction to municipalities in decisions involving water quality and the public health; and the proper strategy that should be followed in order to maximize social welfare in the water area.

The intensity of the dispute over water quality policy can be understood partially as a result of the competition between two professional groups, each trying to impose its standards and values on the field of public health. The dominant group was the physicians, who had founded the public health profession, controlled its professional arm—the American Public Health Association—and headed the local and state boards of health throughout the nation. The challenging group was the sanitary engineers, who had grown out of the nineteenth-century sanitary movement and come to play an increasingly important role in the area of public health. Spokesmen for the sanitary engineering profession believed in the uniqueness of their field and its special role within public health. The term "sanitary engineering," like many other developments in the sanitary area, was first used in England and spread to the United States in the 1870s. By the 1880s, the profession, which had originally included plumbers and other craftsmen, had begun to define itself explicitly in a scientific and disciplinary sense.[25]

Important in this development was the formation in 1887 of the Lawrence Experiment Station as an engineering laboratory under the guidance of civil engineer Hiram Mills. The Lawrence Experiment Station united experts in the areas of engineering, chemistry, and biology and brought their knowledge to bear on the problems of water purification and sewage treatment. The station served as a training ground for many of the great names in sanitary engineering, among them Allen Hazen and George W. Fuller as well as bacteriologist E. O. Jordan.[26]

The consulting biologist to the Lawrence Station, and director of many of its significant investigations, was William T. Sedgwick, head of the Department of Biology at the Massachusetts Institute of Technology. Sedgwick focused his interests on the areas of sanitation and public health and introduced courses in the 1880s in germs and germicides, sanitary biology, water supply and drainage, and hygiene and the public health. In 1889, he first offered a course in sanitary engineering, which, for him, represented a combination of engineering and science that utilized the contributions of each field to solve public health problems.[27]

Sedgwick trained a number of students at MIT, among them George C. Whipple, whom Harvard appointed professor of sanitary engineering in 1911. In 1912, Harvard and MIT joined together to create a school for health

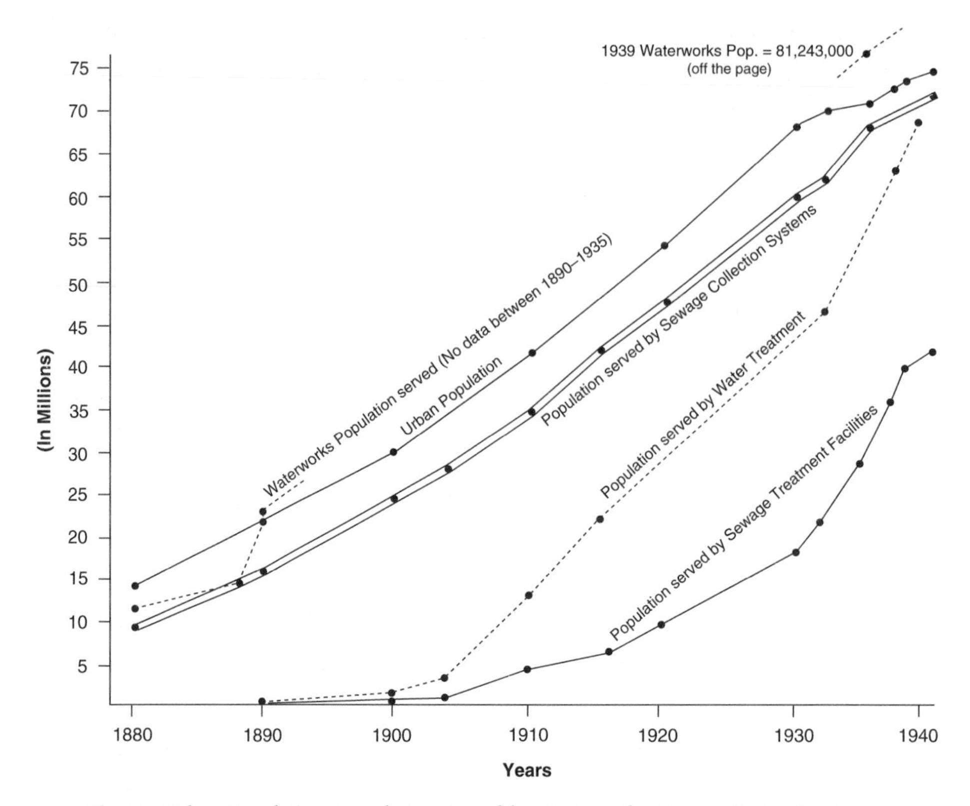

Fig. 6.1. Urban Population, Population Served by Waterworks, Sewers, Water Treatment and Sewage Treatment, 1880–1940.

officers that united the Division of Sanitary Engineering at Harvard, the Department of Preventive Medicine at Harvard Medical School, and the MIT Department of Biology and Public Health. The school was under the direction of William T. Sedgwick. Courses, and in some cases departments, of sanitary engineering appeared at other institutions such as the Universities of California, Illinois, and Pittsburgh. Accompanying the sanitary engineering courses were others in sanitary chemistry, bacteriology, and sanitary and vital statistics, designed to train students for public health careers.[28]

In 1912, George C. Whipple, reflecting the quest of sanitary engineers for disciplinary and professional status, attempted a definition of his field. The sanitary engineer, he said, was "he who adapts the forces of nature to the preservation of public health, through the construction and operation of engineering works." Whipple noted that the profession of sanitary engineering represented "the application of a new science to a new product of civilization. The new science is bacteriology; the new product of civilization is 'The Modern City.'" Sanitary engineers, he added, should be trained primarily in engineering and secondarily in sanitation. Such training would enable the sanitary engineer to interact with students of other disciplines, such as chemistry and bacteriology, to improve the total field of sanitary science.

For Whipple and for other sanitary engineering spokesmen, education in their specialty prepared students far more capably for public health careers than did medicine. Medical training, said Whipple, did not fit men for public health work because the "problem of curing disease is quite different from the problem of preventing disease." Domination of the public health field by physicians, claimed Morris Knowles, director of the Department of Sanitary Engineering at the University of Pittsburgh, resulted in "narrowness of scope; incompleteness of work from the broader sanitary point of view; emphasis on *cure* rather than *prevention;* and insufficient realization of the importance of reliable statistics." The medical "monopoly" of boards of health, said Knowles, had to give way to personnel more properly trained for their tasks.[29]

Another critical element separating the sanitary engineers from physician-dominated boards of public health was their belief that they had a superior conception of the "relative needs and values" of cities in regard to public health. This perception, they claimed, was largely based on their understanding of municipal financial limitations. Within the context of the struggle over water-quality policy, for example, George C. Whipple observed that sewage treatment was a luxury that was important in terms of "standards of comfort and decency" but not critical in regard to public health. If, he continued, the alternative was between spending money to save lives (through other health programs) or for measures providing "comfort and decency," the choice should be obvious. In this context, as *Engineering News* noted in an editorial calling for "Common Sense in State Control of Sewage Disposal," public health physicians such as Dixon of Pennsylvania were "radicals" unwilling to "temper the ideal theories of the

Table 6.1. Population Supplied with Water and Treated Water and with Sewers and Sewage Treatment

Year	*Total Population*	*Urban Population*	*Water*	*Water Treatment*	*Sewered*	*Sewage Treatment*
1880	50,155,783	14,129,735	11,809,231	30,000	9,500,000	5,000
1888			14,857,612			
1890	62,947,714	22,106,265	22,470,608	310,000	16,100,000	100,000
1900	75,994,575	30,159,921		1,860,000	24,500,000	1,000,000
1904				3,160,000	28,000,000	1,100,000
1910	91,972,266	41,998,932		13,264,140	34,700,000	4,455,117
1915				21,998,788	41,800,000	
1916						6,145,117
1920	105,710,620	54,157,973			47,500,000	9,500,000
1930	122,775,046	68,954,823			60,000,000	18,000,000
1932		69,900,000		46,059,000	62,000,000	21,500,000
1935		70,833,000	76,714,000		68,000,000	28,400,000
1937		72,359,000		63,000,000		35,800,000
1938		73,200,000				39,760,000
1939			81,243,480	68,693,196		
1940	131,669,275	74,423,702		74,506,000	70,506,000	40,618,000
1944					71,000,000	42,000,000
1945					74,740,887	46,865,114
1948			93,455,135	83,253,170		48,698,195
1949						50,783,520
1950	150,697,361	96,468,000			91,762,000	56,493,564
1953			106,000,000			
1954			115,000,000			
1957					98,361,396	76,443,731
1958			133,126,310	121,508,296		
1960	179,323,175	125,269,000				
1962					118,371,919	103,684,978
1963			150,602,164	138,292,173		
1968	145,602,000				140,226,049	130,684,771
1970	203,211,926	149,325,000				
1973					162,600,000	159,000,000

scientific laboratory with the practical experience of the engineer and the financier." Sanitary engineers, on the other hand, said the journal, were "the greatest of conservationists, zealous to safeguard health and prolong life, but sparing no pains to see that each dollar is spent to the best advantage."[30]

The third major issue involved in the struggle over water-quality policy in this period pertained to the type of standards that should be utilized to best maximize social welfare in regard to waterway use. The more activist boards of health led by reformer public health physicians, as well as some politicians, often spoke in terms of absolute standards for waterways. A representative of the Pennsylvania Department of Health in 1907 said, for instance, that the policy of the state "is to bring about the abandonment of streams as carriers of sewage. All sewage must finally cease to be discharged untreated into any waters used subsequently for drinking purposes." Or, as the Committee on the Pollution of Streams at the 1908 Conference of State and Provincial Boards of Health maintained, the belief that sewage disposal was a proper use of streams was a "very erroneous idea."[31] These were positions that rejected the concept of utilizing the assimilative power of streams. Underlying them was the argument that society would benefit most from the maintenance of the highest possible water-quality supported by relevant standards.

In contrast to this, sanitary engineers maintained that water-quality policy should maximize the assimilative capacity of waterways and thus balance equities between users. The Committee on Standards of the National Association for Preventing the Pollution of Rivers and Waterways, a group dominated by sanitary engineers, summarized the position in their 1912 report. The committee maintained that it was impossible either to maintain or restore waterways "to their original and natural conditions of purity." Waterway use for waste disposal, they said, was an "economic question," and the "discharge of raw sewage into our streams and waterways should not be universally prohibited by law." In regard to standards, said the committee, no "universal standard" of purity was possible, and local conditions and need should determine water-quality standards.[32]

By the beginning of World War I, the perspective of the sanitary engineers on the question of utilizing the maximum assimilative capacity of streams had triumphed over the "sentimentalists and medical authorities" (sanitary engineer George W. Fuller's phrase) who opposed treatment by dilution. Within the past generation, however, public policy has essentially returned to a position of requiring uniform standards of water purity. Thus, while the events discussed in this paper may be dated, many of the issues today in the water-quality policy area remain the same. Disagreements among professional groups (substitute lawyers for physicians) over proper policy, problems regarding the efficient allocation of funds, and arguments over standards and goals continue to absorb our attention as we seek to resolve our water-quality dilemmas.[33]

Notes

This article is based on research supported by the National Science Foundation under Grant No. ERP 75-08870. Any opinions, findings, and conclusions or recommendations expressed in this publication are those of the author and do not necessarily reflect the views of the National Science Foundation. The author would like to thank Francis Clay McMichael for his comments on an earlier draft of this paper.

1. J. A. Tarr, "The Separate vs. Combined Sewer Problem: A Case Study in Urban Technology Design Choice," *Journal of Urban History* 5 (1979):321–22.

2. S. Galishoff, *Safeguarding the Public Health: Newark, 1895–1918* (Westport, Conn.: Greenwood Press, 1975), 44–51; N. M. Blake, *Water for Cities* (Syracuse, NY: 1956), 262. Chicago constructed the drainage canal to transport its sewage to the Illinois River. The canal was based on the principle of disposal by dilution.

3. "Proposed State Legislation to Prevent Water Pollution," *Engineering News* (1899); B. G. Rosenkrantz, *Public Health and the State: Changing Views in Massachusetts* (Cambridge: Harvard University Press, 1974), 74–112.

4. E. B. Goodell, *Review of Laws Forbidding Pollution of Inland Waters in the United States,* Water Supply and Irrigation Paper 152 (Washington, D.C.: U.S. Geological Survey, 1906). A useful summary of Goodell is "Pollution of Streams," *Municipal Journal of Engineering,* 21: 333–34, 364–65, 384; G. W. Fuller, "Sewage Disposal in America," *Trans ASCE,* 54 (1905): 148.

5. *Engineering News* 42 (1899): 72; "A Judicial Review of American Stream Pollution Decisions," *Engineering Record* 55 (1907): 765–67; and "Sewage Purification and Water Pollution in the United States," ibid. (1902): 276.

6. "Sewage Pollution of Water Supplies," *Engineering Record* 48 (1903): 117; After 1910, the courts awarded damages against municipalities in cases where negligence in the operation of public waterworks resulted in individuals contracting typhoid fever. See James A. Tobey, *Public Health Law,* 2d. ed. (New York: The Commonwealth Fund, 1939), 277–80; E. F. Murphy, *Water Purity: A Study in Legal Control of Natural Resources* (Madison: University of Wisconsin Press, 1961), 66–70.

7. C. V. Chapin, "History of State and Municipal Control of Disease," in Mazyck P. Ravenel, ed., *A Half-Century of Public Health* (New York: American Public Health Association, 1921), 138–39; J. W. Kerr and A. A. Moll, *Organization, Powers, and Duties of Health Authorities,* Public Health Bulletin No. 54 (Washington, D.C.: U.S. Public Health Service, 1912), 14–19. See the untitled editorial in *Engineering News* 41(1899): 169; "The Convention of the American Public Health Association," *Engineering Record* 42 (1900): 423.

8. Rosenkrantz, *Public Health and the State,* 103–6; G. W. Rafter and M. N. Baker, *Sewage Disposal in the United States* (New York, 1894), 150; and "The Water Supply of Large Cities," *Engineering Record* 41 (1900): 73. The Committee on the Pollution of Water Supplies of the APHA, chaired by George W. Fuller, reported at the 1900 convention that "The disposal of sewage offers no difficulties if enough suitable sand is available for filtration. Only when land is scarce trouble may arise." See "Convention of the APHA," *Engineering Record* 42: 423–24.

9. A. Hazen, *Clean Water and How to Get It* (New York, 1907), 68–75; G. C. Whipple, "Fifty Years of Water Purification," in Ravenel, ed., *A Half Century of Public Health,* 161–80. G. A. Johnson, "Present Day Water Filtration Practice," *Journal of the American Water Works Association* 1 (1914): 21–80, contains figures on typhoid death rates for leading cities before and after filtration. Several investigators observed that water filtration reduced mortality and morbidity from other diseases as well as typhoid. See W. T. Sedgwick and S. MacNutt, "On the Mills-Reincke Phenomenon and Hazen's Theorem Concerning the Decrease in Mortality from Diseases Other than

Typhoid Fever Following the Purification of Public Water Supplies," *Journal of Infectious Diseases* 7 (1919): 489–564.

10. "Sewage Pollution of Water Supplies," *Engineering Record* 48:117. See also, "The Water Supply of Large Cities," ibid., 41:73. A further incentive was the apparent limits in the application of intermittent filtration due to the lack of the availability of suitable soils.

11. Hazen, *Clean Water*, 34–37.

12. The literature on progressivism is huge. For insightful surveys, see S. P. Hays, *The Response to Industrialism, 1885–1914* (Chicago: University of Chicago Press, 1957), and R. H. Wiebe, *The Search for Order, 1877–1920* (New York: Hill and Wang, 1967). See also S. P. Hays, *Conservation and the Gospel of Efficiency* (Cambridge: Harvard University Press, 1959), esp. 122–46. The 1899 Rivers and Harbors Act banned refuse disposal in navigable waters but exempted sewage discharges and street runoff.

13. See, for example, Pittsburgh Chamber of Commerce, *Sewage Disposal for Pittsburgh* (Pittsburgh, 1907); G. Soper, "The Sanitary Engineering Problems of Water Supply and Sewage Disposal in New York City" *Science* 25 (1907): 601–5; "Up Stream or Down Stream?" *New York Times*, September 25, 1910; C. D. Leupp, "To the Rescue of New York Harbor," *The Survey* (October 8, 1910), 89–93; *Committee on Pollution of State Waters, Protest against the Bronx River Valley Sewer* (New York: Merchants Association of New York, 1907); and *The Battle of the Microbes: Nature's Fight for Pure Water* (New York, 1908). For New York Harbor and the Passaic Valley Joint Outlet Sewer, see the following: "The Sewage Disposal Problem of Metropolitan New York and New Jersey," *Engineering News* 46 (1906): 25–26; "Hearings on the Proposed New York Harbor Outlet for the Passaic Valley Trunk Sewers," ibid. 60 (1908): 686; "A Scientific Scheme for Sewage Disposal by Dilution in New York Harbor," ibid., 678–79; "The Metropolitan Sewerage Commission against the Discharge of Treated Passaic Valley Sewage into New York Bay," ibid., 63 (1910): 662–63; and G. A. Soper, "Permissible Limits of Sewage Pollution as Related to New York Harbor," *Engineering Record* 66 (1912): 354–55. For Rochester and other upstate New York cities, see "We Blush for Rochester," *New York Times*, August 2, 1910; "A Pure Water Setback," ibid., August 4, 1910; "The Discharge of Sewage into the Hudson River," *Engineering News* 63 (1910): 502–3; "The Rochester Sewage Disposal Case: Sewage Disposal by Dilution Strongly Endorsed," ibid. 64 (1910): 154–55; "The Continued Legislative Defeat of More Stringent Control over Sewage Disposal in New York State," ibid., 282–85; "The Need for a More Rational View of Sewage Disposal," ibid., 304–5; and Palmer C. Ricketts (president, Rensselaer Polytechnic Institute) to the editor, ibid., 396–97. See also "The Pollution of Streams," *Engineering Record* 60 (1909): 157–59.

14. "Mr. Roosevelt and the People," *Outlook* 96 (1910): 1; "The Pollution of Lakes and Rivers," ibid., 144–45.

15. George Courtelyou to President Roosevelt, April 11 and 18, 1908, and William Loeb, Jr., to President Roosevelt and to George Cortelyou, April 13, 1908, in Record Group 90, Public Health Service, General File 1897–1923, National Archives, Washington, D.C.; W. H. Frost, "A Review of the Work of the United States Public Health Service in Investigations of Stream Pollution," *Trans ASCE* 89 (1926): 1332–333; W. G. Smillie, *Public Health: Its Promise for the Future* (New York: Arno Press, 1976), 467; and W. H. Frost, "A Review of the Work of the U.S. Public Health Service," 1332–340.

16. "The Unusual Prevalence of Typhoid Fever in 1903 and 1904," *Engineering News* 51 (1904): 129–30; F. H. Snow, "Administration of Pennsylvania Laws Respecting Stream Pollution," *Proc. Eng. Soc. West. Penna.* 23 (1907): 266–83.

17. Editorial, "The Pollution of Streams," *Engineering Record* 60 (1909): 157–59.

18. M. J. Tierno, "The Search for Pure Water in Pittsburgh: The Urban Response to Water Pollution, 1893–1914," *Western Pennsylvania Historical Magazine* 1 (1977): 23–36.

19. "The Greater Pittsburgh Sewerage and Sewage Purification Orders," *Engineering News* 63 (1910): 179–80; "Pittsburgh Sewage Purification Orders," ibid., 70–71; "The Sewerage Problem of Greater Pittsburgh," *Engineering Record* 61 (1910): 183–84; and G. Gregory, "A Study in Local Decision Making: Pittsburgh and Sewage Treatment," *Western Pennsylvania Historical Magazine* 57 (1974): 25–42.

20. "The Most Important Sewerage and Sewage Disposal Report Made in the United States," *Engineering Record* 65 (1912): 209–12; "Pittsburgh Sewage Disposal Reports," *Engineering News* 67 (1912): 398–402.

21. "Pittsburgh Sewage Disposal Reports," 400–401.

22. "The Most Important Sewerage and Sewage Disposal Report," 209–12; "A Plea for Common Sense in the State Control of Sewage Disposal," *Engineering News* 67 (1912): 412–13. The journal used "engineering" in a broad sense "to include all the teachings of bacteriology, protozoalogy, chemistry, medicine and epidemiology, as regards the relation between sewage disposal, water supply and disease."

23. "The Pittsburg [*sic*] Sewage Purification Order: Letters from Commissioner Dixon and Mayor Magee," ibid., 548–52.

24. Pennsylvania Dept. of Health, *Eighth Annual Report* (Harrisburg, 1913), 901–2; Gregory, "A Study in Local Decision-Making," 41–42.

25. B. G. Rosenkrantz, "Cart Before Horse: Theory, Practice and Professional Image in American Public Health, 1870–1920," *Journal of the History of Medicine* 29 (1974): 62; J. A. Tarr et al, *Retrospective Analysis of Wastewater Technology in the United States: 1850–1972* (Pittsburgh: Carnegie-Mellon University, 1977), V-14–15.

26. Rosenkrantz, *Public Health and the State*, 99–100; E. O. Jordan, G. C. Whipple, and C. E. A. Winslow, *A Pioneer of Public Health: William Thompson Sedgwick* (New Haven: Yale University Press, 1924), 57–60.

27. Jordan, et al., *A Pioneer of Public Health*, 31–41, 57–64, 72.

28. S. C. Prescott , *When MIT was "Boston Tech" 1861–1916* (Cambridge: MIT Press, 1954), 281–83; M. Knowles, "Public Health Service Not a Medical Monopoly," *American City* 7:527, 1912.

29. G. C. Whipple, "The Training of Sanitary Engineers," *Engineering News* 68 (1912): 805–6; "Sanitation More than Medicine," *Literary Digest* 45 (1912): 899; and Knowles, "Public Health Service Not a Medical Monopoly," 527–29. Wilson G. Smillie observed that many physicians who were working in public health left the APHA because its policies were "determined by public health nurses, health educators, statisticians, and sanitary engineers." He appears to be talking about the period from approximately 1910–20. However, up to the 1950s, sanitary engineers had still not achieved what they felt to be equality in the public health field. See Smillie, "Public Health," 310; and Tarr et al., *Retrospective Assessment of Wastewater Technology*, 6:20–21.

30. G. W. Fuller, "Relations between Sewage Disposal and Water Supply Are Changing," *Engineering News Record* 28 (1917): 11–12. See also Fuller, "Is It Practicable to Discontinue the Emptying of Sewage into Streams?" *American City* 7 (1912): 43–45; "A Plea for Common Sense," 412–13; G. C. Whipple, "How to Determine Relative Values in Sanitation," *American City* 11 (1914): 427–32; and G. W. Fuller, "The Problem of Sewage Disposal," ibid., (1911): 343–45.

31. Snow, "Administration of Pennsylvania Laws," 281; H. M. Bracken, "Sewage Pollution Made Compulsory by the Minnesota State Board of Health," *Engineering News* 51: 138, 1904; an unsigned editorial, ibid., 129; R. W. Pratt, "The Work of the Ohio State Board of Health on Water Supply and Sewage Purification," ibid., 57 (1907): 680; and "The Pollution of Streams," *Engineering Record* 60 (1909): 157–59.

32. Standards of Purity for Rivers and Waterways, *Engineering News* 68 (1912): 835–36; "Conference on Pollution of Lakes and Waterways," *Engineering Record* 66 (1912): 485–86. George C. Whipple was chairman of the committee; the president of the association was Calvin W. Hendrick.

33. Fuller, "Relations between Sewage Disposal and Water Supply are Changing," 11–12. An insightful overview linking past and present is E. J. Cleary, "Evolving Social Attitudes on Pollution Control," *Journal of the American Water Works Association* (1972): 405–9.

CHAPTER VII

Water and Wastes

A Retrospective Assessment of Wastewater Technology in the United States, 1800–1932

During the 1970s, to a greater extent than ever before, the federal government attempted to regulate the effects of technology on the society. The environment was a critical area of concern, and environmental quality ranked high on the policy agenda. In December 1970, a presidential order created the Environmental Protection Agency to mount a coordinated attack on environmental problems. The broad legislative authority for EPA programs is based primarily on nine separate acts dealing with air, land, surface and groundwater, and noise pollution passed during the decade. In large part, however, research and policy concerned with these acts have been devoid of a historical dimension that might increase knowledge of the evolution of environmental problems or inform policymakers.

In an attempt to provide such information on one environmental medium, a group of faculty and graduate students at Carnegie-Mellon University conducted a study for the National Science Foundation (NSF) entitled *Retrospective Assessment of Wastewater Technology in the United States, 1800–1972*.[1] This study utilized the technology assessment approach that developed in the late 1960s and early 1970s from concern over the negative effects of technology on society. Technology assessment is generally defined as the "systematic study of the effects on all sectors of society that may occur when a technology is introduced, extended, or modified, with special emphasis on any impacts that are unintended, indirect, or delayed." Retrospective technology assessment attempts to utilize the technology assessment model in regard to past phenomena.[2]

This article derives largely from our NSF report on wastewater technology. The specific objects of the research were to identify the processes and key decision points involved in the evolution of wastewater technology of human-waste disposal, to explore both its primary and secondary social and institutional effects, and to examine the development of public policy toward water pollution problems. In addition, the report dealt with the role played by values in the implementation of technology, as well as the effect of the technology on values. The essential aims of the project were to increase our understanding of water-related environmental problems and to inform policymakers about the larger context of contemporary wastewater questions.

Water Supply and Waste Collection, 1800–1880

The following section considers the reasons for municipal adoption of the water-carriage technology of waste removal in the nineteenth century. Most technology assessments accept the technology under examination as a given, expressing concern only for its effects. In the case of wastewater technology, however, we examined the workings of the previous system used to dispose of human wastes and wastewater—labeled the "cesspool–privy vault" system—in order to gain insight into the reasons for the adoption of the new water-carriage technology.

Until well into the second half of the nineteenth century, most American urbanites depended for their water supplies on local surface sources such as ponds and streams, on rainwater cisterns, or on wells and pumps drawing on groundwater. Water consumption per capita under these supply conditions probably averaged between three and five gallons a day. Householders disposed of wastewater from household functions such as cleaning, cooking, or washing in the most convenient manner. Sometimes this meant simply throwing it on the ground, into a street gutter, or into a dry well or leaching cesspool, a hole lined with broken stone.[3]

Human wastes were occasionally deposited in cesspools but more often in privy vaults, which ranged from shallow holes in the ground to receptacles lined with brick or stone located close by residences, even in cellars.

Theoretically, they could be either permeable, so that the ground could absorb the liquids, or impervious, thus requiring frequent emptying. Often the ground around the privy vaults and cesspools became saturated with wastes, and then they were covered over with dirt and new ones dug. In 1829, it was estimated that each day New Yorkers deposited over one hundred tons of excrement into the city's soil. Most large cities tried to institute periodic cleaning by private scavengers under city contract or by city employees, but services were very inefficient and irregular. The cleaning technology utilized for most of the nineteenth century was labor intensive and rudimentary—dippers, buckets, and wooden casks. The process created both aesthetic nuisances and health problems, primarily through pollution of groundwater and wells. Scavengers collected the wastes in "night soil carts" and disposed of them in nearby watercourses or dumps or on farms, or they sold them to reprocessing plants to be made into fertilizer.[4]

Although both private and public underground sewers existed in the larger cities, such as New York and Boston, they were intended for stormwater drainage rather than human-waste removal. These sewers were usually constructed of stone or brick, in circular or elliptical shapes and were often large enough so that a man could enter them for cleaning. In some of these cities, ordinances prohibited the placing of human wastes in the sewers. The majority of nineteenth-century municipalities, however, had no underground drains. Street gutters of wood or stone, either on the side or in the middle of the roadway, provided for surface stormwater and occasionally for human wastes. Private householders often constructed drains to the street gutter to remove wastewater from cellars.[5]

From approximately 1820 to 1880, demographic and technological factors combined to strain the cesspool–privy vault system and cause its eventual breakdown and replacement. The two most important factors were urban population growth and new urban water-supply systems, with the consequent adoption of household water fixtures. The most critical fixture was the water closet or flush toilet.

By 1860, about 20 percent of the U.S. population was found in communities of over 8,000, and by 1880 the percentage had risen to 28. As cities grew in size, population was also more concentrated, especially in the original central cores. Transportation restricted the distance that population could spread from places of employment and essential urban institutions. Urban density and an explosion of building construction made the existing waste-collection system increasingly inadequate. Overflowing privies and cesspools filled alleys and yards with stagnant water and fecal wastes. Often waste receptacles and their overflows were close to wells and other sources of water supply, causing a serious pollution hazard. In addition, paving reduced the ability of streets to absorb rain and increased the possibility of flooding.[6]

The cesspool–privy vault system of waste collection was further stressed by the adoption of another technology—piped-in water. This technology dates back at least to the ancient Roman aqueducts, with more modern examples from seventeenth-century London and several colonial towns. The movement in nineteenth-century cities away from a localized and labor-intensive water-supply system to a more capital-intensive system that utilized distant sources took place primarily for four reasons in addition to population increase: water from local sources used for household purposes was often contaminated, tasted and smelled bad, and was suspected as a cause of disease; more copious water supplies were required for firefighting; water was needed for street flushing at times of concern over epidemics; and developing industries required a relatively pure and constant water supply. In addition, rising affluence in the nineteenth century undoubtedly increased household demands for water.[7]

Philadelphia was the first city to build a waterworks in 1802. Other municipalities, such as New York, Boston, Detroit, and Cincinnati, followed, and by 1860 the sixteen largest cities in the nation had waterworks. There were then a total of 136 systems in the country; by 1880 this had increased to 598. The availability of a source of constant water in the household caused a rapid expansion in usage, as demand interacted with supply. Chicago, for example, went from 33 gallons per capita per day in 1856 to 144 in 1882; Cleveland increased from 8 gallons per capita per day in 1857 to 55 in 1872; and Detroit from 55 gallons per capita per day in 1856 to 149 in 1882.[8] These figures reflect unmetered usage and include industrial and other nonhousehold uses, yet they indicate greatly increased water consumption over a relatively short span of time.

While hundreds of cities and towns installed waterworks in the first three-quarters of the nineteenth century, few of them simultaneously constructed sewer systems to remove the water because it was believed that the technology was unnecessary, unproved, or too costly. In most cities with waterworks, wastewater was initially diverted into cesspools or existing storm sewers or street gutters; householders often connected their cesspools with the sewers via overflow pipes. The introduction of such large volumes of contaminated water into systems designed to accommodate much smaller amounts caused serious problems of flooding and disposal.[9]

This situation was exacerbated by the widespread adoption of a waste-disposal technology that had not been anticipated by the advocates of piped-in water—the water closet. The water closet actually dated back centuries but was patented in the United States only in 1833. In cities with waterworks, affluent families were quick to install closets and take advantage of their convenient inside location and comparative cleanliness. For example, in Boston in 1863 (population ca. 178,000), there were over 14,000 water closets out of approximately 87,000 water fixtures. In Buffalo in 1874 (population ca. 118,000), there were 5,191 dwellings supplied with water and 3,310 with water closets. By 1880, although the data are imprecise, it can be estimated that approximately one-quarter of urban households had water closets (usually of the pan or hopper type), while the remainder still depended on privy vaults.[10]

Water closets greatly increased both nuisance problems and sanitary hazard in urban areas. They were usually connected with cesspools, which were soon overcharged by the increased flow of waste-bearing water. If connected by overflow pipes with surface gutters or sewers intended as water drains, they contaminated them with fecal matter. Soil became saturated, cellars were "flooded with stagnant and offensive fluids," and the need for frequent emptying of cesspools and vaults greatly increased.[11]

The spreading of feces-polluted water created real and perceived health dangers. During most of the nineteenth century, physicians generally divided into two groups, contagionists and anticontagionists. Contagionists maintained that epidemic disease was transmitted by contact with a diseased person or carrier, while anticontagionists held that vitiated or impure air was the cause. The vitiated air could arise from any number of conditions, including miasmas from putrefying substances such as feces, exhalations from swamps and stagnant pools, or human and animal crowding. By the latter half of the nineteenth century, the majority of physicians were anticontagionists, who believed that filthy conditions accelerated the spread of contagious disease, thus underscoring demands for urban environmental improvements. Public health officials viewed overflowing cesspools with water closet connections as a threat to a healthful environment. As late as 1894, the secretary of the Pennsylvania State Board of Health, Benjamin Lee, complained that householders persisted in connecting water closets to "leaching" cesspools, thereby distributing "fecal pollution over immense areas and . . . constituting a nuisance prejudicial to the public health."[12]

The adoption of two new technologies, therefore—piped-in water and the water closet—combined with higher urban densities to cause the breakdown of the cesspool–privy vault system of waste removal and to generate excessive nuisances and health hazards. Some cities attempted to alleviate these hazards by permitting householders to connect their water closets with existing storm sewers, but the latter were poorly designed for waste removal and merely became "sewers of deposit." Another approach, utilized in more than twenty cities, was the so-called odorless excavator, a vacuum pump that emptied the contents of cesspools and privies into a horse-drawn tank truck for removal. This "interim" technology, adopted between the time that cities turned to piped water systems and the time they installed sewers, was both labor and capital intensive.[13] Engineers, public health officials, and other sanitarians realized that no existing system was capable of meeting the new demands. Increasingly, in the second half of the nineteenth century, they advocated the water-carriage technology of waste removal as a replacement.

Benefits and Costs of Water-Carriage Technology

The water-carriage system of waste removal (or sewerage) was essentially a system that used the wastewater itself as a transporting medium and as a cleansing agent in the pipe. As Jon Peterson notes, it "represented a specialized form of urban planning [that] gradually supplanted long-accepted, piecemeal methods of waste removal, particularly the reliance upon privately-built cesspools and privy vaults and the common municipal practice of constructing sewers without reference to a larger, city-wide plan." In the 1840s, British sanitarian Edwin Chadwick advocated the adoption of a system of self-cleansing earthenware small-pipe sewers that would use the household water supply to dispose of human wastes. A convinced anticontagionist, he believed that odors from decaying organic matter caused the spread of many fatal diseases and that fecal matter had to be swiftly transported from the vicinity of the household. The sewage from this "arterialvenous system," he maintained, could be sold for agricultural purposes. Water-carriage technology, therefore, would provide a system of self-financing health benefits.[14]

Chadwick's system was never implemented as he planned, but his vision stimulated debate in Great Britain about technology and health and strongly influenced American sanitarians concerning the benefits of systematic sewerage. The engineers for the earliest sewerage systems in Brooklyn, Chicago, and Jersey City drew heavily on the English sanitary investigations and debates of the 1840s and 1850s, as well as on the actual experience of London with a system of large brick sewers constructed to remedy the sewage pollution of the Thames River. Throughout the remainder of the nineteenth century, visits to sewerage works in cities in Great Britain and Europe were almost mandatory for American engineers involved in planning new sewerage systems. Thus, water-carriage technology provides a good example of the international interchange and transfer of ideas and experience concerning an urban technology.[15]

The system of water-carriage removal of household wastes had a number of important characteristics that sharply differentiated it from the cesspool–privy vault system: it was capital rather than labor intensive and required the construction of large, planned public works; it utilized continuous rather than individual batch collection; it was automatic, eliminating the need for human decisions and actions to remove wastes from the immediate premises; and, because of its sanitary and health implications and its capital requirements, it became a municipal rather than a private responsibility.

Extensive debates and discussions concerning water-carriage technology were held by professional associations, municipal officials, and citizens' groups. These debates often dragged on for years and involved the preparation of a number of engineering reports that addressed the comparative advantages of various forms or designs of waste-disposal technology.[16] The records of these discussions are marked by a number of forecasts about the water-carriage system, both its benefits and its costs. Benefits forecast by its advocates can be summarized as follows:

1. Capital and maintenance costs of building sewerage systems would represent a savings for municipalities over the annual costs of collection under the cesspool–privy vault scavenger system.
2. Sewerage systems would create improved sanitary conditions and result in lowered morbidity and mortality from disease; such savings could often be translated into financial terms, further stressing the economic benefits to be obtained by adoption of water-carriage technology.
3. Because of improved sanitary conditions, cities that constructed sewerage systems would grow at a faster rate than those without by attracting population and industry.[17]

Opponents countered by enumerating these costs:

1. Water-carriage removal would waste the valuable resources present in human excreta that might otherwise be used for fertilizer.
2. Water-carriage technology would create health hazards, such as contamination of the subsoil by leakage, pollution of waterways with threats to drinking-water supplies and shellfish, and the generation of disease-bearing sewer gas.
3. The costs of sewerage systems would create a heavy tax burden. If financed with bonds, they would impose costs on future generations with no voice in the decision.[18]

It was possible to test the validity of these forecasts only on a very rudimentary basis. Projections of improved local sanitary conditions were accurate, as were those of local health benefits, although the latter often tended to be overstated. The predicted cost savings of sewerage over the annual collection costs of the cesspool–privy vault system probably occurred, although comparisons are difficult; in regard to financial costs, perhaps the most crucial change was the transference of maintenance expenses from the individual householder to the municipality. And forecasts about the risk of wasting the valuable materials contained in human excrement overlooked the difficulty of actually utilizing these wastes on American farms.[19]

The major failure of forecasting concerned the negative effects of sewage disposal by upstream cities on water supplies downstream. While there were a few critics who warned about the health hazards of water-carriage technology, they were largely ignored because of the belief that "running water purifies itself." This hypothesis depended on chemical and physical methods of analyzing water quality, which demonstrated that after sewage had been in a stream for a certain distance its physical elements dissipated.[20] In the late nineteenth and early twentieth centuries, American urbanites found that the promised benefits of the technology outweighed the predicted costs, and cities embarked on a massive implementation of sewerage systems.

Although the first municipal sewer systems intended for human wastes as well as stormwater were built in the 1850s (Brooklyn, 1855; Chicago, 1856; Jersey City, 1859), construction rates accelerated after the 1870s. The first date for which aggregate figures are available is 1890, and in that year the U.S. census recorded 6,005 miles of all types of sewers in cities of over 25,000 population; by 1909, the mileage had increased to 24,972 for cities over 30,000, or from 1,832 persons per mile of sewer to 825 persons per mile. In the latter year, 85 percent of the population of cities with populations of over 300,000 was served by sewers; 71 percent in cities with populations between 100,000 and 300,000; 73 percent in cities with populations between 50,000 and 100,000; and 67 percent in cities with populations from 30,000 to 50,000 (see table 1). Small-diameter sewers were constructed initially almost entirely of vitrified clay pipe, while sewers forty-two inches or more in diameter were made of brick. Beginning around 1905, many sewers were constructed of reinforced concrete.[21]

Of the total mileage in 1909, 18,361 miles, or 74 percent, were combined sewers (human and storm wastes in the same pipe) and 5,258 or 21 percent were separate sanitary sewers; only 1,352 miles were solely storm sewers. The choice between separate or combined sewers was an important design question. For the most part, large cities that needed to remove stormwater from the streets, as well as household wastewater, installed combined sewers, while smaller cities often constructed sanitary sewers alone, leaving the stormwater to run off on the surface. In the 1880s and 1890s, a number of smaller cities installed separate sewers because of the belief, spread primarily by the famous sanitarian Colonel George E. Waring, Jr., that they had superior health benefits compared with the combined sewers. Waring argued, as had Chadwick, that large combined sewers produced sewer gas

Table 7.1. Sewer Mileage by Type and Urban Population Group—1905, 1907, 1909

Population Group	*Sanitary Sewers*	*Storm Sewers*	*Combined Sewers*	*Total Mileage by Population Group*
		1905		
1. >300,000	335.2	157.0	8,229.9	8,722.1
2. 100,000–300,000	809.0	120.5	2,961.0	3,890.5
3. 50,000–100,000	965.8	242.8	2,491.1	3,699.1
4. 30,000–50,000	1,313.4	326.6	1,507.4	3,147.7
Total by type of sewer	3,423.4	846.9	15,189.4	19,459.7
		1907		
1. >300,000	554.8	352.0	9,242.3	10,149.1
2. 100,000–300,000	1,300.0	262.9	3,690.5	5,253.4
3. 50,000–100,000	1,097.1	181.6	2,627.1	3,905.8
4. 30,000–50,000	1,611.3	383.9	1,562.9	3,558.1
Total by type of sewer	4,563.2	1,180.4	17,122.8	22,866.4
		1909		
1. >300,000	789.5	349.9	9,834.3	10,973.7
2. 100,000–300,000	1,404.4	284.2	4,405.8	6,094.4
3. 50,000–100,000	1,831.5	384.2	2,615.5	4,831.2
4. 30,000–50,000	1,232.9	333.8	1,505.9	3,072.6
Total by type of sewer	5,258.3	1,352.1	18,361.5	24,971.9

SOURCES: Moses N. Baker, "Sewerage and Sewage Disposal," Appendix A in U. S. Dept. of Commerce and Labor, Bureau of the Census, *Statistics of Cities Having a Population of over 30,000: 1905* (Washington, D.C., 1907), 342–47; U. S. Department of Commerce and Labor, Bureau of the Census, "Special Reports," *Statistics of Cities Having a Population of over 30,000: 1907* (Washington, 1910), pp. 458-63; and U. S. Department of Commerce, Bureau of the Census, "Special Reports," *General Statistics of Cities with Populations Greater than 30,000: 1909*, (Washington,D.C., 1913), 88–93.

by allowing the accumulation of fecal matter, while small-diameter separate sewers speeded the wastes from the vicinity of the household. By 1900, however, largely because of the work of sanitary engineer Rudolph Hering, most engineers believed that the two sewer designs had equal health benefits, and decisions regarding implementation of one system rather than another were based primarily on cost factors and on the need for subsurface removal of stormwater.[22]

The use of combined rather than separate sewers, given the available treatment technology in the late nineteenth and early twentieth century, made both wastewater treatment and resource recovery more difficult and expensive. The greater volume of wastewater from combined sewers was primarily responsible for the higher costs if treatment at the end of the pipe was planned. Most urban policymakers and their consulting engineers assumed, however, that dumping raw sewage into streams was ade-

Table 7.2. Sewer Mileage and Typhoid Fever Mortality in Fifteen American Cities Located on Streams and Lakes, 1880–1905

	1880 Mileage	*1880 Mortality*	*1890 Mileage*	*1890 Mortality*	*1900 Mileage*	*1900 Mortality*	*1905 Mileage*	*1905 Mortality*	*Response: Filtration (F); Change in Supply or Disposal Site (C)*
Atlanta	—	66.8	24	72	—	74.6	122	70.1	1892, 1904, 1910 (F)
Chicago	337	33.9	525	72.2	1,453	21.1	1,633	16.5	1900 (C)
Louisville	41	65.7	52	75.7	97	64.0	122	49.4	1910 (F)
Nashville	—	133.7	24	64	—	49.5	79	71.2	1908, 1909 (F)
Newark	47	52.7	87	99.5	180	21.1	232	14.1	1889 (C), 1903 (F)
Philadelphia	200	58.7	376	73.6	887	37.2	1,041	51.1	1902, 1913 (F)
Pittsburgh	22.5	134.9	87	127.4	275	144.3	365	107.9	1908 (F)
Richmond	—	61.3	35	61	—	104.6	85	44.9	1909 (F)
Rochester	—	23.4	138	39.6	—	17.2	241	11.5	—
Salt Lake City	0	—	5	—	—	39.2	56	101.8	—
San Francisco	126	35.4	193	55.5	307	30.3	332	23.9	—
Spokane	—	—	3	—	—	45.9	23	86.1	—
Toledo	—	35.9	61	36	156	41.0	191	45.7	1910 (F)
Trenton	0	17.1	4	16	—	32.7	65	24.9	1911, 1914 (F)
Washington, D.C.	169	53.4	266	86.8	405	79.7	484	48.2	1906 (F)

NOTE: Typhoid fever mortality rates expressed per 100,000.
SOURCES: U.S. Department of Commerce and Labor, Bureau of the Census, *Statistics of Cities Having a Population of over 30,000: 1905* (Washington, D.C., 1907), 104–5; ibid., *General Statistics of Cities: 1915* (Washington, D.C., 1916), 152–53.

quate treatment because of the self-purifying nature of running water. Although biologists, chemists, and sanitary engineers were seriously questioning the validity of this hypothesis by the 1890s, as late as 1909, 88 percent of the wastewater of the sewered population was disposed of in waterways without treatment. Where treatment was utilized at the beginning of the twentieth century, it was only to prevent local nuisance rather than to avoid contamination of drinking water downstream.[23]

The consequence of the disposal of untreated sewage in streams and lakes from which other cities drew their water supplies was a large increase in mortality and morbidity from typhoid fever and other infectious waterborne diseases. Bacterial researchers, following the seminal work of Pasteur and Koch, were able to identify the processes involved in such disease. The work of William T. Sedgwick and sanitary engineers, biologists, and chemists at the Massachusetts Board of Health's Lawrence Experiment Station in the early 1890s was critical in clarifying the etiology of typhoid fever and confirming its relationship to sewage-polluted waterways. The irony was clear: cities had adopted water-carriage technology because they expected local health benefits resulting from more rapid and complete collection and removal of wastes, but disposal practices produced serious externalities for downstream or neighboring users. Some of the more striking examples were Atlanta, Pittsburgh, Trenton, and Toledo, cities that constructed sewerage systems between 1880 and 1900 in the expectation of health benefits, all of which experienced substantial rises in typhoid death rates during the same years (see table 2).[24] This, then, was the primary unanticipated impact of sewerage technology—a rise in health costs where health benefits had been predicted. Because these costs were often borne by second parties or downstream users, however, cities continued to build sewerage systems and to dispose of untreated wastes in adjacent waterways.

Policy Options for Dealing with the Unexpected Impacts of Water-Carriage Technology

During the late nineteenth century, municipalities adopted a body of sanitary and plumbing codes in response to the nuisances and health risks of the cesspool–privy vault system and the supposed dangers of sewer gas generated by decaying organic matter. In addition, they created local boards of health because of concern over epidemic disease. Regulations and health boards had been promoted primarily by a sanitary coalition composed of physicians, engineers, plumbers, and civic-minded citizens, and this coalition was instrumental in securing the implementation of the new sewerage technology.[25]

The sanitary coalition also pushed for laws and institutions to deal with the threats to health from urban sewage-disposal practices. The transference of infectious disease carried in waste material from one user to another via sewerage technology, and then through the water medium used for disposal, necessitated an extralocal response. Legal redress for damages was possible in the case of nuisance, but difficulties in specifying the origins of waterborne disease prevented the affected individuals from seeking relief in the courts. The institutional response entailed the creation of state boards of health (beginning with Massachusetts in 1869) and the passage of legislation to protect water quality. In 1905, the U.S. Geological Survey published its *Review of the Laws Forbidding Pollution of Inland Waters in the United States,* listing thirty-six states with some legislation protecting drinking water and eight states with "unusual and stringent" laws. Legislatures had generally passed the stricter statutes in response to severe typhoid epidemics. Supervision of water quality, whether through merely advisory powers or through stricter enforcement provisions, was usually entrusted to the state boards of health.[26]

By 1900, municipal policymakers, public health officials, and sanitary engineers had several alternatives for dealing with the threat of sewage to the water supplies of inland cities. One option was to secure the municipal water supply from a distant and protected watershed, a course followed by cities such as Newark and Jersey City, New Jersey. Another option was sewage treatment or "purification," although this was more effective in preventing nuisance than in protecting drinking-water quality. Sewage-treatment technology was at an early stage of development, and the two methods most commonly used in 1900, sewage farming and intermittent filtration (a physical/chemical microbiological process), were both land intensive and impractical with sewage output from combined systems such as possessed by the great majority of cities. Only one city with a combined system treated its sewage.[27] In addition, while sewage treatment provided benefits for the downstream city, it imposed the costs on the upstream community in an era when there were neither state nor federal programs to subsidize waste treatment.

In the late 1890s, another option became available for municipalities seeking to protect their water supplies—filtration at the intake. Experiments in the 1890s at the Lawrence Experiment Station and at Louisville, Kentucky, showed the effectiveness of both slow-sand and mechanical filters in treating sewage-polluted waters. As a result of these successful demonstrations, many inland cities installed mechanical or sand filters in the years after 1897. The use of water-filtration technology resulted in an impressive decline in morbidity and mortality rates from typhoid fever, as well as other diseases.[28]

The choice confronting municipalities and health authorities faced with polluted water supplies was whether to filter their water *and* treat their sewage, in order to protect both their own water supply and that of downstream cities, or just to filter the water, leaving the downstream user with the responsibility of guarding the safety of its own water supply. This decision involved both public health and municipal financial structures. Debate over the question caused a major rift in the sanitary coalition that had been responsible for the implementation of sewerage systems and the creation of local and state boards of health. This split was most evident in states such as Minnesota, New York, and Pennsylvania, where state boards of health dominated by physicians came into conflict with consulting sanitary engineers and their municipal clients. This dispute continued from about 1905 to 1914; its resolution set the pattern for dealing with water pollution problems and the public health into the 1930s.[29]

In the first years of the twentieth century, sanitary engineers took the position, expressed editorially by the *Engineering Record,* that "it is often more equitable to all concerned for an upper riparian city to discharge its sewage into a stream and a lower riparian city to filter the water of the same stream for a domestic supply, than for the former city to be forced to put in sewage treatment works." Sanitary engineer Allen Hazen expressed the rationale for this position in *Clean Water and How to Get It,* (1907) by noting that "the discharge of crude sewage from the great majority of cities is not locally objectionable in any way to justify the cost of sewage purification." Rather, said Hazen, downstream cities should filter their water to protect the public health, and sewage purification should be utilized only to prevent nuisances such as odors and floating solids.[30]

As sanitary engineers were adopting Hazen's stance on water-quality policy, a considerable body of public and professional opinion was moving in a different direction. A number of business and professional groups, for instance, as well as politicians and representatives of the media, were calling for state and national legislation to protect the purity of waterways. Their demands for legislation protecting water quality reflected the thrust of the Progressive Era movement toward measures for conservation of the nation's natural resources. Foremost among these groups were physicians active in the public health movement who directed boards of health in several states. These officials were influenced by the New Public Health, a movement generated by the bacteriological advances of the 1890s, which stressed the need to control the diseased individual or carrier. In the case of waterborne infectious disease, advocates of the New Public Health argued for restricting or abandoning the use of streams for sewage disposal, especially if they were also the source of a water supply. At the 1908 Conference of State and Provincial Boards of Health, the Committee on the Pollution of Streams, for instance, maintained that both water filtration and sewage treatment were required in order to provide a double safeguard for the public health.[31]

In a number of cases during the years from 1905 to 1914, the most important of which involved the city of Pittsburgh, several state boards of health attempted to compel municipalities to cease discharging untreated sewage into neighboring streams and to convert combined sewer systems into separate ones. In cases where the problem was one of nuisance, municipalities usually constructed sewage treatment plants in order to avoid legal damages. In cases involving assumed threats to water supplies, however, boards of health were unsuccessful in their attempts to change municipal practice. Cities objected, not only because they did not want to adopt costly improvements that primarily benefited downstream communities but also because state constitutional debt limits restricted their capital-improvements capability. The engineering press and engineering

Table 7.3. U.S. Population, Urban Population, and Population with Water Treatment, Sewers, and Sewage Treatment, 1880–1940

Year	*Total Population*	*Urban Population*	*Water Treatment*	*Sewers*	*Sewage Treatment*
1880	50,155,783	14,129,735	30,000	9,500,000	5,000
1890	62,947,714	22,106,265	310,000	16,100,000	100,000
1900	75,994,575	30,159,921	1,860,000	24,500,000	1,000,000
1910	91,972,266	41,998,932	13,264,140	34,700,000	4,455,117
1920	105,710,620	54,157,973	—	47,500,000	9,500,000
1930	122,775,046	68,954,823	46,059,000*	60,000,000	18,000,000
1940	131,669,275	74,423,702	74,308,000	70,506,000	40,618,000

SOURCES: Compiled from miscellaneous federal, state, and professional reports.
*1932 data.

groups gave municipalities their enthusiastic support in these situations. *Engineering News* maintained, for instance, that questions involving water supply and sewage disposal should be decided by engineers rather than by physicians, because engineers had a superior conception of the "relative needs and values" of municipalities. The Committee on Standards of Purity of the National Association for Preventing the Pollution of Rivers and Waterways, a group dominated by sanitary engineers, observed that it was impossible to maintain or restore rivers and waterways to "their original and natural condition of purity." Only if water supplies were not filtered, the committee maintained, should untreated sewage be prohibited from streams.[32]

By the beginning of the First World War, the perspective of the sanitary engineers on the question of the disposal of raw sewage into streams had triumphed over that of the "sentimentalists and medical authorities" who opposed the use of streams for disposal. Essentially, the engineering position was that the dilution power of streams should be utilized to its fullest for sewage disposal, so long as no danger was posed to the public health or to property rights and no nuisance created. Water filtration and/or chlorination could serve to protect the public from waterborne disease.[33]

The practical consequences of this position can be seen in the aggregate figures for sewered population, population served by sewage treatment, and population served by water treatment from 1910 to 1930 (see table 3). In this period, while the population newly served by sewers rose by over 25 million, the additional number whose sewage was treated rose only 13.5 million. At the same time, the increase in the population receiving treated water was approximately 33 million. In 1930, not only did the great majority of urban populations dispose of their untreated sewage by dilution in waterways, but their numbers were actually increasing over those who were treating their sewage before discharge. Because of the successes of water filtration and chlorination, however, waterborne infectious disease had greatly diminished and the earlier crisis atmosphere that had led to the first state legislation had disappeared.

Professional Impacts of Wastewater Technology: Sanitary Engineering

The implementation of any large-scale and capital-intensive technology, such as sewerage, will produce a range of institutional, economic, and social changes. Some of these are logical and predictable, while others are unintended and unanticipated. As was noted earlier, negative health effects were the largest single unanticipated effect of the implementation of sewerage systems and produced attempts to regulate water pollution. This section will deal with the development of the profession of sanitary engineering; the following sections will address the effects of wastewater technology on governmental structure, administration planning, and forecasting. In these realms, changes sometimes were based on perceptions of the needs of the technology and sometimes conceived as adaptations to the impacts of the technology.

The development of a new technology with a set of unique characteristics requiring a special body of knowledge and techniques inevitably produces a community of practitioners. This community, or a more specialized subset of the community, may in time attempt to create a profession—a group of people who profess to hold a body of specialized knowledge that enables them to treat a certain class of problems or phenomena. Although a broad class of practitioners may initially claim to have relevant competence, eventually this group is narrowed down, is institutionalized, sets standards to determine entrance into the group, and acquires professional autonomy.[34] When it achieves that autonomy, it may attempt to extend its domain, as well as prevent others from encroaching on its territory.

This was essentially what happened with sanitary engineering, which originated in England as part of the broader public health and sanitation movements. Its basic concern had been cleansing the environment by means of technology. British civil engineers Baldwin Latham and J. Bailey Denton published the first books with the words "sanitary engineering" in their titles in 1873 and 1877, respectively. In 1881, civil engineer E. S. Philbrick published the first American text with "sanitary engineering" in the title, although the Latham book had been reprinted in Chicago in 1877 as a special supplement to *Engineering News.* At the outset, sanitary engineering was a loosely organized field that included plumbers, plumbing contractors, and sanitarians, as well as engineers. The first journal in the field, founded in 1877, was entitled *The Plumber and Sanitary Engineer;* its name was changed to *The Sanitary Engineer* in 1880. The amorphous nature of sanitary engineering in its early days was noted in 1880 by an English writer: "There is no lack of wisdom in the sanitary world now, for a host of 'sanitary engineers' have sprung up . . . at a moment's notice. It is true they have been following other professions all their life; but a 'fresh door is open here,' and 'right about face!' is the order of the day, which they gladly obey, and turn in to 'fresh fields and pastures new.'"[35]

During the 1880s, sanitary engineering as a field began to define itself more explicitly in a scientific and disciplinary sense. A critical step in the profession's evolution was the formation in 1887 of an institution already noted, the Lawrence Experiment Station, by the Massachusetts State Board of Health. Though under the guidance of civil engineer Hiram Mills, the Lawrence Experiment Station united experts in the areas of engineering, chemistry, and biology and also brought the new knowledge of bacteriology to bear on the problems of water purification and sewage treatment. Researchers at the station made important discoveries in sewage treatment (intermittent filtration) and water filtration (slow sand filtration) in the 1890s, and the station served as a training ground for many of the great names in sanitary engineering and public health of the next thirty years, including Allen Hazen, George W. Fuller, E. O Jordan, Thomas M. Drown, and Ellen H. Richards.[36]

The consulting biologist to the Lawrence Station, and the director of many of its most significant investigations, was William T. Sedgwick, head of the Department of Biology at the Massachusetts Institute of Technology. Sedgwick oriented his department toward sanitation and public health and introduced courses in the 1880s in germs and germicides, sanitary biology, water supply and drainage, and hygiene and the public health. In 1889 he offered his first course in sanitary engineering.[37]

For Sedgwick, however, and for other pioneers in the field, sanitary engineering was more than another engineering discipline—it was a combination of engineering and science that utilized the contributions of each to solve public health problems. Speaking before the Franklin Institute in 1895, William Paul Gerhard, a sanitary engineer with broad interests, articulated this conception of a new breed of professional. Gerhard observed that sanitary engineering was "the art and science of applying the forces of nature in the planning and construction of works pertaining to public or individual health." Sanitary engineers needed to add a knowledge of a wide range of physical and natural sciences to their training in civil engineering. Those who worked in a single branch of sanitation, such as plumbing, said Gerhard, were not entitled to be called sanitary engineers, since the one was a trade and the other a profession. Neither, he added, could physicians engaged in preventive medicine be called sanitary engineers, for they lacked the technical experience and training required to qualify them for executing engineering works.[38]

Gerhard's comments reflected the struggle of a newly emerging profession to free itself from the craft or shop image and to establish its distinctiveness from another emerging profession, public health medicine. Within the universities, the movement toward professionalization proceeded. In 1911, Harvard appointed George C. Whipple, a former student of Sedgwick's at MIT, as professor of sanitary engineering. In 1912, Harvard and MIT joined together to create a school for health officers that united the Division of Sanitary Engineering at Harvard, the Department of Preventative Medicine at the Harvard Medical School, and the MIT Department of Biology and Public Health; the school was directed by William T. Sedgwick.[39]

George C. Whipple's attempt in 1912 to define the field reflected the continued quest of sanitary engineering for disciplinary and professional status. Whipple noted that the new profession represented "the application of a new science to a new product of civilization. The new science is bacteriology; the new product of civilization is 'The Modern City.'" Whipple called for the broad training of sanitary engineers and observed that engineering was "fast coming to be regarded as one of the learned professions, and the education required of one who enters this profession must be not only scientific and technical, but broad and humanitarian." However, he added, sanitary engineers must be trained primarily in engineering and secondarily in sanitation. "The wisely arranged curriculum will . . . cover the field of sanitation in a broad and general way, without taking too much of the student's time from his detailed studies in structures and hydraulics." Such training would enable the sanitary engineer to interact with those educated in other disciplines in order to improve the total field of sanitation.[40]

Concurrent with the institutionalization of sanitary engineering in the universities, the first decades of the twentieth century witnessed the establishment of institutional bases outside. In 1906, the American Society of Sanitary Engineers was founded, although it allied itself to the craft as well as to the engineering tradition by accepting plumbers, plumbing officials, and plumbing contractors for membership. The American Public Health Association formed a sanitary engineering section in 1911, while at approximately the same time the U.S. Public Health Service organized its own staff of engineers and the title "sanitary engineer" was given a civil service classification. In 1920, the Conference of State Sanitary Engineers was formed, composed of the chief sanitary engineer in each of the state departments of health and the chief sanitary engineer in the U.S. Public Health Service. And in 1922 the American Society of Civil Engineers (ASCE) authorized the formation of the Sanitary Engineering Division. In its first year of existence, the division registered a membership of 526; within a decade this figure had tripled. By the 1920s, approximately fifty years after it first emerged as a profession, sanitary engineering had firmly established itself as an engineering discipline.[41]

Leaders in the field, however, had expected that sanitary engineering would become more than an engineering specialty with a focus on sewage, water supply, and refuse collection. In 1924, for example, following in the tradition of Sedgwick and Whipple, sanitary engineer Abel Wolman, then chief engineer of the Maryland State Department of Health, proposed to the American Public Health Association (of which he later became president) that it rename sanitary engineering "public health engineering"; he also called for engineers to move into new areas of environmental control. This attempt to broaden the field was essentially unsuccessful. In 1948, in the second edition of his work *Public Health Engineering,* Earle B. Phelps, professor of sanitary science at Columbia College of Physicians and Surgeons, laid the responsibility for provincialism directly on the sanitary engineers: "Through his textbooks and his professional activities [the sanitary engineer] has defined and limited his field, not as the engineering of sanitary science, but as the engineering of water supply and . . . sewage disposal." Thus, through the immediate postwar period, most engineers were still narrowly linked to the two technologies that had originally led to the growth of sanitary engineering in the nineteenth century—sewerage and water supply.[42]

Governmental Structure and Administration

Sewerage systems have characteristics of economies of continuous collection and of scale that often ignore municipal boundaries and require centralized administration. The same irrelevance of political boundaries is naturally true of the health hazards created by waste disposal and water pollution. Ideally, then, wastewater collection and disposal should be dealt with on a regional basis, and yet many American urbanized areas were (and are) characterized by political fragmentation. In order to secure cost and design advantages as well as efficiency and safety of disposal, sanitary engineers and public health officials pushed for unification of these fragmented districts. As early as the 1870s, the engineering press began urging regional cooperation in sewer and water services, and throughout the late nineteenth and early twentieth centuries it pushed for new regional administrative arrangements. The requirements of sewerage systems for efficient operation offered a powerful argument for overcoming the fragmentation produced by political boundaries that did not conform to environmental needs.[43]

Three means of achieving this unity were actually employed: intermunicipal and interstate cooperation, annexation of or consolidation with suburban areas by a central city, and special district governments. The chief example of intermunicipal cooperation was the Passaic Valley Sewerage and Drainage Commission, formed in 1896, in which seven northern New Jersey municipalities united in construction of a joint outlet sewer; by 1927 seventeen towns were members. There were, however, few other examples of joint action among municipalities because of the difficulty in obtaining agreement on apportionment of responsibilities and costs. An inability to agree on joint responsibilities also inhibited the development of interstate water pollution compacts before 1920, although in the second quarter of the century such pacts were instituted in important regions such as the Ohio River Valley.[44]

A more common method of solving sewerage problems involving several governmental jurisdictions entailed suburban annexation or consolidation. Sewerage and water supply are costly capital systems, and there was a financial incentive for suburban communities with weak tax bases to consolidate with central cities that could supply these services. As a further inducement, the annexed territories often received services at the regular city rate, even though the costs of installation exceeded revenues. Thus, as Jon Teaford observes, the desire for improved service was a "countervailing force for unity" against the forces of fragmentation in the metropolis.[45]

The most readily adopted institutional means to handle sewerage and water-supply projects has been the special district government, a fiscally and administratively independent authority of limited function and extent. Examples include the Chicago Sanitary District (1889), the Boston Metropolitan Sewerage Commission (1889), and the Washington Suburban Sanitary Commission (1918). Sanitary engineers and public health professionals, as well as governmental reformers, pushed for the creation of the special authorities because of the need for a functional structure independent of political boundaries and because of the wish to escape tax or debt limits and be free of municipal political control. Special district governments were an alternative to central city annexation, and they were preferred by suburban authorities for this reason.[46]

Sewerage technology, therefore, with its characteristics of efficiency of continuous collection, scale economies of treatment, and capital intensiveness, as well as its requisite for central administration, was an important factor in facilitating governmental integration. It encouraged consolidation of urban areas and promoted a new governmental form, the special district government. It has also been argued, however, that specialized districts have actually retarded full metropolitan integration.[47] Such institutional innovations may have evolved without wastewater technology, but its requirements undoubtedly accelerated institutional adaptation that provided a model for further innovation.

Planning and Forecasting

When sewerage systems were constructed, engineers had to take into account future urban population growth and changes in city functions in order to avoid constant rebuilding. This required long-range planning and forecasts of population growth. The new technology further required a permanent bureaucracy for day-to-day administration, for data collection, and for efficient planning. In addition, the massive costs of these public works demanded fiscal planning by professionals. Ideally, the works could be constructed and maintained best by experts who could survey the topography scientifically, evaluate alternate materials, plan for probable population change, and keep the system efficiently functioning.

Before the city-planning movement was well begun, major expositions on sanitary reform argued the virtues of planned sewerage. By the 1870s and 1880s, this form of planning was firmly rooted as a major urban art, and sanitary engineering was established as an important branch of civil engineering concerned with sewerage and water supply. The sanitary engineers' view of planning was restricted to sewage collection and disposal and to water supply, but inherent in this view was the concept of the city as a physical container to be organized to provide more efficient delivery of services and disposal of wastes. The engineering ideal was the comprehensively planned city, staffed and managed by disinterested experts such as themselves. Not surprisingly, sanitary engineers played important roles in the emergence of the city-planning profession at the start of the twentieth century, with thirteen of the fifty-two charter members of the American Institute of Planners listed as engineers.[48]

Planning involves making predictions and dealing with probabilities and uncertainties, and sanitary engineers, even more than other engineers, had to be planners. Writing in 1915 and 1916, sanitary engineers Allen

Hazen and George C. Whipple noted the special concern with variation that characterized sanitary engineering and which placed it between the natural and the exact sciences. They also pointed out the importance of probability theory to sanitary engineering problems. Sanitary engineers were concerned with rainfall prediction, water use, and demographic change, all categories that shared the characteristic of high uncertainty. Unique among urban professionals, they attempted long-range urban population predictions, making such calculations considerably before city planners utilized the technique.[49]

The experience of sanitary engineers with population forecasting, social factors such as water use, and public health considerations, as well as their participation in planning large-scale capital works, prepared them for key roles in city government. They adhered to a set of values and procedures that stressed efficiency within a benefit-cost framework, and this appealed to late-nineteenth- and early-twentieth-century reformers attempting to restructure municipal government along lines of professionalism, efficiency, and bureaucratization. Sanitary engineers served in city government not only as municipal engineers but also as administrators and were a principal group from which the majority of city managers were recruited before World War II.[50]

Values and the Technology of Sewerage

This section will consider the relationship between sewerage technology and broader social values. Following Talcott Parsons, we define value as an element of a "shared symbolic system" that serves as a criterion for selection among alternatives. The material is structured to correspond to the model suggested by Theodore J. Gordon in his essay "The Feedback between Technology and Values."[51] This model depicts the mechanism relating values to technology and technology to values as a feedback system in which values help determine the direction of technological development. Once established, the technology, through its effects, influences the formation of new values. Five areas of involvement will be considered: (1.) state of society values leading to sewerage implementation; (2.) technology-specific values encouraging the adoption of sewerage systems; (3.) the effects of sewerage technology on values; (4.) policy adviser and practitioner values; and (5.) user values.

The state of society values that are pertinent to this study are those related to the development of nineteenth-century technology generally and, therefore, to sewerage technology specifically. A belief in progress was widespread in nineteenth-century America, and increasingly throughout the century progress was linked with technology. Modernization and progress were considered synonymous, with the belief that the present was better than the past and that the future would represent an even greater improvement. Sewerage technology was important in the context of urban improvement and boosterism and was strongly linked to city progress by a number of commentators. Sanitary engineer M. N. Baker noted, for instance, that "a village or town without waterworks and sewers is at great disadvantage as compared with communities having these conveniences and safeguards. Industries and population are not so quickly attracted to it." Or, as the engineer for New London, Connecticut, observed, with the building of a sewerage system the "good name" of the city appreciated, "thus attracting population and business, thereby increasing the value of real estate."[52]

Another important value held by most nineteenth-century Americans about technology was a belief in its beneficent nature. As one student of technology and ideas has noted, by the 1840s "the machine"—in this case steamboats, locomotives, and telegraphs—had captured the public imagination and gave evidence that mankind would "realize the dream of abundance."[53] Sewerage technology had strong implications of beneficence because of its associations with health improvements and nuisance abatement. By the late nineteenth century, public health physicians and sanitary engineers, as well as many municipal officials and citizens, viewed sewerage as a life-protecting and life-extending technology.

Americans believed not only that technology was beneficial but also that nature and its resources were meant to be exploited for material benefit. This value originally derived from the frontier experience and was tied to the concept of progress, playing a central role in its formulation.[54] As a value, it was associated with sewerage technology by the assumption that waterways could be utilized almost without restriction for waste disposal. The so-called scientific justification was that running water purified itself. As it turned out, this hypothesis was true only in a limited sense and under special conditions. Before that became clear, however, it seemed that urban progress and a healthful environment would be achieved through the utilization of the natural resource at the city's edge for waste disposal.

Turning to technology-specific values, we can see that a set of values deriving from the sanitary movement had a specific causal reference to the widespread construction of sewerage systems in American cities in the late nineteenth and early twentieth centuries. The sanitary movement began in Great Britain in the 1840s and 1850s with the work of Sir Edwin Chadwick and his followers to promote a healthful urban environment by cleansing the cities. It was essentially a social movement by elites and professionals that aimed to change people's ideas about their own personal habits of cleanliness, to create an enlarged role for government in areas related to health and sanitation, and to promote the construction of urban public works to achieve a healthful city. Chadwick's ideas greatly influenced the pioneer group of American sanitarians and public health reformers. Important among this group were men such as John H. Griscom and Colonel George E. Waring, Jr.; Griscom promoted sewerage as "not only the most economical, but the *only* mode in which the immense amounts of filth generated daily in . . . [large cities] can be effectively removed."[55]

The American sanitary movement had its origins in the pre–Civil War period, and its concerns were reflected in *The Report of the Sanitary Commission of Massachusetts* (the "Shattuck Report") of 1850, in the surveys of the American Medical Association of urban conditions in the late 1840s, and in the work of the Sanitary Commission during the Civil War. After the war, sanitarians and physicians actively propagated the concept of health through cleanliness. Books on personal hygiene, home sanitation, public health, and sewerage flowed from the presses, and articles on sanitation filled the technical and popular magazines. Leypoldt's *American Catalogue of 1875,* for instance, lists only five works on public health ("hygiene") as printed between 1850 and 1865, but fifty-four published between 1865 and 1875. In addition, these years saw the publication of nine works on sewerage, two on drainage, and six on water supply, whereas there had been none listed for the earlier period.[56]

The institutional and organizational embodiments of the sanitary movement were the American Public Health Association (1871), the National Board of Health (1879–83), and the multitude of local and state boards of health that appeared in the late nineteenth century. Municipalities and states passed laws regulating a range of activities relating to sanitation and health, such as cesspool and privy construction and cleaning, sewerage construction, plumbing, and water supply. These laws clearly reflected the acceptance by the informed public of the importance of sanitation. Thus, the sanitary movement helped initiate a value change, convincing many urbanites that filth was not a nuisance to be tolerated but rather a hazard to their health that could be eliminated. And, in this process, sewerage, which one public health historian has called "the most popular sanitary topic of the day" from 1875 to 1895, was a critical element.[57]

As to the effects of the technology on values, we see confirmation of the model of the interaction of values and technology: societal values first encourage the development of the technology, while feedback from the technology to the society reinforces the original values and may also create new ones. Some of these values have already been dealt with in the previous section on governmental and institutional impacts and will only be summarized here. Among the beliefs advanced by the technology were the virtues of planning; the requirement for engineering expertise rather than amateur, popular, or political direction; the desirability of bureaucracy and centralization; the applicability of engineering management to city government; and the need for state and federal regulation to deal with the negative effects of sewerage technology. In addition, the technology, by aiding in the creation of a sanitary and nuisance-free local environment, reaffirmed the American belief in the efficacy of technology to secure desired goals. While water-carriage technology did produce severe health externalities for downstream communities, another technology—water filtration—effectively reduced the risk, further reaffirming the utility of the cycle of technological fixes. Sewerage technology also reinforced the equation between cleanliness and health, with its most concrete embodiment being the modern sanitary bathroom full of devices reflecting the strength of the belief.[58]

As policy advisers, sanitary engineers shared with other professional engineers a set of values that stemmed from their training. They viewed society in problem-solving terms and emphasized efficiency, expertise, and technical solutions. In regard to water use and sanitation, they stressed quantifiable variables rather than nonquantifiable values and other hard-to-measure elements. They differed from other engineers, however, in the broader and interdisciplinary nature of their training, especially in health-related areas. In addition, they frequently worked with public rather than private bodies, serving as city engineers, as engineering representatives on boards of health, and as consultants. As public employees or consultants, they were sensitized to the cost constraints imposed by limited municipal resources and constitutional budgetary restrictions. Their interaction with the political process in city councils and state legislatures often caused them to be skeptical of politics and suspicious of popular causes, although this attitude was also shaped by their professionalism and shared with other engineers.[59]

By the beginning of the twentieth century, sanitary engineers and physicians in public health departments had become the principal professionals involved in decisions about sewerage, water supply, and control of water pollution. (A few cities and states employed bacteriologists and chemists.) In cities and states with an active program to protect the public health, these professionals worked together and shared a number of values in regard to sanitation and health. Both, for instance, largely accepted the germ theory of disease diagnosis and both advocated extensive state measures in disease control. During the second decade of the twentieth century, however, they diverged in their attitude toward risk in regard to the protection of drinking-water quality.[60]

The latter dispute has already been discussed in a policy context and will only be reviewed here in regard to the values component. As noted earlier, followers of the New Public Health argued that utilizing streams for sewage disposal created the risk of exposing populations to typhoid fever and other waterborne disease. The "foremost duty of health of officers," observed Charles V. Chapin, was "the direct control of communicable diseases." This perspective derived from their training in the new methods of bacterial science, from their efforts to distinguish themselves from earlier environmental sanitarians, and from their attempts to capture "the center of action and the criteria for professional identity within the public health movement."[61]

Sanitary engineers, on the other hand, while they agreed on the worth of bacterial science, had a different attitude on the question of water-quality standards: they believed that the risks involved in using streams for sewage disposal were not sufficient to justify the costs of construction of sewage-treatment plants. They argued for fully utilizing the natural dilution power of waterways, for adopting water filtration technology to protect drinking-water quality, and for reserving municipal funds for purposes other than the construction of sewage-treatment plants unless there was a severe nuisance. This position derived not only from their professional training but also from their close relationship with municipalities, which heightened their sensitivity to fiscal limits. In addition, since the sanitary engineering profession was struggling to gain position and prominence within the field of public health, value considerations involving risk were reinforced by considerations of professional prominence. As *Engineering News* boasted in a 1912 editorial, the sanitary engineer was prudent about both health and dollars—he was "a true and the greatest of conservationists, zealous to safeguard health and prolong life, but sparing no pain to see that each dollar is spent to the best advantage."[62]

Finally, user values—the values of municipalities, industries, and the general public. Water users benefited from sewerage technology and also helped create negative externalities from whose consequences they suffered. The extent of benefit or cost depended largely on upstream or downstream location. The most critical values in regard to municipalities and industries seemed to relate to their attitude toward the externalities they created for downstream communities through their waste-disposal practices. Municipalities were likely to construct sewers because they improved local aesthetic and health conditions, increased property values, and helped the city's image. Industries benefited from the ease of disposing their process wastes in sewers. Both, however, resisted the construction of waste-treatment facilities unless necessary to help protect their own water supplies, because these created advantages for downstream populations that paid nothing in return.[63]

As for the articulate public, it had three important values in regard to sewerage technology and water usage. These values favored local conditions of cleanliness and nuisance elimination; drinking water that was clear, free from odor, and potable; and unpolluted waterways for recreational purposes. The latter value gained strength after the turn of the century, when, with growing affluence and a reduced workweek, opportunities for leisure-time recreation increased for members of the middle and working classes. Through the 1920s, however, in the rating system of many sanitary engineers and state agencies concerned with water use, recreation held a lower priority than did other stream uses such as sewage dilution. As the New York Conservation Commission noted in 1923, recreational and industrial uses of streams were incompatible, and recreation "usually causes pollution of the water such as to unfit it for drinking." (Sanitary engineers believed that industrial wastes caused nuisances and impaired the workings of sewage-treatment plants but usually did not result in health problems. In fact, it was argued that wastes such as mine acid drainage neutralized bacterial wastes.) Man could live without recreation, added the commission, but "could not and should not live without work; so, in general, recreational purposes are subordinate to the other uses of streams."[64]

Conclusion

The experience of sewerage technology suggests certain conclusions about both technology and technological systems. The negative effects of new water-supply technology and the unexpected adoption of the water closet illustrate the risk of introducing elements into a balanced technological system without attempting to calculate the impacts of the innovation. Clearly, items that appear to promise only benefits may have severe secondary costs, although these may be difficult to foresee. To anticipate and understand the possible consequences of new technology, it is critical not only to comprehend fully the scope of the innovation but also to understand the operations and interrelationships of the previous system.

When the cesspool–privy vault system was overwhelmed by wastewater from newly installed water closets and other water fixtures, the system itself was replaced by capital-intensive sewers. This decision was based on forecasts of the benefits and costs of the new technology. In the process, forecasts of pollution costs tended to be disregarded because most often they would have to be borne by others, downstream. Municipalities were quite willing to shift the burden of pollution if it meant improving their local environment.

Policy choices in regard to sewer design—specifically the question of separate or combined sewers—and choices involving sewage disposal illustrate the manner in which faulty scientific concepts can effect technology implementation and operation. For instance, advocates of separate sewers in the 1870s and 1880s urged adoption of their system because it supposedly prevented the generation of disease-producing sewer gas. Large cities that built combined systems often discharged their sewage into adjacent waterways on the assumption that the streams were self-purifying, a concept that obtained only under very specific and limited conditions. Thus, in each case, assumptions about disease etiology that later proved faulty encouraged the installation of capital systems that required retrofitting or reconstruction in order to deal with resulting problems.

When municipal sewage-disposal practices had negative health and nuisance effects, especially for downstream communities, public health and engineering groups demanded state regulation of water quality. Legislation was usually enacted in response to crisis situations. The conflict that developed between sanitary engineers and municipalities and state boards of health dominated by physicians over methods of maintaining drinking-water quality illustrates how professionals with varying types of training and perspectives took different attitudes on health risks. The resulting policy reflected the more limited and cost-effective sanitary engineering position rather than the longer-term and more costly public health approach. Water-quality policy, therefore, was shaped by the value perspectives of involved professionals and bounded by the financial limitations of municipalities and an in-place capital technology.

The various characteristics and effects of capital-intensive sewerage technology required governmental and institutional adaptations. Three areas were most important: measures needed to overcome the political fragmentation of urban areas, such as suburban annexation and special district governments; a strengthening of the planning and managerial components of city government; and the development of regulatory bodies on the state, regional, and federal levels to deal with negative consequences. In regard to regulation, there has been a progression to higher and higher governmental levels. Since 1972, the regulatory and standard-setting focus has been at the federal level, and federal grants and subsidies have been required to enable states and municipalities to meet federally mandated standards.

In regard to technology and values, this article has utilized a model emphasizing a two-way feedback. That is, it argues that the nineteenth-century sanitation movement was an important "value change initiator" that facilitated the acceptance of sewerage technology by urban decision makers and voters. By propagating the filth theory of disease, and by convincing urbanites that sewerage technology was a means to improve health as well as eliminate nuisance, it made taxpayers more willing to accept the financial costs of the capital-intensive system. Value change, therefore, was a critical predecessor to technology implementation.

The model also posits that the developing technology itself shaped and reinforced other values: values such as a belief in the need for planning, expertise, bureaucracy, and centralization in government, as well as for an expanded state regulatory role, were all supported by the technology. The profession of sanitary engineering developed around sewerage and water-supply technologies. As a group, sanitary engineers shared with other engineers a belief in technological solutions to problems, in efficiency as a concept, and in the primacy of cost considerations in construction. They differed from many engineers in that they often had public rather than private employers and clients, and they had a professional concern with public health. In its initial decades, the founders of the sanitary engineering profession shared a vision of their discipline as involving more than engineering perspectives. Much of this orientation was lost after World War I, and through the 1950s sanitary engineering remained narrowly wedded to the areas of sewer construction and sewage disposal, water-supply engineering, and municipal refuse collection. Recent decades, however, have witnessed a return of the profession to its early broader and interdisciplinary roots, as symbolized by the replacement of "sanitary engineering" by "environmental engineering."[65]

Ultimately, water-carriage technology, complete with various retrofits, may have been the most efficient and cost-beneficial system the society could have devised to deal with its human-waste removal problems. The system was retrofitted in an incremental manner, often following health crises, and only after considerable damage had been done both to the public health and to the environment. Since its introduction and development, the existence of this capital-intensive system, regulated and governed by a group of special institutions and maintained by a specialized professional group, has been accepted as an unchangeable element in urban America. As a result, little attempt has been made until recently to search for waste-removal and disposal alternatives.[66] The society, therefore, continues to struggle with the problems of a waste-removal technology based on concepts over a century old.

Notes

1. Joel A. Tarr, Francis C. McMichael, James McCurley III, and Terry Yosie, *Retrospective Assessment of Wastewater Technology in the United States, 1800–1972* (Pittsburgh, 1977). The literature on sewerage technology and on related questions of urban technology and planning has grown rapidly in recent years. For two excellent guides, see Suellen M. Hoy and Michael C. Robinson, eds., *Public Works History in the United States: A Guide to the Literature* (Nashville, 1982); and Eugene P. Moehring, "Public Works and Urban History: Recent Trends and New Directions," *Essays in Public Works History* no. 13 (Chicago, 1982). The present article diverges from most articles about sewerage technology by focusing on the range of the technology's effects on society as well as on its development. The most informative articles dealing with the social and institutional effects of sewerage technology are Jon A. Peterson, "The Impact of Sanitary Reform upon American Urban Planning," *Journal of Social History* 13 (Fall 1979): 83–103; and Stanley K. Schultz and Clay McShane, "To Engineer the Metropolis: Sewers, Sanitation, and City Planning in Late-Nineteenth Century America," *Journal of American History* 65 (September 1978): 389–411. For a useful historical case study of special district government, see Louis P. Cain, "The Search for an Optimum Sanitation Jurisdiction: The Metropolitan Sanitary District of Greater Chicago, A Case Study," *Essays in Public Works History* no. 10 (Chicago, July 1980). The critical and definitive work relating sewerage and water pollution to developments in public health is Barbara Gutmann Rosenkrantz, *Public Health and the State: Changing Views in Massachusetts, 1842–1936* (Cambridge, 1972). The relationship between sewers and urbanization is treated in several studies, the most perceptive of which are Stuart Galishoff, "Drainage, Disease, Comfort, and Class: A History of Newark's Sewers," *Societas—A Review of Social History* 6 (Winter 1976): 121–38; Geoffrey Giglierano, "The City and the System: Developing a Municipal Service, 1800–1915," *Cincinnati Historical Society Bulletin* 35 (Winter 1977): 223–47; Roger D. Simon, "The City-Building Process: Housing and Services in New Milwaukee Neighborhoods, 1880–1910," in *Transactions of the American Philosophical Society* 68 (Philadelphia, 1978); and Eugene P. Moehring, *Public Works and the Patterns of Urban Real Estate Growth in Manhattan, 1835–1894* (New York, 1981). A study of one city's struggles to cope with its water supply and wastewater disposal problems is Louis P. Cain, *Sanitation Strategy for a Lakefront Metropolis: The Case of Chicago* (De Kalb, Ill., 1978).

2. For technology assessment, see the articles gathered in Albert H. Teich, ed., *Technology and Man's Future,* 2d ed. (New York: St. Martin's Press, 1977), 229–375; and Edward W. Lawless, *Technology and Social Shock* (New Brunswick, N.J.: Rutgers University Press, 1977), 594–604. For retrospective technology assessment, see Joel A. Tarr, ed., *Retrospective Technology Assessment—1976* (San Francisco: San Francisco Press, 1977); Howard P. Segal, "Assessing Retrospective Technology Assessment: A Review of the Literature," *Technology in Society* 4 (1982): 231–46.

3. Nelson M. Blake, *Water for the Cities* (Syracuse, N.Y., 1956), 12–13; Constance M. Green,

Washington: Village and Capital, 1800–1878 (Princeton, N.J., 1962), 95. The estimates of water usage are based on figures reported for cities without waterworks in John D. Bell, "Report on the Importance and Economy of Sanitary Measures to Cities," *Proceedings and Debates of the Third National Quarantine and Sanitary Convention* (New York, 1859), 576–77.

4. Jon Peterson calls the cesspool–privy vault system the "private-lot waste removal" system ("Impact of Sanitary Reform," 85); B. A. Segur, "Privy-Vaults and Cesspools," *Papers and Reports of the American Public Health Association* (hereafter cited as *APHA*) 3 (1876): 185–87; Mansfield Merriman, *Elements of Sanitary Engineering,* 3d ed. (New York, 1906), 139–42; and Moehring, *Public Works and the Patterns of Urban Real Estate Growth in Manhattan,* 15. There is information on the "municipal cleansing" practices of over one hundred cities in U.S. Department of the Interior, Census Office, *Tenth Census of the United States, 1880, Report of the Social Statistics of Cities,* comp. George E. Waring, Jr., 2 vols. (Washington, D.C., 1887) (hereafter cited as *Social Statistics of Cities, Tenth Census*), and in "Report of Committee and Disposal of Waste and Garbage," *APHA* 17 (1891): 90–119; Joel A. Tarr, "From City to Farm: Urban Wastes and the American Farmer," *Agricultural History* 49 (October 1975): 601–2. In 1880 the wastes of 103 of the 222 U.S. cities listed in the *Social Statistics of Cities, Tenth Census* were used on the land.

5. See regulations cited in *Social Statistics of Cities, Tenth Census;* see also descriptions of sewers in Julius W. Adams, *Report of the Engineers to the Commissioners of Drainage* (Brooklyn, 1857); Henry I. Bowditch, *Public Hygiene in America* (Boston, 1877), 103–9; and Leonard Metcalf and Harrison P. Eddy, *American Sewerage Practice,* 3 vols. (New York, 1914–15), 1:15–19. For private sewers, see Peterson, "Impact of Sanitary Reform," 85; Giglierano, 223–24.

6. U.S. Department of Commerce, Bureau of the Census, *Historical Statistics of the United States . . . to 1970,* 2 vols. (Washington, D.C.: GPO 1975), 1:11–12; George E. Waring, Jr., "The Sanitary Drainage of Houses and Towns, II," *Atlantic Monthly* 36 (October 1875): 427–42, esp. 434; and Clay McShane, "Transforming the Use of Urban Space: A Look at the Revolution in Street Pavements, 1880–1924," *Journal of Urban History* 5 (May 1979): 288.

7. Blake, *Water for the Cities,* 3–17; "Community Water Supply," in *History of Public Works in the United States 1776–1976,* ed. Ellis L. Armstrong, Michael C. Robinson, and Suellen M. Hoy (Chicago, 1976), 217–35; and Moehring, *Public Works and the Patterns of Urban Real Estate Growth in Manhattan,* 23–51.

8. J. T. Fanning, *A Practical Treatise on Hydraulic and Water-Supply Engineering* (New York, 1886), 625.

9. Town of Pawtucket, Committee on Sewers, *Report, 1885* (Pawtucket, R.I., 1885), 15; E. S. Chesbrough, "The Drainage and Sewerage of Chicago," *APHA* 4 (1878): 18–19; and Town Improvement Society of East Orange, *The Sewerage of East Orange* (East Orange, N.J., 1884).

10. May N. Stone, "The Plumbing Paradox: American Attitudes towards Late Nineteenth-Century Domestic Sanitary Arrangements," *Winterthur Portfolio* 14 (1979): 284–85. See Reginald Reynolds, *Cleanliness and Godliness* (New York, 1974), for an amusing description of the evolution of the water closet; see also Lawrence Wright, *Clean and Decent: The Fascinating History of the Bathroom and the Water Closet* (Toronto, 1972); Boston, Cochituate Water Board, *Annual Report for 1863* (Boston, 1864), 43; and City of Buffalo, *Sixth Annual Report of the City Water Works, 1874* (Buffalo, 1875), 47. The estimate for 1880 is based on information in *Social Statistics of Cities, Tenth Census.*

11. William H. Bent, George H. Rhodes, and William Tinkham, *Report of Special Committee on Sewerage for City of Taunton* (Taunton, Mass., 1878), 25–26; E. S. Chesbrough, *Report on Plan of Sewerage for the City of Newport* (Newport, R.I., 1880), 5–6; Rudolph Hering, *Report on a System of Sewerage for the City of Wilmington, Delaware* (Wilmington, Del., 1883), 5–6; and Maryland State Board of Health, "The Sanitation of Cities and Towns and the Agricultural Utilization of Excremental Matter," *Annual Report, 1887* (Baltimore, 1887), 229–30.

12. Bell, "Report on Importance and Economy of Sanitary Measures," 479–575; Charles E. Rosenberg, *The Cholera Years* (Chicago, 1962), 75–81, 117, 202; Charles V. Chapin, "The End of the Filth Theory of Disease," *Popular Science Monthly* 60 (January 1902): 234–39; George E. Waring, Jr., "The Sanitary Drainage of Houses and Towns," *Atlantic Monthly* 36 (November 1875): 535–51;

and Benjamin Lee, "The Cart before the Horse," *APHA* 20 (1895): 34–36. Lee was concerned with the bacterial danger presented by the fecal pollution.

13. Azel Ames, "The Removal and Utilization of Domestic Excreta," *APHA* 4 (1877): 65–70. *Social Statistics of Cities, Tenth Census* listed eleven cities using the odorless excavator in 1880, but this is probably an underestimate.

14. Peterson, "Impact of Sanitary Reform," 84; Francis Sheppard, *London 1808–1870: The Infernal Wen* (Berkeley, 1971), 250–78.

15. Peterson, "Impact of Sanitary Reform," 86–87; Probably the most influential report about European sewage systems by an American engineer was Rudolph Hering, "Report of the Results of an Examination Made in 1880 of Several Sewage Works in Europe," *Annual Report of the National Board of Health 1881* (Washington, 1882). For the background and influence of this report, see Joel A. Tarr, "The Separate vs. Combined Sewer Problem: A Case Study in Urban Technology Design Choice," *Journal of Urban History* 5 (May 1979): 308–33.

16. In some cities, such as Baltimore and Wilmington, opponents of water-carriage technology blocked construction for many years by focusing on costs, design problems, and possible pollution effects. In most cities, however, perceived advantages led to construction within a reasonable amount of time after the water-carriage system was proposed. For information on Wilmington, see Carol Hoffecker, "Water and Sewage Works in Wilmington, Delaware, 1810–1910," *Essays in Public Works History* 12 (Chicago, July 1981); for Baltimore, see Alan D. Anderson, *The Origin and Resolution of an Urban Crisis: Baltimore, 1890–1930* (Baltimore, 1977), 68–72. In Baltimore, also, the politically influential odorless excavator companies were able to delay construction of a sewerage system.

17. John Duffy, *A History of Public Health in New York City, 1625–1866* (New York, 1968), 415; Hering, *Report on a System of Sewerage for the City of Wilmington, Delaware*, 6; Joseph E. Nute, "The Sewerage of Malden" (B.S. thesis, MIT, 1884); Baltimore Sewerage Commission, *Second Report* (Baltimore, 1899), 30; George E. Waring, Jr., *Draining for Health and Draining for Profit* (New York, 1867), 222–23; J. S. Billings, "Sewage Disposal in Cities," *Harper's New Monthly Magazine* 71 (September 1885): 577–84, esp. 580; Bell, "Report on Importance and Economy of Sanitary Measures," 478–83; F. H. Hamilton, "A Plea for Sanitary Engineering," *APHA* 2 (1876): 368–73; Town of Marlborough Sewage Committee, *Report* (Marlborough, Mass., 1885), 7–8; Massachusetts Board of Health, "The Value of Health to the State," *Annual Report, 1875* (Boston, 1875), 57–75; idem, "Political Economy of Health," *Annual Report, 1874*, 335–90; Baldwin Latham, *Sanitary Engineering* (London, 1873), 10–14; Henry E. Sigerist, ed., "The Value of Health to a City: Two Lectures, Delivered in 1873, by Max Von Pettenkofer," *Bulletin of the History of Medicine* 10 (October 1941): 473–503, 593–613; Samuel M. Gray, *Proposed Plan for a Sewerage System for Providence* (Providence, R.I., 1884), 8–11; M. N. Baker, *Sewerage and Sewage Purification* (New York, 1896), 11; and New London Board of Sewer Commissioners, *First Annual Report* (New London, Conn., 1887), 4.

18. Tarr, "From City to Farm," 601–6; C. A. Leas, "A Report upon the Sanitary Care and Utilization of the Refuse of Cities," *APHA* 1 (1875): 456; "The Sewage Question," *Scientific American* 21 (July 24, 1899): 57; Estelle F. Feinstein, *Stamford in the Gilded Age* (Stamford, Conn., 1973), 169; Ernest S. Griffith, *The Conspicuous Failure: A History of American City Government 1870–1900* (New York, 1974), 20; New London Board of Sewer Commissioners, *First Annual Report*, 4; and City of Providence, *Report upon Sewer Assessments* (Providence, R.I., 1877), includes information on assessment practices in sixty-nine cities.

19. For typhoid death-rate figures see U.S. Department of the Interior, Census Office, *Report of the Mortality and Vital Statistics of the U.S., Tenth Census*, pt. 2 (Washington, D.C., 1886), xxvi; U.S. Department of Commerce, Bureau of the Census, *Morality Statistics 1910, Thirteenth Census* (Washington, D.C., 1913), 26–27. Hering, *Report on a System of Sewerage for the City of Wilmington*, 6; Nute, "Sewerage of Malden"; U.S. Department of Commerce, Bureau of the Census, *General Statistics for Cities: 1909* (Washington, D.C., 1913), 20–23; and Tarr, "From City to Farm," 611–12. The leading difficulty was transportation, especially as cities grew larger.

20. William T. Sedgwick, *Principles of Sanitary Science and the Public Health* (New York, 1918), 213, 231–237.

21. See the listing under "Sewers" in U.S. Department of the Interior, Bureau of the Census, *Report on the Social Statistics of Cities, Eleventh Census* (Washington, D.C., 1895), 29–32; and under "Sewers and Sewer Service," in *General Statistics of Cities: 1909* (Washington, D.C., 1913), 20–23. Harold E. Babbitt, *Sewerage and Sewage Treatment* (New York, 1932), 132–33; H. F. Peckworth, *Concrete Pipe Handbook* (Chicago, 1959).

22. Tarr, "Separate vs. Combined Sewer Problem," 308–29, 332. For an appreciation of Waring as an environmentalist, see Martin V. Melosi, *Pragmatic Environmentalist: Sanitary Engineer, Col. George E. Waring, Jr.*, Essays in Public Works History, no. 4 (Washington, April 1977). For a more critical appraisal of Waring, see James H. Cassedy, "The Flamboyant Colonel Waring: An Anti-Contagionist Holds the American Stage in the Age of Pasteur and Koch," *Bulletin of the History of Medicine* 36 (March–April 1962): 163–76.

23. "Sewage Purification and Storm and Ground Water," *Engineering News* 28 (August 25, 1892): 180–81; Sedgwick, *Principles of Sanitary Science*, 231–37; Rudolph Hering, "Notes on the Pollution of Streams," *APHA* 13 (1888): 272–79; and Moses N. Baker, "Sewerage and Sewage Disposal," Appendix A in Department of Commerce and Labor, Bureau of the Census, *Statistics of Cities Having a Population of over 30,000: 1905* (Washington, D.C. 1907), 103–6.

24. Rosenkrantz, *Public Health and the State*, 97–107; George C. Whipple, *Typhoid Fever* (New York, 1908).

25. Howard D. Kramer, "Agitation for Public Health Reform in the 1870s," *Journal of the History of Medicine* 3, 4 (Autumn 1948, Winter 1949): 473–88, esp. 474–76; 75–89; Barbara G. Rosenkrantz, "Cart before Horse: Theory, Practice and Professional Image in American Public Health, 1870-1920," *Journal of the History of Medicine and Allied Sciences* 29 (January 1974): 55–56; idem, *Public Health and the State*, 1–127; Stephen Smith, "The History of Public Health, 1871–1921," in Mazyck P. Ravenel, ed., *A Half Century of Public Health* (New York, 1921), 1–12; and George Rosen, *A History of Public Health* (New York, 1958), 233–50.

26. "Sewage Purification and Water Pollution in the United States," *Engineering News* 47 (April 3, 1902): 276; "Sewage Pollution of Water Supplies," *Engineering Record* 48 (August 1, 1903): 117. After 1910, the courts awarded damages against municipalities in cases where negligence in the operation of public waterworks resulted in individuals' contracting typhoid fever. See James A. Tobey, *Public Health Law*, 2d ed. (New York, 1939), 277–80; Edwin B. Goodell, *Review of Laws Forbidding Pollution of Inland Waters in the United States*, U.S. Geological Survey Water Supply and Irrigation Paper No. 152 (Washington, 1906). A useful summary of Goodell is "Pollution of Streams," *Municipal Journal and Engineer* 21 (October 1906): 333–34, 364–65, 384.

27. Stuart Galishoff, "Triumph and Failure: The American Response to the Urban Water Supply Problem, 1860–1923," in Martin V. Melosi, ed., *Pollution and Reform in American Cities, 1880–1930* (Austin, 1980), 46–47; Metcalf and Eddy, *American Sewage Practice*, 3d ed., 3:190–231; and "Sewage Purification and Storm and Ground Water," 180–81.

28. Allen Hazen, *Clean Water and How to Get It* (New York, 1907), 68–75; George C. Whipple, "Fifty Years of Water Purification," in Ravenel, ed., *A Half Century of Public Health*, 161–80. See George A. Johnson, "Present Day Water Filtration Practice," *Journal of the American Water Works Association* 1 (March 1914): 31–80, for figures on typhoid death rates for leading cities before and after filtration.

29. In some cities, there was heated conflict over whether to filter the water from local polluted rivers or to seek a pure source in a distant locality. For a discussion of this dispute in Boston, see Fern L. Nesson, *Great Waters: A History of Boston's Water Supply* (Hanover, N.H., 1983). This case is treated more fully in Joel A. Tarr, Terry Yosie, and James McCurley III, "Disputes over Water Quality Policy: Professional Cultures in Conflict, 1900–1917," *American Journal of Public Health* 70 (April 1980): 427–35. Sanitary engineers also wanted equal representation with physicians on state boards of health.

30. "Sewage Pollution of Water Supplies," 117. See also "The Water Supply of Large Cities," *Engineering Record* 41 (January 27, 1900): 73; Hazen, *Clean Water and How to Get It*, 34–37.

31. See for example, Pittsburgh Chamber of Commerce, *Sewage Disposal of Pittsburgh* (Pittsburgh, 1907); George Soper, "The Sanitary Engineering Problems of Water Supply and Sewage Disposal in New York City," *Science* 25 (April 19, 1907): 601–5; "Up Stream or Down Stream?" *New York Times*, September 25, 1910; Constance D. Leupp, "To the Rescue of New York Harbor," *Survey*, October 8, 1910, 89–93; Merchants Association of New York, Committee on Pollution of State Waters, *Protest against the Bronx River Valley Sewer* (New York, 1907); idem, *The Battle of the Microbes: Nature's Fight for Pure Water* (New York, 1908); and Samuel P. Hays, *Conservation and the Gospel of Efficiency* (Cambridge, 1959), esp. 122–46; H. W. Hill, "The New Public Health," *Engineering News* 67 (February 29, 1912): 378; idem, "The Relative Values of Different Public-Health Procedures, ibid. 66 (October 12, 1911): 436; Robert W. Bruere, "The New Meaning of Public Health," *Harper's Monthly Magazine* 124 (April 1912): 690–95; and "The Pollution of Streams," *Engineering Record* 60 (August 7, 1909): 157–59.

32. H. M. Bracken, "Sewage Pollution Made Compulsory by the Minnesota State Board of Health," *Engineering News* 51 (February 11, 1904): 138, and "Editorial," 129; R. Winthrop Pratt, "The Work of the Ohio State Board of Health on Water Supply and Sewage Purification," *Engineering News* 57 (June 20, 1907): 680; George Gregory, "A Study in Local Decision Making: Pittsburgh and Sewage Treatment," *Western Pennsylvania Historical Magazine* 57 (January 1974): 23–36; "Standards of Purity of Rivers and Waterways," *Engineering News* 26 (October 31, 1912): 835–36; and "Conference on Pollution of Lakes and Waterways," *Engineering Record* 66 (November 2, 1912): 485–86.

33. George W. Fuller, "Relations between Sewage Disposal and Water Supply Are Changing," *Engineering News Record* (April 5, 1917) 28: 11–12. See also George W. Fuller, "Is It Practicable to Discontinue the Emptying of Sewage into Streams?" *American City* (1912) 7: 43–45.

34. Everett C. Hughes, "Professions," *Daedalus* (Fall 1963) 92: 655–68; George H. Daniels, "The Process of Professionalization in American Science: The Emergent Period, 1820–1860," *Isis* (Summer 1967) 58: 151–66.

35. Henry C. Meyer, *The Story of the Sanitary Engineer* (New York, 1927), 2–7; Rosenkrantz, "Cart before Horse," 55–56. For an excellent discussion of the relationship of sanitary engineering to refuse collection, see Martin V. Melosi, *Garbage in the Cities: Refuse, Reform, and the Environment, 1880–1980* (College Station, Tex., 1981), 79–104. Quoted in Stone, "The Plumbing Paradox," 289.

36. Rosenkrantz, *Public Health and the State*, 99–100; E. O. Jordan, G. C. Whipple, and C. E. A. Winslow, *A Pioneer of Public Health: William Thompson Sedgwick* (New Haven, Conn., 1924), 57–60.

37. Jordan et al., 31–41, 57–64, 72.

38. William Paul Gerhard, "Sanitary Engineering," *Journal of the Franklin Institute* (June, July, August 1895) 139, 140: 457–75, 56–68, 90–105; William Paul Gerhard, "A Half-Century of Sanitation," *American Architect and Building News* (February 25, 1899) 63: 62.

39. Samuel C. Prescott, *When M.I.T. Was "Boston Tech," 1861–1916* (Cambridge, Mass., 1954), 281-83.

40. George C. Whipple, "The Training of Sanitary Engineers," *Engineering News* (October 31, 1912) 68: 805–6; "Sanitation More than Medicine," *Literary Digest* (November 16, 1912) 45: 899.

41. "American Society of Sanitary Engineering," and "Conference of State Sanitary Engineers," in *Encyclopedia of Associations*, 11th ed. (New York, 1977), 1:413; figures on the enrollment in the Sanitary Engineering Division of the ASCE were supplied by the ASCE. See also Frank Woodbury Jones, "The Sanitary Engineering Division of the American Society of Civil Engineers: Its History 1923–1952," a talk delivered before the Centennial of Engineering Meeting of the Sanitary Engineering Division, ASCE, Chicago, September 8, 1952.

42. Abel Wolman, "Values in the Control of the Environment," *American Journal of Public Health* (March 1925) 15: 194; George W. Fuller, "The Place of Sanitary Engineering in Public Health Activities," *American Journal of Public Health* (December 1925) 15: 1069; and Earle B. Phelps, *Public Health Engineering*, 2 vols. (New York, 1948), 1:5. See also three articles by Abel Wolman: "The Engineer and Society," *Johns Hopkins Alumni Magazine* (June 1937) 25: 343; "The

Public Health Engineer and the City Health Officer," *American Journal of Public Health* (May 1941) 31: 435–39; and "Sanitary Engineering Looks Forward," *Journal of the American Water Works Association* (November 1946) 38: 1219–25. For attempts to change both the image and content of sanitary engineering in the 1960s and 1970s, see James W. Patterson, "Environmental Engineering Education: Academia and an Evolving Profession," *Environmental Science and Technology* (May 1980) 14: 524–32; and Gerard A. Rohlich, "Environmental Engineering—A distinct Discipline?" in *Fourth Conference on Environmental Engineering Education, Toronto, June 19–21, 1980*, James W. Patterson and Roger A. Minear (eds.), 21–28.

43. Paul Studenski, *The Government of Metropolitan Areas in the United States* (New York, 1930), 18; Stanley K. Schultz and Clay McShane, "Pollution and Political Reform in Urban America: The Role of Municipal Engineers, 1840–1920," in Melosi, ed., 160–67.

44. "Municipal Cooperation as a Possible Substitute for Consolidation," *Engineering News* (February 16, 1899) 41: 104–6; "Sewerage of the Passaic River Valley," *Engineering Record* (December 28, 1901) 44: 60; and Studenski, 47–48. States such as Ohio and California passed legislation providing for intergovernmental contractual relations. See Jon C. Teaford, *City and Suburb: The Political Fragmentation of Metropolitan America, 1850–1970* (Baltimore, 1979), 81; Edward J. Cleary, *The ORSANCO Story: Water Quality Management in the Ohio Valley under an Interstate Compact* (Baltimore, 1967).

45. Teaford, 39–40, 59–60; Studenski, 166–67.

46. Robert B. Hawkins, Jr., *Self-Government by District: Myth or Reality* (Stanford, 1976), 25; Studenski, 256–62; Teaford, 79–81, 173–74; and Cain, *The Search for an Optimum Sanitation Jurisdiction*, 1–5.

47. For a discussion of this literature, see Cain, *The Search for an Optimum Sanitation Jurisdiction*, 31, n. 4.

48. Peterson, "Impact of Sanitary Reform," 89; Mel Scott, *American City Planning since 1890* (Berkeley: University of California Press, 1969), 163–64; and Nelson Lewis, *The Planning of the Modern City* (New York: J. Wiley, 1916), esp. chapter 21, "The Opportunities and Responsibilities of the Municipal Engineers." William Paul Gerhard was notable among sanitary engineers for directly addressing the question of city planning. See William Paul Gerhard, "The Laying Out of Cities and Towns" *Journal of the Franklin Institute* (August 1895) 140: 90–99.

49. George C. Whipple, "The Element of Chance in Sanitation," pts. 1, 2, *Journal of the Franklin Institute* (July, August 1916) 182: 37–59, 205–27. For examples of population forecasting by sanitary engineers, see Henry N. Ogden, *Sewer Design* (New York, 1899), 93–101, and George W. Rafter and M. N. Baker, *Sewage Disposal in the United States* (New York, 1894), 129–31. Most of the early population forecasts were based on straight-line extrapolation of past trends. They were often faulty. In 1895, e.g., civil engineer Frederic Stearns made 186 population forecasts for twenty-seven cities and towns within 10 miles of Boston. A later check on the accuracy of his predictions showed that he had underestimated growth in twenty-seven cases, overestimated in 156, and made three accurate predictions. Writing in 1928, sanitary engineers Leonard Metcalf and Harrison Eddy noted that "forecasts of population based upon experience of the past are likely to prove somewhat too high in about 85 percent of the cases and too low in the remainder" (2d ed., 1:191–92). For the Stearns estimates, see Paul M. Berthouex, "Some Historical Statistics Related to Future Standards," *Journal of the Environmental Engineering Division, Proceedings of the American Society of Civil Engineers* (April 1974) 100: 423–24.

50. Schultz and McShane, "To Engineer the Metropolis", 409–10; Martin J. Schiesl, *The Politics of Efficiency* (Berkeley, 1977), 171–88. For a discussion of the role of engineers in regard to city government in an earlier period, see Raymond H. Merritt, *Engineering in American Society 1850–1875* (Lexington, Ky., 1969), 136–76.

51. Theodore J. Gordon, "The Feedback between Technology and Values," in *Values and the Future: Impact of Technological Change on American Values*, ed. Kurt Baier and Nicholas Rescher (New York: Free Press, 1969), 148–92; see also Nicholas Rescher, "What Is Value Change? A Framework for Research," in Baier and Rescher, eds., 68–109.

52. M. N. Baker, *Sewerage and Sewage Purification*, 11. New London Board of Sewer Commissioners, *First Annual Report*, 4.

53. Leo Marx, *The Machine in the Garden* (New York: Oxford University Press, 1964), 191–92. Marx discusses intellectuals who question the worth of technological advance.

54. Arthur A. Ekirch, Jr., *Man and Nature in America* (New York: Columbia University Press, 1963), 45.

55. Griscom, quoted in Peterson, "Impact of Sanitary Reform," 86.

56. Kramer, "Agitation for Public Health Reform," 474–76.

57. Ibid., 473–88, 75–89.

58. Reynolds, *Cleanliness and Godliness*, passim.

59. Schultz and McShane, "Pollution and Political Reform in Urban America," 165; Edwin T. Layton, Jr., *The Revolt of the Engineers* (Cleveland: Press of Case Western Reserve University, 1971), 6–8, 64–65; and Charles W. Eliot, "One Remedy for Municipal Government," *Forum* 12: 153–68.

60. Many state and local health departments were headed and staffed by physicians without any particular public health training or orientation. According to sanitary engineer Morris Knowles, such departments were characterized by "narrowness of scope; incompleteness of work from the broader sanitary point of view; emphasis on cure rather than prevention; and insufficient realization of the importance of reliable statistics. . . ." Knowles and other sanitary engineers argued that engineers and trained public health professionals (not necessarily physicians) should be appointed to these boards. See Morris Knowles, "Public Health Service Not a Medical Monopoly," *American City* (December 1912) 7: 527–29.

61. Charles V. Chapin, "History of State and Municipal Control of Disease," in Ravenel, ed., *A Half Century of Public Health*, 136–37, 142; Rosenkrantz, "Cart before Horse," 68.

62. George C. Whipple, "How to Determine Relative Values in Sanitation," *American City* (1914) 11: 427–32; "A Plea for Common Sense in the State Control of Sewage Disposal," *Engineering News* (February 29, 1912) 67: 412–13.

63. As noted, in 1903 the *Engineering Record* suggested that "it is often more equitable to all concerned for an upper riparian city to discharge its sewage into a stream and a lower riparian city to filter the water of the same stream for a domestic supply, than for the former city to be forced to put in sewage treatment works" (*Engineering Record*, "Sewage Pollution of Water Supplies," 117).

64. Quoted in "Discussion on Policy regarding Stream Pollution," following Abel Wolman (chairman), "Domestic and Industrial Wastes in Relation to Public Water Supply: A Symposium," *American Journal of Public Health* (August 1926) 16: 103–7.

65. The Sanitary Engineering Division of the ASCE changed its name to the Environmental Engineering Division in 1972. A convenient summary of developments in the field, although lacking a historical perspective, is Patterson, "Environmental Engineering Education," 524–32.

66. A useful summary of innovative systems including a list of firms manufacturing new systems for human wastes is *Beyond the Pipe Dream: New and Different Ways to Treat Sewage* (Providence, R.I.: Save the Bay, Inc., n.d.). See also U.S. Environmental Protection Agency, *Alternatives for Small Wastewater Treatment*, 2 vols. (Washington, 1977).

PART III
Smoke Pollution

Air pollution—or atmospheric contamination—is caused by a number of factors, some natural, such as volcanic eruptions, but many anthroprogenically produced through processes of inefficient combustion, improperly trained firemakers, or the use of dirty fuel. As a concept, air pollution is relatively new, and before World War II smoke was viewed as the prime cause of atmospheric contamination. Smoke pollution was primarily an urban phenomenon, although there were cases of rural atmospheric contamination from processes such as charcoal burning, copper smelting, and coke-making. Among the leading causes of urban smoke were industries, commercial establishments, residences, railroads, and tug boats—in short, anything that utilized wood or coal as fuel. By the late nineteenth century, however, bituminous coal—the most widely utilized fuel in the United States—was the leading cause of smoke, having supplanted both wood and relatively clean-burning anthracite coal.

Severe smoke pollution persisted in American cities such as Chicago, Pittsburgh, and St. Louis for nearly a century, defying attempts at solution. It posed a major burden for urbanites, blocking the sun, killing vegetation, staining building facades, dirtying clothes, increasing the effort and frequency of household cleaning and washing, raising the costs of goods for department stores, and exacerbating the health difficulties of those with various lung and respiratory complications. Smoke pollution affected with particular intensity those responsible for housecleaning and laundry, women. All urbanites, however, suffered, and only those who were able to flee the city were relatively immune.

Because it affected almost all city-dwellers, support for the control of smoke pollution should have received widespread public backing. Various groups, however, viewed the problem from different perspectives. Engineers, for instance, argued that smoke resulted from inefficient combustion technology and that it could be eliminated by applying a technological fix of the "best available technology." Other professionals maintained that smoke was the result of human factors and could be reduced through the effective training of firemakers and stokers. And urban reformers saw smoke regulation as mainly a matter of determination on the part of the courts and the legislatures to compel polluters to control their effluents. Compounding the difficulty was a powerful association between smoke and prosperity. To many urbanites, smoke signified prosperity and jobs, not nuisances or health hazards. Thus, values entered strongly into the equation.

The first essay in this section deals with a successful attempt by Pittsburgh—the nation's smokiest city—to control its smoke. The second deals with attempts to regulate one of the most difficult-to-control sources of smoke—railroads. In each case, the problem was eventually solved by fuel and technological substitutions, with changes in values reflected in legislation and legal action playing important but difficult-to-weigh roles. The focus on smoke as the main atmospheric contaminant shifted in the postwar decades to automobile fumes and to air pollution in general. Smoke pollution, however, raises a number of interesting and provocative questions about how a society protects the environment when such protection appears to raise costs for different groups, especially those with lower incomes, how values and value change enter into these considerations, and how technological change to reduce pollution can be induced.

Photo III.1. Smoke from the "Metals District," Pittsburgh, 1914. Source: Mellon Smoke Investigation, Pittsburgh, 1914.

Photo III.2. A railroad marshaling yard in Pittsburgh, 1914. Source: Mellon Smoke Investigation, Pittsburgh, 1914.

Photo III.3. "Smoke-blighted" trees in Munhall Hollow, above the Homestead Steel Works of the Carnegie Steel Company. Source: Margaret Byington, *Homestead: the Household of a Mill Town* (New York, 1910), reprinted, University Center for International Studies, University of Pittsburgh, 1974.

Photo III.4. "Darkness at Noon in Downtown Pittsburgh, 1920." Photo by W. S. Brown. Source: Carnegie Library of Pittsburgh.

Photo III.5. Pittsburgh Smoke Control Campaign, 1941. Source: Allegheny Conference on Community Development.

Photo III.6. The partially cleaned Carnegie Library of Pittsburgh, 1990. Source: Author's photograph.

Photo III.7. "Steam Locomotive Going through Oakland, 1953." Photo by Clyde Hare. This Baltimore and Ohio Railroad train is proceeding along Junction Hollow, a cut located in Pittsburgh's Oakland district. The picture was taken just before the B & O converted to diesel-electric. The buildings on the hill in the background belong to Carnegie Institute of Technology. Source: Carnegie Library of Pittsburgh.

CHAPTER VIII

Changing Fuel Use Behavior and Energy Transitions

Pittsburgh Smoke Control Movement, 1940–1950

A Case Study in Historical Analogy

The United States today is in the beginning of a transition involving a shift from an overwhelming dependence on cheap and abundant supplies of oil and natural gas to a variety of other fuels, most of which bear higher costs in regard to both production and environmental controls. Transitions from one energy source to another are not new in the United States. The nineteenth century witnessed, for instance, a slow shift from renewable energy sources such as wood and water power to fossil fuels, while in the twentieth century oil and natural gas have displaced coal for many uses. What is new about the current situation is that it is the first time, aside from wartime periods, that the federal government has assumed responsibility for energy policy on such a sweeping front.[1]

Federal energy policy, however, has had limited success, especially in areas such as household conservation, which potentially involves changes in both capital investment and energy-related behavior for millions of consumers. It has recently been suggested that since energy policy decisions are inextricably bound up with processes of social change, utilizing social science perspectives on energy behavior would increase the likelihood of devising successful policy. Such research is now underway with a focus on the information that can be derived from social psychological and sociological investigations. But since energy transitions are not a new phenomenon, important insights should be available from the examination of past cases involving attempts to alter energy-using behavior. This paper presents one such case in an attempt to provide energy policymakers with a "device of anticipation" that will aid them in successful policy formulation.[2]

The particular example of energy transition to be analyzed in this paper derived from municipal policy to control smoke in Pittsburgh in the 1940–50 period. The city succeeded in eliminating dense smoke in the late 1940s, after over half a century of failure, by generating a new policy that required domestic consumers, as well as industries and transportation companies, to change their fuel type and/or combustion equipment. The policy was originally based on the expectation that consumers would use smokeless coal and/or smokeless heating equipment burning bituminous coal in order to meet clean air standards. In the years from approximately 1945–50, however, over half the households in Pittsburgh shifted their fuel from bituminous coal to natural gas (see figure 8.1). This change also required either the retrofitting or replacement of existing combustion equipment. While other cities were moving from coal to natural gas or oil as domestic fuel, there was no other city in the nation where the rate of change was as rapid (see figures 8.2, 8.3).

The case of Pittsburgh, therefore, offers an example of energy transition accelerated by environmental policy based on control of fuel use and combustion equipment. This paper will focus particularly on the generation and implementation of policy in regard to domestic fuel consumers. Achieving successful smoke control in Pittsburgh required both creating public support for a law controlling the energy-consuming behavior of individuals in the name of a larger social good, and devising a strategy for implementing this law under conditions of higher fuel costs and equipment uncertainties. An analysis of this case suggests that it contains insights on a number of important issues pertinent to energy policy today. These insights derive primarily from questions concerned with organizational and individual roles in policy development and implementation; means of individual behavior modification; the quality and diffusion of information regarding fuel supply, technological capabilities, and policy impacts on consumers; and the equity implications of policy. These questions will be examined in the context of Pittsburgh smoke control.

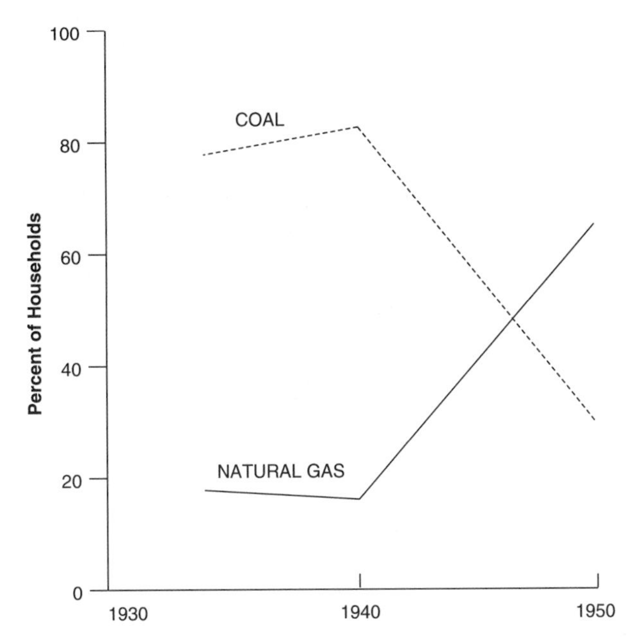

Fig. 8.1. Coal and Natural Gas Use in Pittsburgh, 1934, 1940, 1950—Domestic Heating. Source: 1934: Bureau of Business Research, University of Pittsburgh, *Real Property Inventory of Allegheny County* (Pittsburgh, 1937), p. 148; 1940 & 1950: Bureau of the Census, *Population and Housing for Census Tracts.*

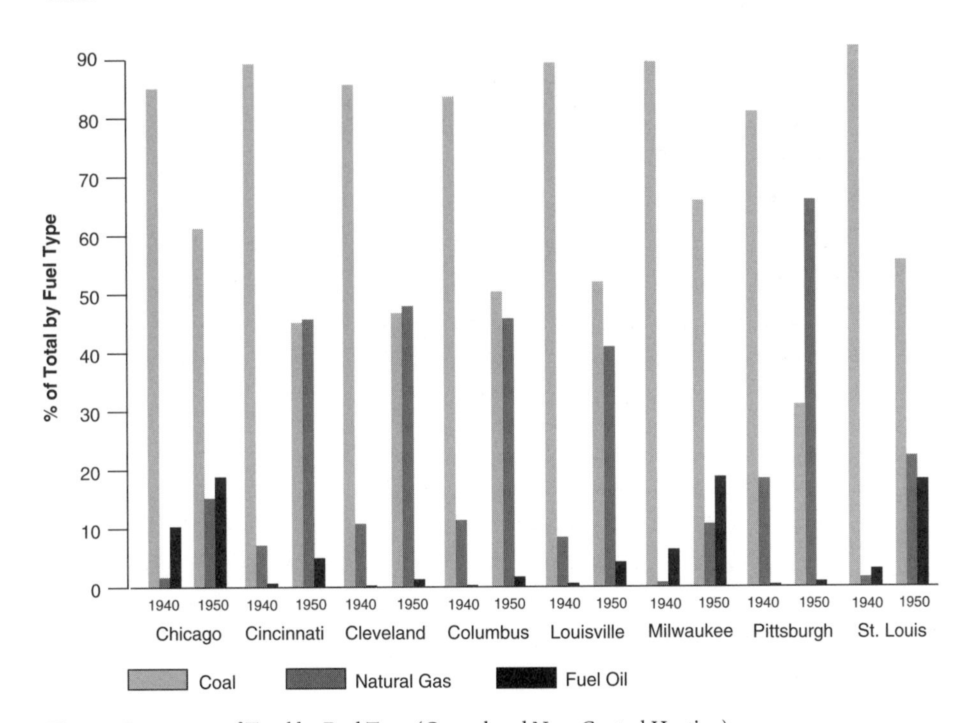

Fig. 8.2. Percentage of Total by Fuel Type (Central and Non-Central Heating).

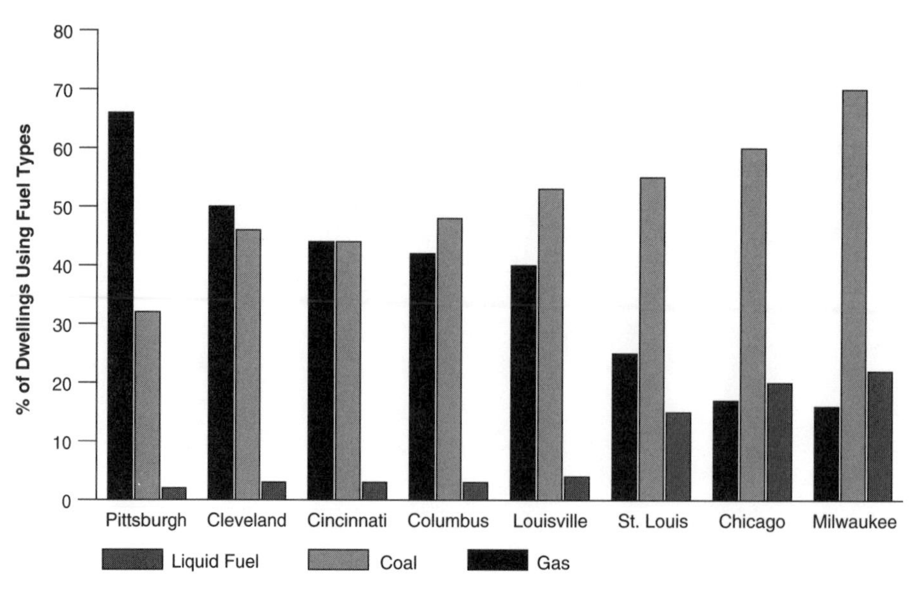

Fig. 8.3. Eight Cities—Percentages of Homes Heated with Fuel Type, 1950. Source: U.S. Census Bureau, *Population and Housing Statistics for Census Tracts* (1940 and 1950).

Pre–Smoke Control Pittsburgh

The problem of smoke pollution in Pittsburgh resulted from a conjunction of the factors of urbanization, industrialization, topography, and the availability at low cost of large sources of high-volatile bituminous coal. While air pollution is not only a problem of cities, it is normally more critical under urban conditions because of population density and increased economic activity. As Lawrence Tilly has observed, the "metropolis is a dependent ecosystem . . . [and] all ecosystems require a continuous supply of energy to power their activities." In Pittsburgh this source of energy—bituminous coal—served for domestic and commercial heating purposes, for processing raw materials and manufacturing goods, and for providing fuel for transportation systems. Heavy usage and local topographical and climatic factors that often produced temperature inversions combined to give Pittsburgh a reputation as "the smoky city" early in the nineteenth century.[3]

Smoke pollution grew progressively worse as Pittsburgh industrialized, and public authorities made some gestures at controlling it. In 1868, the city council passed a statute banning the use of bituminous coal or wood by railroads within the city limits, and in 1869 they forbade the construction of beehive coke ovens. Neither statute, however, was enforced. During the 1880s, the discovery and exploitation of local supplies of natural gas provided the city with approximately six years of clean air. Exhaustion of the local gas supply, however, caused industries and residences to resume the use of soft coal for fuel, with a consequent return of heavy smoke. The resumption of severe smoke pollution resulted in agitation by various elite and professional groups for smoke control. The city council responded by the passage of a series of ordinances in the 1890s and the beginning of the twentieth century that regulated the emission of dense smoke from industrial, commercial, and transportation sources but did not attempt to control domestic heating plants. A Bureau of Smoke Regulation was created in 1911 for enforcement purposes. The bureau's director believed that education and persuasion rather than legal sanctions would persuade polluters to stop producing smoke. The methods to achieve this were educational—training firemen to operate furnaces and boilers more efficiently—and implemented through retrofitting with various smoke-consuming devices. Fuel efficiency, it was argued, provided an incentive to take these actions, since smoke was a sign of incorrect firing and fuel waste.[4]

However, the smoke control movement failed to control the smoke nuisance to any appreciable degree during the first third of the century. During the 1920s and 1930s, therefore, smoke and fuel researchers and regulators set about to redefine the problem. There was general agreement among them that industries and railroads had made advances in the elimination of dense smoke. These improvements had been achieved through technological and fuel improvements, by care in firing methods, and through cooperation with smoke bureaus. The smoke problem persisted, smoke investigators held, because of a failure to control domestic furnaces. Experts argued that smoke from household furnaces was especially objectionable because "the amount of black smoke produced by a pound of coal is greatest when fired in a domestic furnace and that domestic smoke is dirtier and far more harmful than industrial smoke."[5]

Domestic furnaces had not been regulated for several reasons, the most important of which were the political and administrative problems involved in controlling the heating habits of a multitude of householders. In

1940 there were 175,163 dwellings in the city, 141,788 of which burned coal and 30,507 consumed natural gas; 53,388 of those burning coal had no central heating plant and used stoves to heat their homes. Smoke regulators lacked an effective administrative mechanism to control domestic smoke without employing hundreds of smoke inspectors. Politically, the issue was difficult because control threatened to impose higher costs for capital equipment and fuel on householders. And, because of a historical equation between smoke and prosperity in Pittsburgh and other industrial cities, it was difficult to develop a public consensus for stringent controls.[6] In short, the problem was one of devising a strategy to change individual behavior in regard to fuel use in the name of the collective social goal of clean air.

Organizing the Smoke-Control Campaign

The climate of opinion in Pittsburgh in the late 1930s made discussion of smoke control extremely difficult. A city dependent on heavy industry, Pittsburgh was badly scarred by the Great Depression; clear skies suggested closed factories and unemployed workers. In addition, many local businesses were related to the coal-mining industry, which also suffered severely from the depression. As a sign of its belief that smoke equaled prosperity and its relief at the return of full employment, in 1939 the Pittsburgh City Council actually eliminated the Bureau of Smoke Regulation. "You'll never get elected again," said one politician to a member of the City Council who supported antismoke legislation. "Don't you know, the poor people, they don't want smoke control. It's going to cost them more money." "We like to see smoke," added another politician; "it means prosperity."[7] Opinions such as these were representative of those held by many working-class people. Although they found smoke a nuisance and an annoyance, they were concerned that they would have to pay a substantial proportion of their incomes for smoke control. Pittsburgh, therefore, appeared an unlikely environment for the passage of substantive legislation controlling smoke emissions from either industries or homes.

Changing this attitude required a sustained citywide campaign that would convince a substantial body of the public that the benefits of smoke regulation would outweigh the costs. (The last such drive, which had resulted in the formation of the Bureau of Smoke Regulation, had occurred in the 1911–17 period.) Organization of this campaign actually began in 1936 when a small group of individuals were instrumental in securing a grant from the federal Works Progress Administration to do a survey of the impacts of smoke pollution on the community. The survey updated the reports of the important six-volume Mellon Smoke Investigations (1912–14) and provided the basis for a renewal of interest in eliminating the smoke nuisance by several voluntary organizations that had been involved in past control efforts. The Civic Club and the chamber of commerce, which were concerned with the city's image and feared that the lack of clean air would cause population and industrial loss, were most important.[8]

Historically, however, campaigns against smoke led by such voluntary organizations had never been effective. Distinguishing the situation in 1940–41 from the past were two elements: the active involvement of three key individuals, who cooperated with the voluntary organizations but who represented other critical elements in the decision-making field, and the example of a city that appeared to have successfully solved its smoke problem.

The critical individuals, who might appropriately be called "entrepreneurs" for the collective social good of smoke control, were Abraham Wolk, a lawyer and city council member; Edward T. Leech, editor of the city's most influential newspaper, the *Pittsburgh Press;* and Dr. I. Hope Alexander, director of the Pittsburgh Department of Public Health. Wolk, described by a colleague as a "monomaniac" on smoke control, appears to have become involved in the campaign because of the effect of the smoke on the health of his son. He organized the political coalition necessary for passage of the 1941 ordinance, convincing Mayor Cornelius D. Scully, who was up for reelection in 1942, that it would be to his advantage to support the legislation. Ed Leech, a crusading editor, furnished critical media leadership, using his paper "like a war club" to advance smoke control. And Alexander, a physician, made smoke control into a public health crusade, emphasizing the damage that smoke was doing to the lungs and health of all Pittsburghers, regardless of class.[9]

The model city was St. Louis, also an industrial center dependent on bituminous coal. While Pittsburgh had been faltering in its fight against smoke in the middle and late 1930s, St. Louis was making inroads into its smoke pollution problem. The important role in St. Louis was played by Raymond R. Tucker, a combustion engineer who had been a member of the Department of Mechanical Engineering at Washington University of St. Louis. In 1934 he was appointed secretary to the mayor with instructions to devise a strategy to "clarify the air." Tucker realized that the method by which St. Louis and other cities had solved their water pollution problem—filtration at the source—furnished a model that could also be applied to the air. That is, if water could be rendered potable by removing impurities before distribution, then the air could be cleansed by controlling the quality of the fuel before consumption. "Smokeless air" would result from a law that required the burning of "smokeless" fuel or the use of smokeless equipment. Tucker fought an uphill battle attempting to secure legislation, but in April 1940, after a heating season characterized by heavy smoke palls, the St. Louis Board of Aldermen finally approved an ordinance requiring the use of smokeless fuel or smokeless mechanical equipment by fuel consumers, including homeowners. The essential control mechanism was city licensing of fuel dealers in order to control the quality of fuel at the source. The result of the first test of the ordinance in the 1940–41 heating season was a series of smokeless days that city officials claimed were the result of the smoke ordinance.[10]

St. Louis offered an example of a municipality that had successfully controlled its smoke problem, and it had used a mechanism that was transferable to other cities. The Pittsburgh antismoke forces were in contact with St. Louis, and in February 1941 the *Press* began a concerted series of articles and editorials pointing to St. Louis's success and asserting that Pittsburgh could also achieve clean air. Most effective in mobilizing opinion were the vivid cognitive images created by two pictures published on the *Press*'s front page showing a smoke-darkened St. Louis street before smoke regulation and the same street sunlit after the control ordinances had become operative. Egged on by the *Press,* readers—especially irate housewives—began bombarding Pittsburgh mayor Cornelius D. Scully with over two hundred letters a day demanding action. During the same month, a delegation of Pittsburghers led by Dr. Alexander, and including several members of the Smoke Elimination Committee of the Civic Club of Allegheny County, visited St. Louis and returned with glowing reports of the city's achievement. Shortly thereafter, the city council also visited St. Louis on a "civic pilgrimage" to examine the administrative machinery of smoke control and to assess its potential political costs. While some members of the council raised questions about the impacts of smoke control on the coal miners and on the poor, most returned convinced of its technical feasibility. Its political practicality, however, was the critical question, and aside from Wolk, the other councilmen maintained a cautious attitude, waiting to see the public response.[11]

The media were essential in generating a positive public attitude. And in Pittsburgh in 1941, "the media" meant the newspapers; radio played a subsidiary role. The leader among the papers in advocating a smoke-control ordinance was Leech's *Pittsburgh Press;* of the city's other newspapers, the *Post-Gazette* generally followed the *Press*'s lead in support of smoke control, while the *Sun Telegraph* played a more neutral role, raising questions about possible costs as well as benefits. Leech effectively declared war on smoke, bombarding the public with editorials and articles about its evils and the advantages of control. Investigative reporter Gilbert Love, assigned the role by Leech of "knowledgeable community informant," translated medical information and technical reports into concise and informative articles. Items detailing the costs of smoke to the city in terms of damage to health, property, vegetation, and cleaning expenses and information concerning the support of various community groups for control appeared almost daily. The fact that the 1940–41 winter was marked by a number of extremely smoggy days accentuated the urgency of the need for legislation. The *Press* played on the adverse atmospheric conditions by publishing photographs contrasting air-quality conditions in the city with those in the suburbs. Letters to the editor calling for smoke control were published almost daily, although there was also a scattering of letters expressing concern over the possible loss of jobs.[12]

The Mayor's Commission for the Elimination of Smoke

The antismoke campaign attempted to win support from all organized groups within the community. The campaign directors stressed that the benefits of smoke control would outweigh the costs both for the community and for individuals within it. The communitywide character of the campaign was reflected in the membership of the Mayor's Commission for the Elimination of Smoke, appointed on February 18, 1941. In his charge to the commission, Mayor Scully declared that "Pittsburgh must, in the interest of its economy, its reputation and the health of its citizens, curb the smoke and smog which has made this season, and many others before it, the winter of our discontent." The commission was chaired by Councilman Wolk and included Dr. Alexander; A. K. Oliver, chairman of the board of the Pittsburgh Coal Company, the district's largest coal company; H. Marie Dermitt, secretary of the Civic Club; Ed Leech; Ralph C. Fletcher, research director of the Federation of Social Agencies; William N. Duff, chairman of the Smoke Abatement Committee of the chamber of commerce; Patrick T. Fagan, president of District 5 of the United Mine Workers; Mrs. H. K. Breckenridge, a "prominent" club woman; and Mrs. W. C. Ridge, member of the Pittsburgh Board of Education.[13] Thus, the commission included representatives of business, labor, government, the media, the health professions, and voluntary associations with a civic and a welfare orientation; the inclusion of three women reflected the campaign leadership's perception of the importance of women in achieving smoke control.

In addition to these members, a technical advisory group was appointed to the commission, consisting of M. A. Mayers, Carnegie Institute of Technology Research Laboratories; Edward Weidlein, director of Mellon Institute of Science; Sumner B. Ely, consulting engineer and former Carnegie Tech faculty member; Ragnar Berg, chief engineer at Koppers Company; and Henry F. Hebley, control manager at Pittsburgh Coal Company. The role of the technical committee was to present recommendations concerning control of specific sources of smoke, such as railroads and metallurgical companies, and to gather information on questions such as the availability of smokeless fuel and smokeless equipment.[14]

While the commission was holding its hearings, the Civic Club and the League of Women Voters conducted a countywide campaign of public arousal and education. The dissemination system was a network of voluntary associations that could be utilized to communicate information about the benefits of smoke control throughout the community and also to serve as a political pressure group on the city council if necessary. While voluntary organizations of all types were represented in the network, women's groups were the most numerous, reflecting the deep involvement of women in the smoke-elimination campaign. As homemakers, women of all classes knew how much extra cleaning smoke necessitated, with the burden falling most severely on working-class women who lived close to the mills. Middle- and upper-class women in the Civic Club and the League of Women Voters coordinated luncheons and lectures and provided speakers to interested groups. These events were reported in the newspapers, especially the *Press* and *Post-Gazette,* keeping the issue before the public.[15]

Eventually the Smoke Elimination Commission held twelve closed meetings and four public meetings. The purpose of the public meetings was to give interested groups a chance to be heard, but even more to "get across to the public something which . . . still needs more hammering,—the need for smoke control. . . . It gives the papers and the Commission a show. It gives the people a chance to be part of [the] . . . meetings."[16] The first three public meetings were carefully orchestrated for maximum impact and were held soon after the Commission's formation. The first hearing presented the testimony of ten local physicians concerning the negative health effects of smoke; the second heard the impassioned complaints by representatives of a number of women's clubs concerning the negative impacts of smoke on the family, on physical and mental health, and on the enslavement of women to the drudgery of constant housework and cleaning; the third, arranged by the Civic Club Smoke Committee, involved the testimony of householders and other fuel users about the advantages of mechanical stokers and processed fuel as opposed to high-volatile bituminous coal. The fourth public meeting was called at the request of the Western Pennsylvania Coal Operators Association and took place near the end of the commission's proceedings.

While the public meetings served the function of public arousal and information transmission about the negative effects of smoke, the closed meetings provided information for the commission on the more policy-relevant questions: smokeless fuel and technology supply; costs and characteristics; possible effects of the policy on the coal industry, on coal miners, and on the poor; and administrative procedures and timetables for enforcement. It became clear early in the proceedings that the members of the commission shared a consensus on the following points:

1. Smoke should be eliminated not merely abated.
2. Households and commercial firms contributed more to the smoke nuisance than did large industries and posed the greatest enforcement problem.
3. Any policy adopted should not damage any Pittsburgh industry by either reducing coal production or consumption or by imposing excessive control costs.
4. Special concern should be paid to the possibility that the poor might pay higher fuel costs as a result of smoke control.

The discussion within the commission clearly reflected the conviction of its members that, with the right policy, the socially desirable end of smoke elimination could be achieved without excessive costs to individuals or to industry. As Chairman Wolk noted early in the commission's proceedings, "We want to make this city smokeless without hurting anybody."[17]

The costs of a smoke-control policy based on the use of smokeless fuels and mechanical equipment, however, promised to impact the coal miners and low-income consumers more than other groups and was therefore a frequent topic of discussion. The Steel City Industrial Union Council, which represented over 100,000 CIO members in Allegheny County, including the United Mine Workers, had unanimously endorsed the principle of smoke control early in March. A representative of the mine workers' union sat on the commission to watch over their interests and those of labor in general. During the hearings, he often asked witnesses questions about fuel costs and possible job losses by miners, but he appeared to accept the position that smoke control would not substantially impact mine employment. This stand was based on the argument that the need for smokeless coal would actually result in the mining of larger amounts of bituminous.[18]

The issue of how the poor would pay for higher fuel costs or for new combustion equipment was also an important subject. Speaking before the commission, Mayor Scully noted that "our problem is a psychological propaganda problem to keep somebody from attacking the program . . .

on the theory that it's going to hurt a little fellow and do great harm to people who are underprivileged." Generally the commission members, including the labor representative, agreed that smoke control would bring more benefits than costs to working-class and low-income people. Because the poor suffered the most from the effects of smoke pollution, commission members maintained, they would reap the most benefits from a strong law. Specific mechanisms for subsidizing or aiding the poor, however, were seldom discussed aside from the argument that implementation of the law against domestic consumers should be delayed until cheap and adequate supplies of smokeless coal were available. While both a subsidy plan and a revolving fund were briefly mentioned as options, they were never pursued by the commission.[19]

Many of the critical questions facing the commission were technical, and here the members depended on the advice of their technical committee and the testimony of experts from the various relevant industries. During the month of May, close to the end of the commission's hearings, the Technical Advisory Subcommittee presented its reports. These dealt with domestic combustion technology, tonnage of coal used for domestic purposes, and control of smoke and dust from railroads, steamboats, and stationary and metallurgical sources. The most important section of the subcommittee's recommendations dealt with methods of regulation and enforcement dates. The subcommittee recommended that domestic fuel supply be regulated so that householders would either have to use smokeless fuel or smokeless equipment. However, in order to ensure that adequate supplies of both fuel and equipment were available, they suggested that enforcement against domestic consumers be delayed until after the commercial and industrial sectors had complied with the smoke-control law. The subcommittee chairman recommended a deferment for a period of at least three years. The long delay was necessary not only to secure compliance by the nondomestic sectors but also to permit the construction of plants to produce smokeless fuel.[20]

While some members of the Smoke Elimination Commission and the subcommittee were associated with coal-mining firms, the industry was still concerned about securing full representation of their views both before the commission and the public. The coal producers were under pressure from cleaner competing fuels, such as oil and natural gas, as well as from antismoke campaigns in other cities besides Pittsburgh, and were concerned about the loss of a critical market. The Western Pennsylvania Coal Operators Association, representing seventy-six firms that produced over 80 percent of the region's coal, created their own committee to investigate the problem. In mid-May, at the same time as the commission's Technical Advisory Subcommittee was presenting its recommendations, the Coal Operators Association requested a special public hearing to present their technical committee's report. In addition, they ran full-page advertisements in the Pittsburgh newspapers detailing their plan.[21]

The Coal Operators Association took the position that it was "in the long range interest of coal producers—and coal users—to help find a way by which smokeless and efficient heating can be economically obtained." They believed, however, that a smoke-control policy based on the use of processed coal was "wishful thinking," since they doubted that the capital required for its production could be obtained. The coal producers feared that such a policy would result in the capture of the Pittsburgh domestic market by the producers of other cleaner fuels such as natural gas or oil or low-volatile coal from other regions. The result would be severe harm to the industry, the unemployment of over two thousand Allegheny County miners, and higher consumer fuel prices.

The operators proposed the adoption of "improved" equipment designed for the combustion of bituminous coal as an alternative to the use of processed coal or other smokeless fuels. The coal industry was already supporting research and development of such equipment at Bituminous Coal Research, the Battale Memorial Institute, and various universities. In order to allow time for the commercial development of the smokeless technology, said the coal producers, smoke control should be implemented in a series of stages, with industry first and domestic consumers last. In this manner, clean air would be attained without the destruction of the markets for Western Pennsylvania bituminous.[22]

The Commission's Report and the Passage of the Ordinance

At the conclusion of the commission's hearings, John P. Robin, the mayor's executive secretary, and Gilbert Love, of the *Pittsburgh Press*, prepared a final report to the mayor.[23] The report was signed by all members of the commission, including coal industry and labor representatives, and listed the names of two hundred voluntary organizations—including fifty-six women's clubs, twenty-four business organizations, and many labor and civic groups—that supported smoke control. The report held that smoke elimination in Pittsburgh was feasible and that it would "bring about a new era of growth, prosperity and well-being" for the city. Clean air, it maintained, would improve conditions for the city's population, halt population movement to the suburbs, attract light industry, and create new industrial activity and employment, especially in the area of processed fuels. These benefits would be the result of a policy requiring the use of smokeless fuels and/or smokeless mechanical equipment for combustion purposes.

The report dismissed the concerns of the Coal Operators Association that such a policy would negatively effect the coal industry and maintained that requiring the burning of smokeless fuels would actually increase regional production. "Pittsburgh might well become the location of a new coal processing industry, selling its products not only in this city and the suburbs . . . , but in distant communities."[24] While the Western Pennsylvania Coal Operators Association had disagreed with this position, it reflected the belief of commission member A. K. Oliver, chairman of the board of Pittsburgh's largest coal company, who signed the final report.

The commission dismissed the possible negative impacts of the policy on low-income families in the same manner as it had treated the concerns of the coal industry. That is, the report focused on the benefits of smoke control for the poor and deemphasized the possible costs, noting that "the smoke elimination plan would impose little or no additional burden on the low-income groups of the city."[25] Smokeless fuel might cost more than high-volatile bituminous, but it would produce more heat. In addition, the commission predicted that many low-income families would purchase new smokeless stoves when their present equipment wore out, making possible a return to cheap high-volatile fuel. In order to control the price and quality of smokeless fuel, the commission recommended the creation of a fuel consumer's council that would protect the low-income consumer against short weight and other fraud and help extend the fuel dollar. John P. Bussarello, representative of the United Mine Workers, signed the report, indicating that the unions were willing to accept the position that smoke control would not place undue burden on the miners or the working class.

The enforcement procedures recommended by the commission rested on the concept of control of smoke at the source. Over a staged two-year period, all fuel users would have to either burn smokeless fuels or utilize smokeless mechanical equipment. The timetable for implementation was as follows: industries, office buildings, hotels, apartment houses and commercial establishments by October 1, 1941; railroads by October 1, 1942; domestic users and all other fuel consumers by October 1, 1943. Specific standards in regard to the emission of dense smoke as defined by the Ringelmann chart were issued for different classes of users. In addition to smoke, the suggested ordinance provided for enforcement against other air pollutants such as fly ash, noxious acids, gases, and fumes.[26]

The commission recommended the creation of a Bureau of Smoke Prevention housed in the Department of Health and headed by a "qualified engineer." Permits were required for new installations and reconstruction of existing heating plants, and the bureau had the right of inspection. The superintendent could impose fines of $25 to $100 and seal equipment in violation of the law. Appeals from his action would be heard by a board of five citizens. Ultimately, however, the report maintained, public opinion would determine if the city would become smoke free: "Public cooperation is the vital ingredient of any civic improvement, and the bureau of smoke prevention by mobilizing the force of public determination will be able to see to it that the law will be enforced right up to the hilt."[27]

Chairman Wolk introduced the ordinance prepared by the Smoke Elimination Commission in the city council. The railroad and steamboat interests and Jones and Laughlin Steel Corporation attempted to secure modifications of certain provisions, but without success. Early in July 1941, the council voted eight to one to approve the smoke ordinance essentially as submitted. Edward Leonard, a councilman active in labor circles, cast the only dissenting vote, protesting that the ordinance would be a burden to both the poor and to industry.[28]

The success of the smoke-control advocates in obtaining the 1941 legislation is explained by several different factors. The experience of St. Louis showed Pittsburgh elites and political influentials that smoke control could be a reality in an industrial city. St. Louis, therefore, served both as an incentive and a model for emulation. The managers of the campaign were able to convince most Pittsburghers that the benefits of smoke control would outweigh the costs and that the policy would have minimal distributional effects. They accomplished this end through a media blitz that emphasized the evils of smoke and the benefits that would ensue from its eradication and by organizing a citywide network of voluntary institutions for educational and political purposes. The media and communitywide campaign generated an enthusiastic response, indicating a strong latent public desire for eliminating smoke. While there was opposition to the ordinance, it was unorganized and fragmented. Two groups whose combined opposition might have blocked or delayed its passage—organized labor and the coal industry—were co-opted by the argument that they would also benefit from smoke control and were involved in the decision-making process. The power of the coalition backing the ordinance was most clearly revealed in the lack of serious political debate concerning its provisions. All significant power groups within the city either supported or were neutral on its passage.

Implementing the Smoke-Control Program

The smoke-control ordinance as passed in July 1941, included three stages of implementation. Stages one and two, directed primarily against industries, commercial establishments and railroads, were implemented as planned, although the Bureau of Smoke Prevention only engaged in limited enforcement during the war. Dr. Sumner B. Ely, head of the bureau, maintained that industry and the railroads were doing their best to meet the standards considering wartime conditions, and he stressed the need for cooperation and understanding in seeking compliance. The smoke inspectors, he maintained, should supply "advice and assistance" rather than playing the role of policemen. The original implementation date for domestic consumers—October 1, 1943—was waived, with the agreement that implementation would occur six months after the war's end.[29]

The intense focus on productivity, the limited enforcement against industry and railroads, and the lack of regulation of domestic heating equipment created very poor air quality in Pittsburgh during the war. The city suffered a number of severe smoke palls, some requiring that streetlights be turned on in the downtown at midday. The need for strict enforcement of the smoke-control ordinance after the war appeared urgent to the antismoke forces. The most critical question in terms of implementation continued to be enforcement against domestic consumers. In order to maintain the impetus that had resulted in the ordinance's passage, the Pittsburgh Civic Club created a new organization—the United Smoke Council (USC), consisting of eighty allied organizations from Pittsburgh and Allegheny County. The slogan of the council was "Now More Than Ever," and it organized a committee to study the availability of new smoke-elimination devices and smokeless fuels; an enforcement committee to aid the Smoke Bureau; and a committee to encourage antismoke laws throughout the county. The council's most critical function was to continue public educational efforts about the need for smoke control. M. Jay Ream, chairman of the Civic Club's Smoke Elimination Committee, was appointed the council's first head.[30]

In addition to the United Smoke Council, a second important civic association entered the field of smoke control. This was the Allegheny Conference on Community Development (ACCD), formed in 1943. The ACCD had as its mission the development of "an over all community improvement program" for Pittsburgh. Primarily concerned with revitalization of the Pittsburgh Central Business District and ultimately the regional economy, it understood that environmental improvements like smoke and flood control were essential to community revitalization. The ACCD was especially critical to the smoke-control campaign because of its concentration of corporate power. The key figure behind the ACCD was Richard King Mellon, who exercised enormous economic power through his leadership of the Mellon banking interests and linked corporations such as Gulf Oil, Koppers Coke, and Pittsburgh Consolidation Coal. Mellon indicated his interest in smoke control in 1943 in response to a request for help from M. Jay Ream of the United Smoke Council and noted that it was an issue with which his father had been concerned. He suggested that Ream make contact with the officials at the ACCD, and a merger between the two groups became a reality at the end of 1945.[31]

The merger of the United Smoke Council into the conference was important because it supplied the smoke-control group with funding, access to professional and technical staff, and the backing of powerful corporate leadership. It also made its single-issue focus of smoke control part of a larger campaign for city improvement. The USC, with its wide net of community organizations, enlarged the scope of the conference program and increased public support. Thus the United Smoke Council and the Allegheny Conference joined to provide the bureaucratic organization and planning essential for policy implementation.[32]

In the postwar period, as in 1941, political support was necessary as well as that of voluntary associations. In the spring of 1945, David Lawrence was elected mayor of Pittsburgh. Chairman of the Allegheny County Democratic Committee, Lawrence was concerned with strengthening his organization's hold on the city. In a move often described as politically courageous, Lawrence allied himself with the Allegheny Conference plan for civic reconstruction, including a commitment to implement the smoke-control law at an early date. Lawrence later explained that although this decision displeased many Pittsburgh Democrats, he "resisted all pressure and temptation to ease up and take it slowly."[33]

Setting an effective implementation date for smoke control depended on the accuracy of the available technical information. The important questions continued to be the availability of low-volatile fuel and of mechanical equipment to make possible public compliance with the law. The ACCD hired C. K. Harvey, an engineering consultant, to study the problem and provide it with recommendations. The coal companies again initiated their own study group under the auspices of the newly formed Western Pennsylvania Conference on Air Pollution. Each technical committee investigated the following major questions relating to a proposed implementation date of October 1, 1946:

1. Would low-volatile fuel be available for consumers? Could it be supplied by local producers or would it have to be imported?
2. What would the costs of low-volatile fuel be in comparison to high volatile?
3. Would smokeless combustion equipment be available at a low enough cost to enable consumers to burn high-volatile fuel smokelessly?
4. What effect would the requirement for low-volatile fuel have on the mining industry in the Pittsburgh region?

5. What would be the financial impact of the requirement for low-volatile fuel or smokeless equipment on low income consumers?

In his report to the ACCD, Harvey concluded that while there would not be sufficient smokeless equipment available to permit the use of high-volatile coal by October 1, there would be adequate supplies of low-volatile fuels to meet the smoke-control ordinance's objectives. Such fuels would have to be obtained from central Pennsylvania. The technical report noted that the 500,000 tons of coal that needed to be imported equaled only 3 percent of the county's production and would, therefore, not discernibly affect either the coal companies or labor. In a separate letter to M. Jay Ream, Harvey repeated the argument made in the 1941 Smoke Elimination Commission report that the demand for low-volatile coal would stimulate the creation of a processed-fuel industry based on local bituminous.[34]

In regard to the impact of the policy on consumers, Harvey also repeated the 1941 argument that the higher heating value of smokeless coal might conceivably lower "the average fuel bill." By using smokeless coal in their present equipment, he added, consumers would be able to salvage the "recovery life" of their furnaces before converting to smokeless equipment. "Thus the Ordinance can be put into effect without further delay and with a minimum of capital investment and cost." Harvey argued that delaying implementation until sufficient smokeless equipment was available would be an error because "the public may become discouraged and lose its present enthusiasm for smoke pollution abatement."[35]

Satisfied with the accuracy of the Harvey report, on February 25 the ACCD Executive Committee approved the October 1, 1946, date for enforcing the smoke-control law against domestic users. Within several weeks, however, members of the ACCD Technical Committee raised questions concerning the accuracy of Harvey's claim that there would be adequate supplies of smokeless coal available to meet domestic needs. Simultaneously, President George Love, of Pittsburgh Consolidation Coal, and Vice President J. D. Morrow requested a special meeting of the ACCD Executive Committee to present the bituminous coal producers' arguments for a delay of the implementation date.[36]

The coal producers maintained that it would be a mistake to implement the smoke-control law on October 1, 1946, because of inadequate supplies of smokeless coal. Enforcement, they warned, would cause great hardship throughout the city. The smoke problem would ultimately be solved through the utilization of the smokeless stoves and furnaces under development by Bituminous Coal Research and other research groups, not by smokeless coal. Since the commercial production of this equipment had been delayed by postwar shortages of capital and materials, implementation concerning domestic users should be put off. The representatives of the coal interests urged that time and research, not emotional arguments, were required to solve the complex problem of air pollution without harming both consumers and the industry.[37]

The continued concern of the coal industry that enforcement of smoke control without the availability of technology to burn coal smokelessly would drive consumers to cleaner competing fuels underlay this position. "Coal producers," said the technical report of the Western Pennsylvania Conference on Air Pollution, were "opposed to smoke not only from the civic or humanitarian standpoint, but because it results in dissatisfaction with their product." If Pittsburgh acted, the report continued, other cities would follow, affecting thousands of miners and workers in related industries. The end result would be "the progressive crippling and possible destruction of a large segment of the Western Pennsylvania bituminous coal industry."[38]

The conflicting data presented by the two technical committees on the availability of smokeless coal, and the potential costs to domestic users, the coal industry, and miners of acting in 1946, led to a compromise by the ACCD: the enforcement date against domestic users would be delayed one year to October 1, 1947. Strict enforcement against industrial and commercial coal consumers and institutions and apartment houses would begin on October 1, 1946. In addition, the ACCD, the Coal Operators Association, and the city agreed to cooperate in maintaining public interest in smoke abatement. While some members of the United Smoke Council's technical committee opposed the delay, the executive committee of the ACCD accepted it unanimously. This was understandable since the compromise was prepared by Arthur B. Van Buskirk, the representative of Richard King Mellon on the executive committee. Mayor David L. Lawrence also indicated his crucial agreement.[39]

In late April, the city council held hearings on the revised smoke ordinance providing for the new implementation date. Here, representatives of the Allegheny Conference and the United Smoke Council presented a petition outlining the agreement reached earlier between the ACCD, the coal operators association, and the city; it was signed by the leaders of various civic and labor groups. Representatives of several coal corporations, the Retail Coal Merchants Association, the United Mine Workers, the Teamsters, and the General Laborer's Union all made statements appealing for delay. Presentations by engineers representing the coal companies and Bituminous Coal Research provided technical information confirming the lack of availability of low-volatile fuel and smokeless equipment. The League of Women Voters and several women's clubs, the Board of Trade of an elite Pittsburgh neighborhood, and several disaffected members of the United Smoke Council were the only voices calling for retention of the 1946 date. On April 29, 1946, the Pittsburgh City Council unanimously passed the amended ordinance.[40]

Enforcement and Response to the Smoke-Control Ordinance

With the passage of the ordinance secured, the ACCD and the USC, in coordination with the Pittsburgh Bureau of Smoke Prevention and the Lawrence administration, proceeded with efforts to ensure effective implementation and to secure a countywide law. From the perspective of this paper, the important questions concern enforcement of the ordinance by the Bureau of Smoke Prevention and the citizens' response to the law.

By requiring domestic users either to install smokeless equipment or to utilize smokeless fuels in order to meet the standards imposed by the ordinance, the Bureau of Smoke Prevention aimed at altering the behavior of consumers in heating their homes. Individual householders could choose among several alternatives in order to meet the law's requirements. One option was to purchase smokeless combustion equipment that would permit the continued use of high-volatile fuels. A second alternative was to buy combustion equipment using alternative fuels such as natural gas or oil or to retrofit existing equipment to burn these fuels. And a third alternative was to utilize smokeless coal in existing stoves and furnaces. Because the third option required the least capital outlay, most low-income households preferred it. From the perspective of enforcement, however, this alternative posed the most difficulties.

There were approximately 100,000 homes using hand-fired, coal-burning stoves and furnaces in Pittsburgh, including about 69,000 that depended on small coal stoves for cooking as well as heating. To enforce against the individual domestic consumer was an impossible task for the Bureau of Smoke Prevention's twelve inspectors. The bureau solved the enforcement problem by focusing on the coal distribution yards (approximately thirty) and the coal truckers. It forbade yards to sell high-volatile coal for use in hand-fired equipment and truckers to deliver it. Truckers hauling coal for consumption in the city had to be licensed and have license numbers painted on the side of the truck for easy identification. Those caught hauling illegal high-volatile (or bootleg) coal were subject to fines, as were dealers who made illegal sales.[41]

The winter of 1947–48 was critical for the new statute. Some of the more dire predictions of the law's critics were borne out. Supplies of smokeless fuels were inadequate, prices were inflated (smokeless coal was usually approximately 25 percent costlier than high-volatile bituminous, but in some cases it sold for considerably more), and inferior grades of coal were peddled to consumers. Because of a lack of familiarity with the characteristics of low-volatile coal, as well as the sale of low-grade mixtures, many citizens had serious problems obtaining sufficient heat. Complaints about the statute poured into city hall, newspaper offices, and radio stations, and several city councilmen attempted to have it suspended. "Undoubtedly," an important city official later admitted, "some very real difficulties were imposed on many people . . . perhaps if we had waited a year, it could have been done smoothly and with less inconvenience and cost and hardship to some people."[42]

The ordinance fell hardest on poor families, because fuel costs composed a larger percentage of their budget than higher income groups' and because of their fuel-buying habits. Working-class families often purchased their coal by the week in bushel lots from itinerant truckers, since they had neither the cash nor the storage space to buy larger amounts. In addition, the poor commonly used older, inefficient stoves for heating and cooking. While there had been considerable discussion in both the 1941 and 1946 hearings about special provisions for the poor, cash subsidies of a limited amount were provided only for public assistance recipients. The Bureau of Smoke Prevention attempted to educate low-income consumers as to the proper measures of firing smokeless coal rather than fining people for smoke violations. In 1948, for instance, only fifteen people were summoned before magistrates for smoke violations, as compared with 250 trucker violations. As one lawyer involved in the coal trade noted, the inspectors "deliberately went real easy on the smoke in poor neighborhoods and thought that they would get rid of that by eliminating the trucking of illegal fuel rather than by moving against poor Mrs. Murphy."[43]

In spite of the many difficulties with fuel supply, the heating season of 1947–48 showed a considerable improvement in air quality compared to previous years. An unusually mild winter aided in reducing the smoke palls. "PITTSBURGH IS CLEANER" reported the *Press* on February 21—the worst smogs were gone, homes were cleaner, and white shirts did not develop black rings around the collars. In statistical terms, Pittsburgh received 39 percent more of the available sunshine in the winter of 1947–48 compared with 1946–47; the hours of "moderate" and "heavy" smoke decreased approximately 50 percent during the same period; and dust-fall measurements also showed a slight decline as compared with 1938–39. In a publication entitled "The New Look in Pittsburgh," the United Smoke Council boasted that Pittsburgh was losing its reputation as "the Smoky City." During the next few years, heavy smoke nearly disappeared from the Pittsburgh atmosphere. In 1955, for instance, the Bureau of Smoke Prevention reported only 10 hours of "heavy" smoke and 113 hours of "moderate" smoke, as compared with 298 hours of "heavy" smoke and 1,005 hours of "moderate" smoke in 1946. Smoke elimination caused the public and the regulatory bodies to reorient themselves toward other constituents of air pollution such as odors, noxious gas, and fumes.[44]

Many of the benefits of smoke control in terms of improved air quality, more sunshine, and health were difficult to quantify, but the Bureau of Smoke Prevention did attempt to put a dollar figure on savings resulting from "greater cleanliness." The Bureau calculated the total savings as $26,808,000, or $41 per capita. This included savings on cleaning costs, reduced building depreciation, laundry bills, injury to vegetation, and fuel costs; it compared to a $20 per-capita figure utilized by the 1941 Smoke Commission as their estimate of the cost of the smoke nuisance to the average Pittsburgh resident. In addition, the bureau observed that domestic consumers would save from improved fuel combustion caused by the installation of "more efficient burning equipment." (The bureau, however, did not attempt to factor in the costs of new capital equipment.) Finally, the report noted that the range of other civic improvements underway in the Pittsburgh "renaissance" would not have taken place without successful smoke control.[45]

Because of the visible improvement in Pittsburgh air quality, public opinion shifted from limited to strong approval of the law. Public opinion surveys taken by a local advertising agency before and after the 1947–48 winter showed marked increases in favorable opinion. A poll in the summer of 1947 reported that 63.2 percent of those questioned (n=460) favored smoke control. A follow-up poll taken in August 1948 showed 75 percent in favor (no n is given). There were positive changes across all income levels, but low-income groups ($2,000 yearly or less) showed the largest increases, from 35.7 percent favorable to 62 percent. Of the total polled in 1948, 32 percent said they found conditions "much better" in terms of the amount of smoke and dirt in Pittsburgh; 40 percent found conditions "somewhat better"; 22 percent found them "about the same"; and 2 percent found them "worse." The high approval rate for the ordinance is noteworthy given that 57 percent of those using fuel other than natural gas (mostly low-volatile processed coal) complained about the fuel, mostly in regard to price. The public-opinion data are fragmentary and incomplete, but they do suggest that many Pittsburghers were prepared to pay higher prices for environmental quality.[46]

The Shift to Natural Gas

The improvements in Pittsburgh air quality that occurred after the implementation of the smoke-control ordinance in 1947–48 were not necessarily the result of the type of fuel and equipment substitutions projected by the 1941 policymakers. In 1941, and also to an extent in 1946–47, those involved in formulating and implementing the ordinance assumed that coal would continue to be the city's dominant domestic heating fuel for some years. The price of natural gas and oil was higher than that of coal through World War II and supplies were erratic. "Lower cost to consumers and availability at all times," noted the Bureau of Mines in 1943, "are the principal factors favoring the use of coal." Clean air would thus result from the use of smokeless coal produced from local bituminous or the use of equipment permitting smokeless combustion of bituminous. This strategy protected the interests of the local coal industry and miners who were concerned about inroads from competing fuels such as natural gas or oil. Repeated assurances by policymakers that the smoke problem could be solved without damaging the coal industry were clearly intended to secure the acquiescence of the industry and the United Mine Workers in the law's passage and implementation.[47]

While the use of low-volatile and processed coal (Disco) and smokeless coal-burning equipment did play a role in reducing smoke in 1947–48, they steadily declined in significance. Increasingly, low-priced natural gas, furnished by pipelines from the Southwest and stored in underground storage pools, became the dominant fuel used for Pittsburgh domestic heating. The rates of change for the city are striking. In 1940, 81 percent of Pittsburgh households burned coal and 17.4 percent natural gas (from Appalachian fields); by 1950, the figures were 31.6 percent for coal and 66 percent for natural gas. This represented a change in fuel type and combustion equipment by almost half the city households, most of which took place after 1945 (see figure 8.1).[48]

The shift to natural gas in Pittsburgh would undoubtedly have occurred without the smoke-control law because of price and convenience factors. Union contracts calling for higher wages and improved working conditions drove coal prices upward in 1947–49, while numerous strikes caused supply difficulties. Coal thus lost the advantages of lower cost and constant availability that had caused the Bureau of Mines in 1943 to predict its continued supremacy over competing fuels. Moreover, heating with gas rather than coal was much more convenient. Price and convenience, therefore, drove a fuel and equipment transition. The data reviewed here, however, also suggest that the Pittsburgh smoke-control ordinance accelerated the rate of change. Price and supply factors were operative in other cities as well as Pittsburgh, but in no other city in the nation was the rate of fuel change so rapid (see figures 8.2, 8.3). A definitive evaluation of the exact role of the ordinance in the transition, however, requires careful examination and comparison of fuel price and supply factors and smoke-control policy in other cities that are beyond the province of this chapter. The evidence does suggest that the coal producers and miners were correct that the smoke-control law would hasten the loss of their domestic markets and their jobs, although even they did not anticipate the rapidity of the change in Pittsburgh.[49]

Conclusions

This chapter has maintained that an examination of the passage and implementation of Pittsburgh's smoke-control laws from 1940–50 would yield insights helpful to those formulating energy policy today in regard to the energy-using behavior of individuals. The Pittsburgh ordinance required either fuel or combustion-equipment change in order to reach the goal of smoke-free air. Policymakers originally believed that processed coal, smokeless stoves, and mechanical stokers would provide the fuel and equipment necessary to bring clean air to the city, but natural gas rapidly became the preferred fuel in the Pittsburgh area in the postwar period. The availability of clean natural gas at a reasonable price ensured the success of the smoke-control policy, but the goals set by smoke-control policymakers would undoubtedly have been met without it. From the perspective of policy development and implementation, the critical questions involve the passage of the 1941 ordinance and its implementation in regard to domestic fuel consumers in 1946–48. These events took place under conditions of uncertainty as to fuel and equipment supplies and represented a clear value statement by private and public decision makers supported by numerous communitywide organizations for smoke-free air.

Campaign Roles and Strategies

The passage of the Pittsburgh smoke-control ordinance of 1941 must be understood in the context of the city's past policy failures to control smoke and St. Louis's achievement in devising a strategy to deal with the problem. St. Louis furnished both an example for emulation and a policy model that could be transferred. It is doubtful if Pittsburgh would or could have acted successfully in 1941 without St. Louis's success in controlling smoke.

The media, especially the *Pittsburgh Press*, played a crucial role in disseminating the image of a smoke-free St. Louis and insisting that Pittsburgh could also clean its skies.[50] Throughout the campaign, the *Press* consistently gave front-page coverage to antismoke activities and disseminated information concerning the costs of smoke and the public and private benefits that would ensue from control. Aside from one newspaper, the media did not seriously consider the costs of regulation or how they would be distributed.

Leadership in the campaign was provided by figures from both the private and public sectors ("entrepreneurs for the public good")—that is, by representatives from city government, from the media, and from voluntary organizations. The voluntary organizations established a network throughout the city for the dissemination of antismoke information. The Committee for the Elimination of Smoke provided a forum for discussing the issues, supplied "expert" testimony on technical questions, and kept the problem before the public. Unlike many other "study" commissions, it produced a report that actually became the basis for the ultimate legislation.

The heavy involvement of the public in the campaign, acting largely through the voluntary organizations, reflected a deep concern over the nuisance and health effects of smoke and a belief that a strong law could solve the problem. It also suggested a public willingness to pay the costs of smoke controls, although these costs may not have been well understood.

Finally, the statute became law in 1941 because it was supported by the leading organized groups in the community—government, business, and labor—as well as the media. The opposition to smoke control, whatever its extent, lacked a forum and an organization to present its position. The defense of smoke was not a popular stance.

Control and Implementation Strategies

The smoke-control ordinance set the policy goal of the elimination of dense smoke, as well as other constituents of air pollution, such as fly ash. It operated through the policy innovation of regulating the fuel and combustion equipment used by domestic consumers, a group never before regulated in Pittsburgh. While consumers experienced the higher costs of fuel and new combustion equipment directly, the Bureau of Smoke Prevention exercised control at a distance through coal dealers and truckers. In its relations with householders, the bureau stressed education and cooperation and only used legal proceedings against distributors who violated the ordinance. The network of voluntary organizations developed to pass the ordinance aided in obtaining public support for enforcement and to help "police" against offenders. Cooperation rather than conflict, however, marked the implementation.

Equity Questions

Because it required switching to a higher priced fuel or investment in new capital combustion equipment, the smoke-control ordinance created financing and equity problems.[51] While aware of these difficulties, policymakers who formulated the terms of the ordinance and made decisions about implementation never seriously considered subsidy mechanisms to cushion impacts on the poor. They assumed that the public and private benefits that would result from smoke control would overshadow costs and inequities. Public opinion polls taken in 1947 and 1948 suggest that public approval of the act was high and cut across income classes. Low-income groups, however, paid a higher percentage of their income for clean air than did middle- and upper-income groups, because they no longer could purchase cheap but dirty fuel. Also, coal miners suffered more than other labor groups because their employment was directly affected.

The evidence suggests, however, that labor and working-class groups were unclear as to their exact interests in the situation. The argument that smoke control would come through the use of processed fuel suggested that miners would retain their jobs, although fuel costs would be initially higher. More pertinent is that while lower income groups might have to pay a disproportionate share of their income for clean air, they also suffered more from the smoke because of the location of working-class residential areas in low-lying sections of the city. Workers, and especially their wives, therefore, approached the question of smoke control as victims of smoke as well as potential payers of higher costs. While several leaders of the United Mine Workers objected to the ordinance in 1947–48 as opening the way for natural gas to capture coal markets, representatives of other unions supported implementation.[52]

Roles of Technical Advisers and Technical Information

Technical advice was important in both the formation of a smoke-control policy and its implementation because the problem was viewed as susceptible to a technological solution. The key questions had to do with the adequacy of supplies of smokeless coal and of smokeless combustion equipment. There tended to be divergences of opinion between the technical committees representing the coal industry and those representing the public; there were also splits within the committees themselves over these questions. In both the policy formation and implementation phases, policymakers were willing to make limited concessions following the technical recommendations of industry but insisted on holding to their goal of smoke elimination. Ultimately, policymakers used the technical information and advice that suited their larger policy goals and discarded that which conflicted; the value choice for smoke-free air superseded the technical details.[53]

Impacts of the Policy

When implemented, smoke-control policy resulted in a number of public and private benefits and some costs. The aesthetic benefits were obvious and some were actually quantified by the Bureau of Smoke Prevention. The cleaner air also brought health improvements (although difficult to quantify). Many of these benefits had been predicted by the advocates of a new smoke-control ordinance in 1941. Smoke control was also extremely important as a preliminary step toward the urban renewal program known as the Pittsburgh Renaissance. Without the elimination of smoke, it would not have taken place.

On the consumer side, the costs of the policy fell most heavily on householders who had difficulty in obtaining and utilizing smokeless fuels or paid higher prices for fuel in the 1947–48 heating season. These were mostly lower income consumers. Thus, the policy had regressive distributional effects in that the poor paid a larger share of their income for smoke-free air than did other income groups. Heavy costs were also borne on the supply side by the coal industry and the miners. While natural gas would ultimately have replaced bituminous coal as the preferred domestic fuel in Pittsburgh, the smoke-control ordinance accelerated the rate of transition. The rapidity of this shift had not been predicted, although it ensured the success of the ordinance.

The case of Pittsburgh smoke control suggests that changing energy-using behavior and accomplishing environmental goals is often the result of a long complex process of trial and error. A successful model that could be copied, heavy media support, capable powerful leadership from the private and public sectors, the utilization of voluntary organizations to generate public support, and strong public concern were all required. While economic factors were considered in policy formulation and implementation, they were not more important than, and were perhaps subsidiary to, values factors and political elements. Ultimately, economic forces in the form of cheap natural gas made the policy a complete success, but this development came after rather than before the determination to rid Pittsburgh of smoke.

Notes

This chapter is based on research supported by the National Science Foundation, Program in Ethics and Values in Science and Technology, Grant No. SS78-17308. Any opinions, findings, and conclusions or recommendations expressed in this publication are those of the author and do not necessarily reflect the views of the National Science Foundation. The authors would like to thank Cliff Davidson, Samuel P. Hays, Sara Kiesler, Steven Klepper, Pat Larkey, Richard Smith, and Lee Sproull for helpful criticisms and G. David Goodman, Kenneth Koons, and Todd Shallat for aid in collecting and analyzing materials.

1. For discussions of energy transitions, see Sam H. Schurr and Bruce Nershert, *Energy in the American Economy, 1850–1975: An Economic Study of its History and Prospects* (Baltimore, 1960); Lewis J. Perelman (ed.), *Energy Transitions: Long-Term Perspectives* (Boulder, 1981); and National Academy of Sciences, *Energy in Transition 1985–2000* (San Francisco, 1979). For energy policy, see Robert Stobaugh and Daniel Yergin (eds.), *Energy Future: Report of the Energy Project at the Harvard Business School*, 2nd ed. (New York, 1979), and Sam H. Schurr, et.al., *Energy in America's Future: The Choices Before Us* (Baltimore, 1979). Paul C. Stern, "Social Science Perspectives on Energy Conservation: A Working Paper," prepared for the National Academy of Sciences Committee on Energy Conservation, 1980.

2. For a discussion of the uses of historical analogy, see Bruce Mazlish (ed.), *The Railroad and the Space Program: An Exploration in Historical Analogy* (Cambridge, 1965), especially 1–18.

3. Lawrence Tilly, "Metropolis as Ecosystem," in Charles Tilly (ed.), *An Urban World* (Boston, 1974), 466–473. Comments were made about smoke in Pittsburgh as early as 1800. For a review of nineteenth century developments see John O'Connor, Jr., "The History of the Smoke Nuisance and of Smoke Abatement in Pittsburgh," *Industrial World* (March 24, 1913).

4. *Ibid.*, 2. See, for example, J.W. Henderson, Bureau Chief, Pittsburgh Bureau of Smoke Regulation, "Smoke Abatement Means Economy," *Power* (July 24, 1917); R. Dale Grinder, "The Battle for Clean Air: The Smoke Problem in Post-Civil War America," in Martin V. Melosi (ed.), *Pollution and Reform in American Cities, 1870–1930* (Austin, 1980), 89. An excellent analysis of attempts to regulate Pittsburgh and Allegheny County air pollution with an emphasis on events since 1967 is Charles O. Jones, *Clean Air: The Policies and Politics of Pollution Control* (Pittsburgh, 1975).

5. Victor J. Azbe, "Rationalizing Smoke Abatement," in *Proceedings of the Third International Conference on Bituminous Coal* (Pittsburgh, 1931) II, 603.

6. Osborn Monnett, *Smoke Abatement, Technical Paper 273, Bureau of Mines* (Washington, 1923); H.B. Meller, "Smoke Abatement, Its Effects and Its Limitations," *Mechanical Engineering* (November 1926) V. 48, 1275–1283.

7. Interview with Abraham Wolk, April 26, 1973, in "An Oral History of the Pittsburgh Renaissance," Archives of Industrial Society, University of Pittsburgh, hereafter referred to as "Oral History." Letters defending air pollution still appear in the Pittsburgh newspapers at times when tighter air pollution controls are discussed for industry.

8. *Pittsburgh Press*, February 9, 1941, hereafter referred to as *Press*; Jennie Herron to Dr. Edward Weidlein, September 26, 1939, in Edward Weidlein Papers, Mellon Institute of Research, Pittsburgh.

9. David L. Lawrence, "Rebirth," in Stefan Lorant, *Pittsburgh: The Story of an American City*, 2nd ed. (Lenox, MA, 1975). Lawrence himself was the boss of the county Democratic party. Wolk noted that in 1941 "Dave Lawrence never said a word to me against it . . . ." Noted in Interview with Abraham Wolk, April 26, 1973, "Oral History." One commentator observed that it was important that Alexander, a physician, was prominently involved because "he was able to generate some real fear on the part of the people that maybe after all this was dangerous to health as well as being an economic proposition." Interview with William Willis, April 16, 1973, "Oral History." See, also, Dr. I. Hope Alexander, "Smoke and Health," United Smoke Council of Pittsburgh, 1941. Alexander noted that "The medical profession is in accord with the statement that smoke is a health menace of major proportions . . . For the present, however, there are few scientific facts to definitely establish a case of cause and effect."

10. Oscar H. Allison, "Raymond R. Tucker: The Smoke Elimination Years, 1934–1950" (unpublished dissertation, St. Louis University, 1978). "Cities Fight Smoke," *Business Week*, April 6, 1940, 33–34.

11. *Press*, February 3, 4, 12, 1941. Also, see *Pittsburgh Post-Gazette* for the same dates. *Press*, February 16, 19, 20, 21, 1941; *Business Week*, February 22, 1941, 30; "Smoke Meeting," February 20, 1941, Civic Club Records, Archives of Industrial Society, University of Pittsburgh. *Press*, February 19, 21, 22, 1941; and *Pittsburgh Sun Telegraph*, February 20, 22, 1941.

12. Interview with Gilbert Love, December 19, 1972, "Oral History." See, for instance, *Press*, February 3–5, 8, 9, 14 and March 1, 2, 13, 20, 21, 1941.

13. *Report of the Mayor's Commission for the Elimination of Smoke* (Pittsburgh, 1941), in Civic Club Records, hereafter referred to as *Report of the Mayor's Commission; Press*, February 19, 1941.

14. *Ibid.*, March 13, 1941.

15. See, for example, *Press* and *Post-Gazette*, March 13, April 30, June 24, 1941; M. Jay Ream to __________, April 3, 1941, Civic Club Records; "Notes Taken on Smoke at Annual Meeting," May 8, 1941, Civic Club Records. See, also, "List of Organizations Co-Operating with Civic Club on Smoke Elimination," *ibid.*

16. The comment is that of John Robin, the Mayor's Executive Secretary, in "Fourth Meeting of the Mayor's Commission for the Elimination of Smoke," March 12, 1941, 74, Pittsburgh City Council Archives.

17. "Hearings Before the Pittsburgh Smoke Commission," April 15, 1941, 49, City Council Archives.

18. *Report of the Mayor's Commission*, 22; "Hearings Before the Pittsburgh Smoke Commission," April 15, 1941, 38–39, City Council Archives.

19. Proceedings of the Tenth Meeting of the Pittsburgh Smoke Commission," April 24, 1941,

57, City Council Archives; "Proceedings of the Fifth Meeting of the Pittsburgh Smoke Commission," March 31, 1941, 23, 69–78; "Proceedings of the Sixth Meeting," April 7, 1941, 38–39; "Proceedings of the Fifth Meeting of the Pittsburgh Smoke Commission," March 31, 1941, 69; "Hearings Before the Pittsburgh Smoke Commission," April 15, 1941, 49–50; and "Proceedings of the Twelfth Meeting," May 14, 1941, 22–32.

20. "Proceedings of the Twelfth Meeting of the Pittsburgh Smoke Commission," May 14, 1941, 13–32.

21. *Press*, May 15, 1941.

22. See "Hearings of the Coal Operators Association of Western Pennsylvania Before the Mayor's Commission for Elimination of Smoke," May 13, 1941, Pittsburgh City Council Archives; "The Western Pennsylvania Coal Operators Association Reports on a Plan to Reduce Air Pollution in Greater Pittsburgh," in Civic Club Records, Archives of Industrial Society.

23. Lorant, *Pittsburgh*, 377, 380.

24. *Report of the Mayor's Commission*, 18.

25. *Ibid.*, 19.

26. The Ringelmann Chart was a method of estimating smoke density through a visual comparison of a column of smoke with a shaded chart supplying different ratings to various densities. For a discussion of the chart and its weaknesses, see L.S. Marks, "Inadequacy of the Ringelmann Chart," *Mechanical Engineering*, 59 (September 1937): 681–685. This provision was also included in the St. Louis ordinance, and represented an advance in air pollution control as compared to earlier ordinances that had focused on dense smoke only.

27. *Report of the Mayor's Commission*, 37.

28. The railroads wanted an alteration in the amount of time they were allowed to produce dense smoke. Jones & Laughlin Steel Corporation opposed the provisions in regard to fly ash and fumes on the grounds that it would be "a burden to the steel business." Their representative argued that fly ash and fumes involved "air pollution" and not smoke. See Committee on Hearings, Pittsburgh City Council Records, June 24–25, 1941. *Press*, July 7, 1941.

29. *Press*, October 25, 1942 and April 25, 1943; "Memo of the Meeting of the Associated Municipalities," November 19, 1942, Civic Club Records. *Press*, January 20, and March 11, 1942.

30. See, for instance, *Press*, Nov. 5, 7, 1945. In its 1944 *Report for Stationary Stack Conditions*, the Bureau of Smoke Prevention noted that 75 percent of the heating equipment of non-domestic structures was "more or less antiquated." 5. See "The United Smoke Council of the Allegheny Conference," in United Smoke Council Records, and "Minutes," Meeting of the United Smoke Council of Pittsburgh and Allegheny County, October 18, 1945, United Smoke Council Records.

31. Roy Lubove, *Twentieth Century Pittsburgh* (New York, 1969), 106–121. M. Jay Ream to Richard King Mellon, June 9, 1943 and Mellon to Ream, n.d., in Allegheny Conference on Community Development files, Pittsburgh. In 1945, Mellon noted that smoke control was "the most important project in Allegheny County."

32. "Minutes, Meeting of the United Smoke Council of Pittsburgh and Allegheny County." Nov. 3, 1945.

33. Lorant, *Pittsburgh*, 390.

34. See C.K. Harvey, "Report to the Technical Committee on Smoke Abatement," February, 1946, Allegheny Conference on Community Development, in Abraham Wolk Papers, Archives of Industrial Society; C.K. Harvey to M. Jay Ream, January 17, 1946, United Smoke Council files, Allegheny Conference on Community Development.

35. Harvey, "Report to the Technical Committee on Smoke Abatement," 1–2, 15. Park Martin, Director of the Allegheny Conference, supported the October 1, 1946 date with the comment that "there probably never will be a 'just right time.'" "Minutes," Special Meeting of the United Smoke Council of the Allegheny Conference, February 23, 1946, Civic Club Records.

36. Dr. Robert Doherty to Park Martin, March 29, 1946, copy in Miscellaneous Correspondence of Dr. Edward Weidlein, Mellon Institute.

37. See, for instance, "Conference on Smoke Before the County Commissioners," Dec. 20, 1945, in Civic Club Records; "Minutes," Meeting of the United Smoke Council," Dec. 13, 1945.

38. The Western Pennsylvania Conference on Air Pollution, *The Role of Coal in Smoke Abatement* (April 9, 1946), copy in Wolk Papers; Mrs. Gregg to Members of the Executive Committee, March 29, 1946, Weidlein Papers.

39. "Proposed Solution of Smoke Abatement Problem," April 2, 1946, draft in papers of Dr. Edward Weidlein, Mellon Institute. See, also, Executive Minutes of the A.C.C.D., April 2, 1946, Allegheny Conference.

40. *Press*, April 17–18, 1946. The various petitions are available in the City Council Archives.

41. There are discussions of methods of enforcement by Dr. Sumner B. Ely in "Minutes," United Smoke Council, December 13, 1945 and in *Report of Stationary Stacks*, Bureau of Smoke Prevention, Department of Public Health, Pittsburgh, 1948, 9–10. See, also, Interview with Albert Brandon, July 21, 1972, Pittsburgh Renaissance Project. Brandon was Assistant City Solicitor in charge of the enforcement of smoke control.

42. Interview with John Robin, September 20, 1972, "Oral History." *Press*, December 16, 1947, January 13, 1948. For comparative costs of competing fuels, see Peoples Natural Gas Company, *Operating and Financial Statistics Years 1943–1972* (Pittsburgh, 1973), 23a. This source shows that in 1948 bituminous coal was $10.70 a ton while "Disco" or treated coal was $17.10.

43. The U.S. Department of Labor reported a tabulation of data from 42 cities in 1940 showing the percentage spent by different income groups for fuel and electricity. Noted in "The Western Pennsylvania Coal Operators Association Report on a Plan to Reduce Air Pollution in Greater Pittsburgh," 15, in Civic Club Archives. For discussion of these issues, see A.P.L. Turner to David L. Lawrence, October 31, 1947; T.B. Lappen, Spiritual Director, Society of St. Vincent De Paul, to Junction Coal & Coke Company, April 17, 1946; and Howard D. Gibbs, Executive Secretary, Retail Coal Merchants Association, March 24, 1946, all in Wolk Papers. Interview with Ralph German, February 27, 1979; Statement of John J. Grove, Public Relations Director, A.C.C.D., in "Oral History." Bureau of Smoke Prevention, Department of Public Health, *Report on Stationary Stacks 1948* (Pittsburgh, 1948), 11.

44. *Press*, February 2, 1948; United Smoke Council, *That New Look in Pittsburgh*, (n.p., n.d.). City of Pittsburgh, Department of Public Health, Bureau of Smoke Prevention, *Report 1955* (Pittsburgh, 1955), 6. Further efforts by industry to control smoke emissions and railroad conversion from coal-burning to diesel-electric locomotives also played an important role.

45. *Report of the Mayor's Commission*, 10; Bureau of Smoke Prevention, Department of Public Health, *Report on Stationary Stacks 1951* (Pittsburgh, 1951), 13. For new equipment, see *ibid.*, 15. The following chart supplies estimates for the costs of equipment (located in the Harvey Report on Smoke Abatement, 15):

Equipment	Life Maintenance	Cost Installed	Depreciation	Interest		
Gas Furnace	$20	$550	$27.50	$33	1%	$5.50
Coal Furnace	$20	$433	$21.65	$24.98	3%	$13.00
Gas Conversion Units	$15	$163	$10.87	$ 9.78	1%	$ 1.63
Stoker	$ 7	$283	$40.57	$17.04	5%	$14.20

46. *Press*, June 22, 1947 and August 29, 1948. The Bureau of Smoke Prevention also reported the results in its 1948 report. Smoke control appeared as an issue in the 1949 Democratic primary election. Edward J. Leonard, who had voted against the ordinance in 1941 on the grounds that it would hurt the little man, ran against David Lawrence. Lawrence was renominated by a vote of approximately 77,000 to 50,000 in a hard campaign, with Leonard winning many of the labor wards. Lawrence later claimed that the primary "was the final climax for smoke control." See Lorant, *Pittsburgh*, 398–402. However, smoke control was not the only issue in the election. The *Press* suggested in an editorial on March 28, 1946, that Pittsburghers would have to "Make small financial sacrifices" for smoke control.

47. Arno C. Fieldner, *Recent Developments in Fuel Supply and Demand*, U.S., Department of the Interior, Bureau of Mines (November 1943), 11. For discussions of the high price and erratic supply conditions of natural gas in 1941, see transcripts of the Mayor's Commission for Smoke Elimination, Mar. 12, 1941, 17–19, 60; for March 28, 1941, 7–8; and March 31, 1941, 74–76. On April

19, 1946, the *Press* cited higher prices for home heating by natural gas as compared to coal. *Ibid.*, April 7, 1946, Feb. 15, 1947.

48. See "Jobs for Inches," *Business Week*, Dec. 29, 1945, 19; "Natural Gas is on the Up," *ibid.*, March 13, 1948, 26; and *Annual Reports* of the Philadelphia Company (Equitable Gas Company), for 1946, 15; 1948, 16–17; 1949, 21–22. Data on household fuel use are available from the 1940 and 1950 Censuses of Housing. See U.S. Department of Commerce, Bureau of the Census, *Housing: Characteristics by Type of Structure, 16th Census of the U.S.* (Washington, 1945), and *Census of Housing: 1950* Vol. I *General Characteristics 17th Census* (Washington, 1953).

49. An updated and unsigned memo in the Wolk papers (probably from 1949), notes: "The cost of heating by natural gas is so close to that of coal that hundreds are converting from coal to gas. If the price of coal is forced up, the miner will lose a large part of his market. A six room house, stoker heated, costs $100 per year ($40 overhead, $60 for coal). The gas companies can heat it for $105 to $110 per year." The conversion to gas was also accelerated by coal strikes in 1946–49 which caused uncertainty in regard to bituminous supplies and price. The coal operators and the *Press* charged that the strikes and higher prices were causing coal to lose its markets to competing fuels. See *Press*, September 29, October 4, 1949 and March 8, 22, 1950. Even in St. Louis the rate of change to natural gas and oil was slower, but here a "maze of contracts of affiliated utilities" caused problems on the supply side. See Allison, "Raymond Tucker," 73–76. Representatives of the United Mine Workers charged in the post-war period that there was a conspiracy in Pittsburgh to drive out coal in favor of natural gas. See *Press*, December 21, 1945; March 8, 9, April 23, December 2, 5, 1947; and July 3, 1948. In its 1950 *Annual Report*, Pittsburgh Consolidation Coal noted that, "It was obvious five years ago that in many of the areas into which our coal normally moved there would be a gradual decrease in the amount of coal used for domestic heating as natural gas, with its greater convenience, found its way into these markets." In its 1949 *Annual Report*, the Equitable Gas Company, which served more households in Pittsburgh than any other gas utility, observed that smoke abatement legislation and the high cost of coal were increasing their domestic sales.

50. For the importance of cognitive images, see Herbert A. Simon, "The Birth of an Organization: the Economic Cooperation Administration," *Public Administration Review*, 13:227–236.

51. For a concise discussion of equity questions in environmental and energy policy, see Lester C. Thurow, *The Zero-Sum Society: Distribution and the Possibilities for Economic Change* (New York, 1980), chaps. 2 and 5.

52. For discussions of the split among labor unions in connection with support of the smoke control ordinance, see *Press*, Mar. 9, September 9, and December 2, 1947.

53. For discussions of the role of "experts" in science and technological policy, see A. Teich (ed.), *Science and Public Affairs* (New York, 1974); J. Primack and Von Hippel, *Advice and Dissent: Science in the Political Arena* (New York, 1974); and Dorothy Nelkin (ed.), *Controversy: Politics of Technical Decisions* (Beverly Hills, 1979).

CHAPTER IX

Railroad Smoke Control

The Regulation of a Mobile Pollution Source

Introduction

The last two generations have witnessed significant progress in the improvement of air quality in the United States. Control of pollutants from mobile sources such as internal combustion engines used in automobiles and trucks, however, have shown less reduction than emissions from other sources. The persistence of relatively high pollution rates from vehicles highlights the special problems encountered in reducing emissions from mobile as compared to stationary sources. The automobile and motor truck are not the first mobile sources of air pollution to affect urban air quality adversely.[1] The steam locomotive, a technology utilized to transport many millions of passengers and tons of freight for more than a century, also seriously affected the quality of urban air through its production of smoke.

The problem of controlling smoke from the steam locomotive included many of the elements present in attempts to regulate emissions from vehicles powered by the internal-combustion engine. There were technical questions involving technology and fuel, as well as policy questions involving emission standards and enforcement. Regulations, however, were primarily municipal statutes rather than national policy.

Fuel, of course, was a critical element. American railroads had originally burned wood, reflecting the cheapness and wide availability of this resource. This contrasted with the British experience, where the scarcity of wood required early reliance on mineral fuels. As wood supplies diminished, and as the railroads penetrated regions with plentiful coal deposits, fuel substitutions gradually took place. The initial mineral fuel used by the railroads was clean-burning anthracite, but as the tracks moved further west, away from the anthracite mines (northeastern Pennsylvania), they adopted the more readily available (and hence cheaper) high-volatile bituminous coal. During the 1860s and 1870s, most of the technical problems associated with the burning of coal had been solved, and by 1880 coal, mostly bituminous, constituted more than 90 percent of locomotive fuel.[2]

The dominance of bituminous coal as a locomotive fuel was based on its wide distribution, its easy accessibility for mining, and its high average heat content. During the first third of the twentieth century, commercial mining of bituminous coal was carried on in twenty-nine states, with over 90 percent of the total coming from seven states east of the Mississippi River. Next to labor, fuel constituted the most expensive single material purchased and used by the railroads, ranging between a high of 11.2¢ (1920) and a low of 5.6¢ (1933) per revenue dollar between 1911 and 1950.[3] When fuel costs were high, the railroads paid special attention to factors that would help them achieve the maximum amount of transportation at a minimum fuel expense.

Coal was important to the railroads not only as their principal fuel, but also because it constituted their largest single item of revenue tonnage. In 1919, the first year for which exact figures are available, Class 1 railways hauled over ten million cars of coal out of a total of 45 million cars, or 22 percent. In 1940, the ratio dropped to about 19 percent. (Tonnage figures were actually larger because coal can be loaded into freight cars more heavily than any other commodity except iron ore and pig iron.) Fuel also constituted more than 50 percent of the non-revenue tonnage hauled by the railroads. Some railroads—such as the Norfolk and Western, the Chesapeake and Ohio, the Pennsylvania, and the Baltimore and Ohio—specialized in coal shipping.[4] The fact that coal was the largest tonnage item carried by the railroads and, as fuel, constituted a critical item in their cost picture, gave them a vital interest in both their own efficiency of operation and factors affecting coal use in other markets.

Railroad locomotives often generated copious amounts of dense smoke and cinders, not only because they burned high-volatile bituminous coal but also because they were inefficient fuel consumers compared to stationary steam plants. The restricted space in the locomotive cab for the firebox made extremely high rates of combustion necessary, requiring a strong draft. These conditions resulted in the loss of a considerable amount of the coal heat at the stack and the emergence of cinders and gases from the stack as smoke. Furthermore, locomotive smoke, as compared to smoke produced from other industrial sources, was particularly offensive because it was discharged at relatively low levels and often dispersed over a wide area by moving trains.[5]

Railroad locomotives actually varied considerably by size and design, depending on their function. Intercity passenger trains differed from those used for suburban runs, and both of these differed in traction requirements from freight locomotives. Switching engines, intended for use in railroad yards, created a special type of problem, with their frequent stops and starts under heavy loads. Railroad terminals, where locomotives clustered with banked fires or for firing-up, were a particular nuisance because they were often located in densely populated areas. Some authorities considered railroad-terminal smoke to be the most acute smoke abatement problem in the city.[6]

Incentives for the Railroads to Control Locomotive Smoke Emissions

The primary incentives for the railroads to control locomotive smoke can be divided into internal and external factors.[7] Internal conditions involved factors that drove the railroad companies to make changes that reduced smoke in order to improve their efficiency and economy of operation and to meet special operating situations. External elements involved pressures from the outside environment—primarily municipal smoke-control regulations that forced the railroads to undertake smoke-control measures that they would not have voluntarily adopted because they were perceived as adding to rather than diminishing the costs of operation. Actually, the cost effectiveness of a particular innovation or action was not always clear, and in practice there was often considerable overlap between "internal" and "external" situations.

Internal Conditions

SPECIAL OPERATING SITUATIONS

The most obvious special situation requiring action by the railroads to deal with the smoke problem involved the need for tunnels to secure access to city centers. Steam locomotives were unacceptable in subaqueous tunnels, especially with grades, because the subterranean passages could not be adequately ventilated. Several fatalities had occurred, for instance, in subaqueous tunnels where steam was used. The solution used to deal with the problem was electrification of the lines. In 1895, in what may have been the first case where electricity supplanted steam as motive power in main-line railroad service, the Baltimore and Ohio Railroad (B&O) electrified a 3.6-mile portion of its Baltimore belt line, including the route through a 7,000-foot tunnel. This gave the B&O direct access to the city and allowed it to compete with the Pennsylvania Railroad (PRR) for the Baltimore trade.[8]

Another important example of a technology substitution that occurred because of the smoke problem concerned the attempts of the PRR to secure direct access to Manhattan Island. Here, after considering construction of a bridge, the PRR management decided that the experience of the B&O in electrifying part of its Baltimore belt line, including a long tunnel, justified its construction of subaqueous tunnels. Construction of the tunnels, an electrified line, and a new station took nine years, and in 1910 the PRR began hauling passengers into its new midtown Manhattan station using electric power.[9]

FUEL CONSERVATION

The production of dense ("black") smoke is a sign of inefficient fuel consumption, and the railroads were sensitive to "the intimate relation between smoke abatement and fuel economy," especially when fuel costs were high. The "economical utilization" of fuel, railroad managers maintained, could only be secured by continual producer supervision over coal preparation, careful coal handling in the transition from cars to locomotive tenders, the fitting of coal types to particular service requirements, the maintenance of good locomotive operating conditions, and the education of engineers and firemen in proper firing.[10]

In order to secure fuel savings, the railroad companies, which were masters of bureaucratic organization, often created central fuel bureaus that exercised supervision over all states of fuel distribution and consumption. Beginning after 1900, instructors in fuel economy were assigned to the different rail districts "to ride and inspect locomotives, observe the handling and firing of the locomotives, instruct the engine crews where necessary in correct methods of handling and firing, report power conditions, give attention to the methods of cleaning, banking and preparing fires in locomotives and the coaling and handling of the locomotives while in the charge of engine terminal employees, and in general to endeavor to bring about every practicable improvement in the detail conditions which effect economy in the use of locomotive fuel."[11] The fuel consumption bureaus usually included smoke inspectors, whose duty it was to report locomotives emitting black smoke.

Industry self-regulation of smoke emissions had primarily two motivations. One motivation related to the fact that dense smoke was a sign of poor combustion and fuel waste. A second motivation involved the desire to avoid regulation and interference with operations by municipal smoke inspectors. Both motives combined to produce relatively large staffs of railroad smoke inspectors. In Pittsburgh, for instance, in the 1920s, the city's five railroads maintained a corps of twenty inspectors to observe locomotives and to correct conditions responsible for smoke. The number of railroad inspectors contrasted with only four municipal smoke inspectors. In 1943, the railroad companies employed twelve inspectors who reported to the Pittsburgh Bureau of Smoke Control on a daily basis. Similar cooperation between the railroad companies and bureaus of smoke control also occurred in Chicago, Kansas City, and Hudson County, New Jersey.[12]

Railroad smoke inspectors were vulnerable in times of financial stringency. Often hired in response to public agitation over smoke pollution, they might disappear when agitation subsided and regulation was deemphasized. In addition, fuel savings became less critical when prices were low, reducing the need to maintain vigilance in the search for fuel-wasting and smoke-producing firemen and engineers. As one Pennsylvania Railroad Division Manager remarked in 1909, the cost of supervising railroad smoke "is a very serious burden on the cost of operation, and while the railways would not provide such supervision but for the belief that it will yield adequate return or from realization of the duties which the railways owe the public, there must be a limit to the amount of money which they can so expend."[13]

PROTECTING PASSENGER COMFORT

Railway management wanted to reduce locomotive smoke from passenger trains in order to protect passenger comfort as well as conserve fuel and avoid regulation. Competition for passenger traffic was intense, and the reduction of smoke and cinders was one means to enhance travel enjoyment. As early as the 1850s, the B&O and the PRR were experimenting with smoke-control devices on their passenger trains. In 1868, the Illinois Central Railroad placed special instructions in their passenger locomotive cabs concerning smoke prevention. The list concluded with the statement that "much of the annoyance from smoke and coal dust will be prevented and a large saving in fuel effected by attention to the above rules."[14] Similar concern over comfort, however, did not apply to the freight and switching trains that comprised the bulk of railroad traffic.

Pressures from the External Environment

The main external pressures on the railroad corporations to reduce smoke came from the municipalities. The smoke-control movement in American cities extended from approximately the late nineteenth century to the 1950s. Smoke pollution was particularly heavy in industrial cities, such as Pittsburgh and St. Louis, that depended on bituminous coal for their industrial and domestic fuel. In many of these cities, topographical and climatic features produced inversions that trapped the smoke and particulates and created smoke palls. As these cities grew in industry and population, air-quality conditions worsened. While smoke signified economic progress to some urbanites, others became concerned with its negative affects on the quality of urban life. Smoke-control advocates listed damage to buildings, higher costs of cleaning, loss of sunlight, and chronic health impairment as costs of smoke pollution. Concerned about these negative impacts and their effect on the ability of the cities to continue to attract population and business, civic associations, women's clubs, and engineering associations spearheaded a drive for smoke regulation in the early twentieth century.[15]

In 1912, the U.S. Bureau of Mines, which had conducted several studies of smoke pollution, polled a number of cities concerning their smoke control regulations. While only twenty-nine of three hundred cities with under 200,000 inhabitants had smoke ordinances, all twenty-eight cities with over 200,000 had regulations. These laws, however, differed greatly in terms of specificity and sources regulated. No city aside from Los Angeles, for instance, mentioned private residences in their ordinance. Several ordinances prohibited the emission of "dense smoke" within the city limits as a nuisance but did not attempt to define standards or time limits. Regulations in other cities, such as Baltimore, Chicago, and Cincinnati, prohibited the emission of dense smoke for more than a specified number of minutes in an hour (usually six to eight minutes). A few cities attempted to supply a scientific grading of smoke density utilizing the Ringlemann chart, a visual grading mechanism.[16]

The industrial cities with bad smoke conditions were also railroad centers, and urban reformers often pinpointed locomotives as especially offensive smoke generators. Estimates of the amount of the total smoke problem created by the railroads ranged between 20 and 50 percent in cities like Chicago and Pittsburgh. The Committee of Investigation on Smoke Abatement and Electrification of Railway Terminals of the Chicago Association of Commerce, for instance, reported in 1915 that while steam locomotives burned 11.9 percent of the total fuel consumed within Chicago, they were responsible for 22 percent of the city's visible smoke and 7.4 percent of the solid constituents in the atmosphere. Locomotives engaged in yard freight service were the worst offenders.[17]

In the regulations reviewed by the Bureau of Mines in 1912, railroads were usually dealt with under the general smoke ordinances. In those cases where they received specific mention, they were either exempted from regulation under certain conditions or were subjected to more stringent controls than other classes of smoke producers. In Louisville and Rochester, for instance, the smoke ordinance did not apply to locomotives entering, leaving, or in transit through the city. In contrast, Jersey City, which possessed many railroad marshaling yards, had an ordinance that prohibited the emission of dense smoke from any engine or locomotive within the city limits and provided for fines in cases of injury to health or property.[18]

During the years between 1912 and 1941, the number of cities with smoke-control ordinances increased, as did the severity of the standards. In 1939, the Bureau of Mines issued another survey of smoke ordinances, this time of the nation's eighty largest cities. Of the eighty, twenty-five followed a model ordinance prepared by a group of engineering societies in 1924 calling for the emission of dense smoke for no more than six minutes an hour. Fifty cities used the Ringlemann chart as a means to measure smoke density. Other cities had a variety of standards, with several allowing locomotive smoke for one out of every fifteen minutes. More significantly, nine cities exempted railroads from regulation, reflecting the political power of railroads within those communities.[19]

Beginning in 1940 with St. Louis, and followed by Pittsburgh in 1941, a number of cities adopted ordinances with more stringent regulations that required the use of smokeless fuels or technology and controlled domestic sources of smoke, as well as smoke from industries, commercial establishments, and railroads. The Pittsburgh Smoke Control Ordinance of 1941 was the most stringent in the nation in the 1940s and permitted railroads to produce dense smoke (defined as No. 2 on the Ringlemann chart) for no more than one minute an hour. It was adopted over railroad protests that the ordinance made it impossible for them to operate.[20]

Whatever the terms of the various ordinances, enforcement tended to be sporadic, depending on the orientation of the administration in power. Regulatory bureaus were often understaffed and underfinanced and lacked substantial powers of implementation. At times of popular outcry against the smoke nuisance, smoke inspectors would become more active and occasionally haul railroad companies into court for violations of the ordinances. The railroads, in turn, would increase the number of smoke inspectors on their staffs and make efforts to reduce locomotive smoke. Conditions generally returned to their original state, however, once the crusade had ended, although occasionally there were some permanent improvements because of retrofits. Municipal smoke inspectors often tended to be sympathetic to railroad problems in reducing smoke and cooperated with them in seeking solutions, rather than resorting to the sanction of legal proceedings.[21]

There are several possible explanations for the behavior of municipal smoke bureaus. One explanation derives from the common professional training and associations of municipal smoke inspectors and railroad personnel. In the twentieth century, regulatory bureaus were usually headed by mechanical engineers who viewed themselves as primarily technological problem-solvers. As such, they shared a professional interest with railroad managers in seeking to improve railroad operating procedures in order to bring about efficiency and economy of operation. "Smoke abatement," they argued, "was a problem for the Engineer and not the Legislator." Municipal smoke inspectors and railroad personnel also shared membership in the Smoke Control Association of America. The latter frequently delivered papers at the association's annual meetings concerning progress in locomotive smoke control, while papers given by city inspectors at these meetings often cited the railroads for their cooperative attitude.[22]

Another set of plausible explanations for the actions of these regulators derives from the various theories involving regulatory behavior. The evidence suggests, for instance, that the regulated corporations—the railroads—had actually "captured" the regulatory bureaus in regard to their normal operations.[23] Given the importance of the railroads to urban economic vitality in the 1900–1945 period, and given their political power as a major industry, municipalities were reluctant to burden them with regulatory procedures and fines that hampered their operation. On the other hand, municipal administrations were also susceptible to political pressures and agitation from various groups interested in smoke control, such as businessmen's associations and women's clubs, for more stringent law enforcement. The administrations were often forced to respond to these demands, resulting in a period of tougher law enforcement. Normally, however, given the power and importance of the railroads, it was rational for smoke-control bureaus to follow a policy of cooperation and education in enforcement. At the same time, it was in the interests of the railroads to appear to be making reasonable attempts at smoke control.

Whenever possible, the railroads attempted to keep disciplinary action against the personnel responsible for smoke violations in their own hands. When city officials used legal force in enforcing ordinances, the railroads considered the action "overzealous" behavior on the part of the municipality. On the other hand, some railroad officials held that municipal action reflected a failure of the railroads to police themselves. When discipline was required, noted the chief smoke inspector of the Chicago and Northwestern Railway, the railroad itself must provide it. Such discipline could take the form of reprimands, layoffs, or even discharges from the service. In 1949, for instance, there were 102 suspensions and 254 reprimands delivered by the railroads against personnel involved in violations of the Pittsburgh Smoke Control Law.[24]

The railroads were particularly concerned about laws that might take decisions regarding major capital expenditures out of their hands. This actually happened in 1903 in a case involving the New York Central Railroad. The company used a two-mile tunnel that was often filled with smoke (a partially covered cut) to reach their Manhattan terminal. It was considering the electrification of the line but had not made a final decision by 1902. In that year, however, seventeen people died in a crash in the tunnel caused by smoke obscuring the signal lights. In 1903, in reaction to the crash, the New York legislature passed a law prohibiting the use of steam locomotives south of the Harlem River after July 1, 1908. The New York Central faced the choice of either electrifying its line or abandoning its Manhattan terminal. The former had to be its choice, although the timing was not of their choosing and was forced by the legislative action. Similarly, in Pittsburgh and Allegheny County in 1946 and 1947, passage of strict smoke-control regulations added to the pressure on the Pittsburgh-area railroads to convert to diesel-electric locomotives.[25]

Railroad Approaches to Dealing with the Smoke Problem

There are primarily four methods by which society can deal with the problems of technology-induced air pollution, besides ceasing to use that technology completely. These four approaches involve training human operators to use the technology differently than they had previously (the "human fix"); retrofitting existing equipment to reduce emissions of pollutants; switching from a dirty fuel to a cleaner fuel ("fuel switching"); and substituting a different and cleaner technology for the polluting technology. All of these approaches were tried by the railroads in their attempts to deal with the smoke problem. All involved costs; and the railroads, many of which had financial difficulties in the early twentieth century, usually chose those options that were least costly in terms of immediate capital outlay.

The Human Fix: Educating the Engineer and Fireman

Throughout the articles and conference discussions about methods of reducing smoke, there is a consistent theme of the necessity for intelligent and well-trained firemen and enginemen. "When a railroad is in service," noted one railroad official, "no factor has so important an influence over the control of smoke as the manner in which it is fired by the fireman and operated by the engineman." In order to deal with the human element, railroad managers and smoke inspectors advocated a program of education accompanied by constant supervision. Many railroads had short courses in smoke prevention for their operating personnel. Railroad smoke inspectors were expected to teach proper methods of firing as well as report bad smoke conditions. Cartoons, posters, and other literature provided regular reminders about the need to prevent smoke in order to save fuel and avoid violations of city ordinances.[26]

Even though technological innovations during the twentieth century reduced the need for human involvement in the firing process, it could not be totally eliminated. "No matter how good the design is," observed a well-known smoke-control expert in 1944, "you still have to have men to operate this equipment. You are not going to do away with any of the smoke laws by the best designed locomotive and the best designed boilers of any kind, because we will still have the human element to contend with."[27] In short, the human element was responsible for smoke and, given the nature of the steam locomotive, there were limitations to the extent to which technology could replace human beings.

Technological Retrofits and Technological Forcing

Various technological retrofits to reduce locomotive smoke emissions were available during the late nineteenth and early twentieth centuries. In fact, the basic retrofitting devices remained the same for about fifty years;

they improved in quality and efficiency but not in concept. These devices were the automatic stoker, the brick arch, the pneumatic firebox door closer, and the steam-air jet.

Technological changes in regard to the steam locomotive itself were directed toward power enhancement, but within the same basic structure. Considerable improvements were made within the framework of the boiler, the frame structure, driving wheels and trucks; but no radical changes occurred. Design improvements, therefore, involved the inclusion of additional known elements. The primary changes involved the manner in which the generated steam energy was utilized in the cylinders and transmitted from the cylinders to the rails. Between 1900 and 1950, the maximum pounds of tractive effort available increased from 49,700 pounds of tractive effort to 135,375 pounds for a passenger locomotive. Simultaneously, the overall efficiency of the average high-speed locomotive increased from about 3.5 percent to 7.5 percent (1940).[28]

To a large extent, engine-horsepower output depended on the replacement of the human element with mechanical factors. For instance, mechanical stokers were able to feed coal at a much more efficient rate than firemen. Efficient fuel feeding and consumption also meant reduced smoke. As the managing editor of *Railway Age* noted in 1946, "because . . . the human element has been unable to perform the job as effectively as the public interest demands, science is stepping in to complete the task." It might be noted that this statement was made at a time when the steam locomotive was about to be replaced by the diesel-electric. Rates of adoption of innovation, therefore, were critical in smoke reduction, but there were important limitations (such as costs and technical restraints) on both the pace and extent to which human factors could be eliminated in the context of the steam locomotive.[29]

Retrofits were installed to satisfy both internal motivations of fuel efficiency and power generation and to satisfy regulatory pressures. Sometimes internal and external factors overlapped. When both motivations were involved, or when there were clear advantages in regard to fuel savings, innovations were more rapidly adopted. When retrofits were adopted primarily to meet smoke-control regulations, the pace was considerably slower.

Mechanical stokers, brick arches, and steam jets were retrofits or mechanical improvements that responded to both internal and external pressures. Mechanical stokers, for instance, both improved steam locomotive performance and reduced smoke. Railroad and locomotive manufacturers began working to develop automatic stokers in the late nineteenth century, and by 1911–12 several types had been perfected. One authority called stokers the "greatest single mechanical contribution to the coal-burning steam locomotive," because by vastly increasing fuel input they permitted large increases in power production. However, firemen were still required along with mechanical stokers in order to make manual adjustments. In addition, the stokers reportedly reduced the firemen's incentive to fine-tune the fire for efficiency. Even with mechanical stokers, locomotives could produce black smoke.[30]

The brick arch was also developed in the late nineteenth century, and it both improved fuel consumption and reduced smoke production. It worked by providing a longer flame passage for burning gases, thus enhancing their ignition rather than allowing them to escape as smoke. The arches often became a source of difficulty, however, because they required constant maintenance and had short lives, needing frequent replacement. Arches enhanced the ability of the firemen to reduce smoke and improve power production but did not eliminate the necessity for intelligent firing.[31] The human element remained.

The air or steam jet was a technology that provided a uniform and constant flow of air to support the combustion process and therefore improved the efficiency of fuel consumption and prevented smoke emissions. Experiments with steam jets began in the early twentieth century and continued through the 1940s. The early jets were crude and often introduced excess air into the firebox, reducing the amount of heat and steam produced for a given amount of fuel. Most jets were operated manually, and crews often used them at the wrong time or, because they generated a large noise volume, did not use them at all. There was no consensus among railroad men as to the efficiency or utility of the steam jets, and some held that they were actually unnecessary.[32]

In the 1930s and 1940s, as the coal industry faced increased competition from competing fuels, it accelerated research efforts to retain its existing markets. In 1934, the National Coal Association formed Bituminous Coal Research (BCR) to begin exploring other coal uses. By 1946, BCR had a membership of 250 coal companies and ten railroads. It pushed the development of the relatively smokeless, coal-burning gas-turbine locomotive and advocated the adoption of a new steam-air jet as cheap, effective, and quiet. Most important, it was supposedly fully automatic and reduced the need for the human element.[33]

Under pressure from smoke-control ordinances in various cities, a number of railroads installed the BCR jet. By 1949, for instance, almost all of the locomotives in the Pittsburgh area had steam-air jets. The results, however, were mixed, and violations continued. The superintendent of the Pittsburgh Bureau of Smoke Prevention reissued the old complaint that "the real problem is to control the human element. . . . It makes little difference whether the locomotive is equipped with stokers, steam-air jets or what the equipment is. So long as a valve must be turned or a firedoor cracked, we are sure to have man failures."[34]

In addition to retrofitting locomotives in order to reduce smoke, the railroads also installed smoke-control devices at their terminals and roundhouses. Although there was some internal motivation to make these improvements in order to improve work conditions, the pressure was primarily external. Terminals and roundhouses were especially objectionable in terms of smoke because it was more difficult to control smoke production from a sitting than a moving locomotive. These railroad facilities were often located in congested districts, and the companies faced pressure from citizens and smoke-control bureaus to regulate the nuisance. One step the railroad companies could have taken was to relocate their facilities outside of populated areas, but this was often impracticable in regard to access. Instead, the railroad companies installed smoke collectors and washers, high chimneys, and direct steaming systems. An advantage of the latter was that they prepared locomotives quickly for runs as well as reducing smoke.[35]

Fuel Switching

Fuel switching was a method used by the railroads to control bad smoke emissions in cities. The fuel used was usually coke or low-volatile coal. Ideally, the universal use of coke or low-volatile fuel such as anthracite would have almost eradicated the railroad smoke problem, but the railroad corporations were unwilling to bear the higher costs of these cleaner fuels. Therefore, they were only used in regard to the worst cases—pusher locomotives working on steep grades or shifting heavy loads on work trains and in locomotives without mechanical stokers.[36]

Technological Substitutions

No matter which retrofits were applied to steam locomotives burning bituminous coal, it was nearly impossible to entirely prevent smoke. The Pittsburgh Smoke Control Ordinance of 1941 was the nation's strictest, only allowing a locomotive to make one minute of dense smoke (defined as No. 2 smoke or greater on the Ringlemann chart) an hour, but it did not eliminate the railroad-smoke nuisance. As the superintendent of the Pittsburgh Bureau of Smoke Prevention noted, "if dense smoke is blown from a steam locomotive stack for only 15 or 20 seconds, it makes a very objectionable black cloud that may hang in the atmosphere for some time."[37] Such a condition, however, did not violate the ordinance. In addition, a group of locomotives could legally produce No. 1 smoke that formed a heavy cloud without breaking the law. Harsher standards, the superintendent noted, would have placed a heavy burden on the railroads and made it exceedingly expensive to operate.

If making the steam locomotive smokeless was almost impossible, there existed the option of replacing it with a smokeless alternative. The electric locomotive, which was technologically superior to the steam engine in terms of traction, reliability, safety, riding quality, cleanliness, and ease of operation, was first used on American railroads in 1895. Its technological advantages, plus its smokeless operation, made electrification a favorite cause of civic and business associations concerned with railroad smoke. Yet electrification also required large amounts of capital, and financially pressed railroads hesitated to make the expenditures. In Chicago, for instance, which was one of the world's great railroad centers, reform and business organizations advocated electrification as an alternative to steam. In 1915, the Chicago Association of Commerce assembled a committee of experts to examine the practicality of electrification. The committee concluded that improvements in the steam locomotive were still possible, and that electrification of the city's railroad terminals was "financially impracticable." The high capital costs of electrification, combined with railroad financial difficulties, prevented all but a few companies from adopting the system in spite of its attractive technological and environmental qualities.[38] Aside from New York in 1903, no governmental body enacted and enforced regulations so strict that a railroad was forced to electrify a portion of its system in order to comply with the law.

A second technological substitution of the steam locomotive became available in the 1920s and 1930s. This was the diesel-electric, which, while involving initial large capital costs, did not require the central power sources and extensive building that catenary electrification necessitated. Railroads substituted diesels for steam locomotives on some long-distance passenger runs in the 1930s and also used them for switching engines in a number of cities. Railroad adoption of the diesel-electric was slowed by the depression, and massive changes did not come until after World War II.[39]

Railroad adoption of the diesel locomotive was inhibited in some areas not only by financial limitations but also by the position of the railroad companies as heavy coal haulers. In the Pittsburgh region, for instance, the coal industry used its power as a large shipper to pressure the railroads to continue to use steam locomotives. The coal industry also demanded that the railroads help them block a strong 1949 Allegheny County Smoke Control law that threatened to drive coal users to alternative fuels and force the railroads to retrofit their steam engines or shift to diesel locomotives. Railroad lobbyists were active in the state capital opposing the law, but unsuccessfully.[40]

Throughout the nation during the postwar years, the diesel-electric rapidly displaced the steam locomotive as the chief form of railroad motive power, especially as its operating efficiencies became clear. In 1946 there were 39,592 steam locomotives in service, 867 electric, and 5,008 diesels. By 1951 there were almost as many diesel locomotives (19,014) in service as steam (22,590), while electrics had begun a slow decline (817). In 1960 there were 30,240 diesels, 498 electrics, and only 374 steam locomotives.[41] The substitution of an environmentally and technically superior and cost-effective technology for the steam locomotive had finally solved the railroad-smoke problem.

Conclusion

The problem of railroad-smoke pollution must be put in the context of environmental pollution in general. Historically, the free-market pricing system has failed to provide incentives to protect the environment—that is, to guard the society's common property in clean air, water, and undegraded land resources.[42] The true social costs of air pollution by smoke from the burning of bituminous coal were not adequately valued. The incentive structure for reducing smoke came either from the internal cost-savings available to the railroads that might be generated by fuel conservation measures, management concern over public response to the smoke nuisance that might decrease passenger revenues, or public policy that threatened legal sanctions. Ultimately, none of these approaches brought results that satisfied public demands for the reduction of smoke pollution and the improvement of air quality.

Not surprisingly, the railroads themselves undertook few initiatives to reduce smoke without the expectation of improving performance. From the point of view of maximizing profits, their action was rational; environmental considerations did not enter into their considerations. Retrofits such as brick arches or mechanical stokers, whose installation management could justify as cost reducing and power enhancing, were more readily adopted than other retrofits (such as the steam-air jet) that threatened to reduce power supply. Steam-air jets were not widely installed until the 1940s, when the steam railroads were faced with intense competition from other, cleaner modes of transport, such as the automobile and the motor bus, or when municipal regulations and regulators became tougher.

In addition to the retrofits, the railroads attempted to educate their personnel (the "human fix") to reduce smoke, especially when such training promised fuel savings or was needed to meet municipal standards and avoid penalties. (It can be argued that the "human fix" was cheaper and more flexible for the railroads than retrofits or technological substitutions.) Without such incentives, however, the railroads generally ignored the problem. In 1948, for instance, when the city of Cumberland, Maryland, enacted its first smoke-control ordinance, the representatives of the Baltimore & Ohio Railroad hastened to promise their cooperation. In apologizing for delays in inculcating their personnel with "a sense of smoke consciousness," they observed that Cumberland's ordinance was new. The implication was clear that before the passage of the ordinance, the railroad had made little or no attempt to train its men to prevent smoke.[43]

The function of public policy from the perspective of the public interest was to persuade or force the railroads to reduce their smoke emissions to conform to municipal smoke standards. The political power of the railroads and their economic primacy in the cities, however, often resulted in limited implementation by the smoke-control bureaus. Smoke-control inspectors shared professional interests with railroad personnel and were often sympathetic to the technological difficulties of controlling railroad smoke. They therefore preferred to follow a policy of cooperation and education rather than one of strong sanctions. Often, they allowed the railroads themselves to undertake disciplinary action against firemen and engineers responsible for breaking the law.

Thus, public policy toward locomotive smoke followed a path of what James E. Krier and Edmund Ursin call "least steps along the path of least resistance,"[44] except in cases of crisis, such as New York City in 1905, or of strong public concern, such as in Pittsburgh in 1941 and 1947. "Policy-by-least steps" permitted the railroads to make gradual adjustments at their own pace and according to the industry's own rate of technological progress. Municipal smoke-control policy did develop tougher standards over time, but cooperation with the railroads usually avoided a last resort of using the courts to compel technological retrofit or technological substitutions to meet new standards.

Most engineers involved in locomotive smoke regulation realized that the ultimate answer to the railroad-smoke problem was to replace the steam engine. Both the electric and diesel locomotives were options, but conservative management and concern over high capital costs, as well as pressure from coal producers, prevented all but a few lines from electrifying and slowed the rate of adoption of the diesel engine. Eventually, in the post–World War II period, the diesel replaced the steam locomotive and cities no longer faced a railroad-smoke problem.

The issue of railroad-smoke abatement presented a difficult problem of environmental regulation. The problem lay in the nature of the technology itself, the principal fuel consumed, and the political power and economic importance of the regulated industry. Complicating the situation were coal industry pressures applied to the coal carrier railroads to continue using the steam locomotive. The railroad-smoke problem finally disappeared due to the substitution of a new and superior technology, but environmental considerations played a very small role in the change. Economic factors were eventually dominant, but Thomas G. Marx has shown that railroad managers were motivated by considerations aside from pure profit maximization in their attachment to the steam engine and resisted adoption of the diesel-electric after its many advantages over the steam locomotive were obvious.[45]

Given this conservatism on the part of the private-sector managers, the ideal public policy posture for the purpose of controlling railroad smoke would have been to force the railroads to speed up their rate of substituting a smokeless technology for the steam locomotive.[46] But, given the political and economic power configurations in American cities and the limited strength of environmental values, such an approach never received serious consideration.

Notes

1. U.S. Environmental Protection Agency, *Trends in the Quality of the Nation's Air—A Report to the People* (Washington, 1980). The horse was also a mobile pollution source. It was eliminated as a pollution problem by the automobile. See Joel A. Tarr, "Urban pollution—many long years ago," *American Heritage* 22 (October 1971): 65–69, 106.

2. John H. White, Jr., *American Locomotives: An Engineering History, 1830–1880* (Baltimore, 1968), 83–90; Frederick Moore Binder, *Coal Age Empire: Pennsylvania Coal and its Utilization to 1860* (Harrisburg, 1974), 111–132.

3. W. L. Robinson, "Locomotive fuel," *Proceedings of the Second International Conference on Bituminous Coal* (Pittsburgh, 1928), 366. Association of American Railroads, Bureau of Railway Economics, *Railroad Transportation: A Statistical Record 1911–1951* (Washington, 1953), 37.

4. Ibid., 23; Robinson, "Locomotive fuel," 367. For a discussion of the coal shipping railroads, see Joseph T. Lambie, *From Mine to Market: The History of Coal Transportation on the Norfolk and Western Railway* (New York, 1954).

5. John O'Connor, *Some Engineering Phases of Pittsburgh's Smoke Problem*, Bulletin No. 8, Smoke Investigation, Mellon Institute of Industrial Research and School of Specific Industries (Pittsburgh, 1914), 71–81; Benjamin Linsky, ed., *A Different Air: The Chicago Report on Smoke Abatement* (Elmsford, NY, 1971), 175–178.

6. J. B. Irwin, "Railway smoke abatement," *Transactions of the American Society of Mechanical Engineers* (1927–1928) Vols. 49–50, Pt. 156; and I. A. Deutch and S. Radner, "Abating locomotive

smoke in Chicago," in Smoke Prevention Association of America (hereafter referred to as SPAA), *Manual of Ordinances and Requirements* (1941).

7. The question can also be viewed from the perspective of technology forcing. For instance, in regard to the internal combustion engine, technological advances were improving engine efficiency, but not necessarily fast enough to meet the clean air standards desired by policy makers. The 1970 Amendments to the Clean Air Act attempted to speed up the development of a cleaner and more efficient engine through various sanctions (fines). For a discussion of technology forcing, see Eugene Seskin, "Automobile air pollution policy," in Paul R. Portney, ed., *Current Issues in U.S. Environmental Policy* (Baltimore, 1978), 83–90.

8. In the early stages of railroad development, steam locomotives had actually been banned from some city centers because of the hazard and nuisance they presented and because they frightened horses. For New York see George Rogers Taylor, "The beginnings of mass transportation in urban American: Part 1," *Smithsonian Journal of History* (Fall 1962), 31–38; for Pittsburgh, see Joel A. Tarr, *Transportation Innovation and Changing Spatial Patterns in Pittsburgh, 1850–1934* (Chicago, 1978), 56, note 23; and Carl W. Condit, *The Pioneer Stage of Railroad Electrification—Transactions of the American Philosophical Society* (Philadelphia, 1977), Vol. 67, Pt. 7, 10–17.

9. Michael Bezilla, *Electric Traction on the Pennsylvania Railroad 1895–1968* (University Park, PA, 1980), 9–55.

10. D. F. Crawford, "The abatement of locomotive smoke," *Railway Age Gazette* 53 (Dec. 24, 1913) 762–765; Robinson, "Locomotive fuel," 372–377.

11. Robinson, "Locomotive fuel," 373–374; A. W. Gibbs, "The smoke nuisance in cities," *Railroad Age Gazette* 5 (Feb. 26, 1909): 415. For railroad organization, see Alfred D. Chandler, Jr. (ed.), *The Railroads: The Nation's First Big Business* (New York, 1965).

12. H. B. Meller, "Smoke abatement, its effects and its limitations," *Mechanical Engineering* 48 (November 1926): 1279; Sumner B. Ely, "Steam-air jet application to locomotives," *Proceedings*, SPAA 38th Annual Meetings, June 6–9, 1944, 81; Roy V. Wright, "Railroad smoke abatement," *Proceedings*, SPAA, 1946, 27; and Linsky, ed., *A Different Air*, 52.

13. Gibbs, "The smoke nuisance in cities," 415.

14. White, *American Locomotives*, 89–90; Binder, *Coal Age Empire*, 127–130.

15. R. Dale Grinder, "The battle for clean air: the smoke problem in post-Civil War America," in Martin V. Melosi, ed., *Pollution & Reform in American Cities, 1870–1930* (Austin, 1980), 83–104; Joel A. Tarr, "Changing fuel use behavior and energy transitions: the Pittsburgh smoke control movement, 1940–1950," *Journal of Social History* 14 (Summer 1981): 561–588.

16. Samuel B. Flagg, *City Smoke Ordinances and Smoke Abatement*, Bull. 49, Bureau of Mines (Washington, DC, 1912).

17. O'Connor, *Some Engineering Phases of Pittsburgh's Smoke Problem*, 71–81; Linsky, ed., *A Different Air*, 175–178.

18. Flagg, *City Smoke Ordinances*, 10–25, gives a review of the terms of smoke control ordinances in cities of over 200,000 residents.

19. J. S. Barkley, "Some fundamentals of smoke abatement," Information Circular 7090, *Bureau of the Mines*, 1939.

20. "Hearings before the Pittsburgh City Council on the smoke control ordinance," June 25, 1941, Pittsburgh City Council Records.

21. For a discussion of the debate over strict and nominal enforcement in municipal smoke control, see Grinder, "The battle for clean air," 95–99. For policy shifts in regard to railroads, see Irwin, "Railway smoke abatement," 153–154. Descriptions of alterations in railroad smoke conditions can be found in the *Annual Reports* of most municipal smoke control bureaus. For examples of papers by municipal smoke inspectors discussing the necessity of cooperation rather than legal sanctions, see Meller, "Smoke abatement," 1279, and W. H. Kimberly, "Railroad smoke control in Pittsburgh," *Proceedings*, SPAA, 1943, 113.

22. For a discussion of how professional training can influence positions taken on public questions involving environmental regulations, see Joel A. Tarr, Terry Yosie, and James McCurley III,

"Disputes over water quality policy: professional cultures in conflict, 1900–1917," *American Journal of Public Health* 70 (April 1980): 427–435. Also see Edward T. Layton, Jr., *The Revolt of the Engineers: Social Responsibility and the American Engineering Profession,* especially 53–79. Quoted in Grinder, "The battle for clean air," 99; see *Annual Reports,* SPAA.

23. For a discussion of the various regulatory models, see R. A. Posner, "Theories of economic regulation," *The Bell Journal of Economics and Management Science* 5 (Autumn, 1974): 335–358.

24. Irwin, "Railway smoke abatement," 153. Pittsburgh Department of Public Health, Bureau of Smoke Prevention, *Report on Railroad Smoke Conditions* (Pittsburgh, 1949), 4.

25. Bezilla, *Electric Traction,* 28. The railroads opposed the emission standards in the Pittsburgh Smoke Control Ordinance in 1941 and fought passage of county enabling legislation in 1947. See *Pittsburgh Press,* Aug. 26, 1949.

26. Irwin, "Railway smoke abatement," 153; see Wright, "Railroad smoke abatement," 27; R. E. Howe, "Fuel conservation and smoke abatement," *Proceedings,* SPAA, 1947, discussion after 16; and S. E. Back, "Action taken by the Pennsylvania Railroad to eliminate smoke through educational activities and the application of smoke eliminating appurtenances to locomotives," *Proceedings,* SPAA, 1947, 105–110.

27. William Christy, commenting in discussion after paper by Sumner B. Ely, "Steam-air jet application to locomotives," *Proceedings,* SPAA, 1944, 87. See also William G. Christy, "The human side of smoke abatement," *Mechanical Engineering* (June, 1933), 350.

28. Alfred W. Bruce, *The Steam Locomotive in America: Its Development in the Twentieth Century* (New York, 1952), 387–393; Harold Barger, *The Transportation Industries, 1889–1946: A Study of Output, Employment, and Productivity* (New York, 1951), 104–105.

29. Wright, "Railroad smoke abatement," 28; Bruce, *The Steam Locomotive,* 387–388. For discussions of the rate of innovation in the railroad industry, see Edwin Mansfield, "Innovation and technical change in the railroad industry," National Bureau of Economic Research, *Transportation Economics* (New York, 1965), 171–197, and Jacob Schmookler, *Invention and Economic Growth* (Cambridge, 1966), 104–130.

30. Bruce, *The Steam Locomotive in America,* 159–163; W. S. Bartholomew, "Locomotive stokers and smoke prevention," *Railway Age Gazette* (Sept. 6, 1918) 65:451–453.

31. "Pennsylvania locomotive brick arch tests," *Railway Age Gazette* 62 (May 4, 1917): 926, 933–935; Bruce, *The Steam Locomotive,* 149–150.

32. For early development of the jets, see D. F. Crawford, "The abatement of locomotive smoke," *Railway Age Gazette* 55 (Dec. 24, 1913): 763. For criticisms of their performance, see Ely, "Steam-air jet application," 78–82, 85.

33. Harold J. Rose, "BCR developments to prevent smoke," *Proceedings,* SPAA, 1946, 56.

34. Back, "Action taken by the Pennsylvania Railroad," 105; Pittsburgh Department of Public Health, Bureau of Smoke Prevention, *Report on Railroad Smoke Conditions,* 1947, 4. Ibid.

35. Linsky, ed., *A Different Air,* 175–178; M. D. Franey, "Washing locomotive smoke," *Railway Age Gazette* 59 (Sept. 24, 1915): 538–540; Irwin, "Railway smoke abatement," 156; Meller, "Smoke abatement," 1279; and Deutch and Radner, "Abating locomotive smoke in Chicago."

36. Irwin, "Smoke abatement," 155–156; O'Connor, *Some Engineering Phases of Pittsburgh's Smoke Problem,* 75.

37. For a discussion of the Pittsburgh law, see J. C. Kuhn and David Stahl (eds.), *Public Health Laws of the City of Pittsburgh* (Pittsburgh, 1950), 302–344; Pittsburgh Department of Public Health, Bureau of Smoke Prevention, *Report on Railroad Smoke Conditions* (Pittsburgh, 1949), 1.

38. For the electric locomotive, see Bezilla, *Electric Traction of the Pennsylvania Railroad,* and Condit, *The Pioneer Stage of Railroad Electrification.* Linsky, ed., *A Different Air,* 1046–1050.

39. For the diesel-electric, see Charles F. Foell and M. E. Thompson, *Diesel-Electric Locomotive* (New York, 1946); and Thomas G. Marx, "Technological change and the theory of the firm: the American locomotive industry, 1920–1955," *Business History Review* (Spring 1976) 50: 4–6.

40. Pittsburgh Department of Pubic Health, Bureau of Smoke Prevention, *Report* (Pittsburgh, 1954), 20. Dr. Edward Weidlein, the first director of the Allegheny County Air Pollution Control Bureau as well as director of the Mellon Institute of Research, told the author that the presidents

of the Pittsburgh area railroads had confided to him that they opposed the county law because of pressure from the coal companies. When the powerful financier and civic leader Richard King Mellon heard of the actions of the railroad lobbyists, he threatened to remove all the business of the Pittsburgh corporations he controlled from the Pennsylvania railroad. Interview with Dr. Edward Weidlein, Nov. 14, 1978.

41. U.S. Bureau of the Census, *Historical Statistics of the U.S.* (Washington, DC, 1975) Vol. 11, 727–728.

42. See, for instance, Gerald Garvey, *Energy, Ecology, Economy: A Framework for Environmental Policy* (New York, 1972), 33–35.

43. See George M. Hitchcock, Department of Smoke Abatement, Cumberland, Maryland, "Annual Report" (1948), n.p. The B & O had a Department of Fuel Conservation with trained inspectors in fuel economy and smoke abatement. The railroad cooperated to a certain extent with the city's regulations, but there were still violations. In 1951, the city hired Raymond R. Tucker, a noted mechanical engineer and the man responsible for controlling smoke in St. Louis, as a consultant. Tucker reported that "the railroads appeared to be flagrant violators as far as the emission of dense smoke." See Raymond R. Tucker, "Report to the Mayor and City Council of the City of Cumberland on the results of the survey made for the elimination of smoke and dust," (unpublished manuscript in Raymond R. Tucker Papers, Washington University Library, St. Louis). See also Charles Z. Heskett to The Honorable Thomas S. Post (Mayor, City of Cumberland), July 24, 1951, ibid. Heskett noted that although the railroads had had three years in which to install over-fire jets, many engines operating in and through Cumberland were not equipped with them.

44. The concept of "Policy-by-least-steps" is that of James E. Krier and Edmund Ursin, in *Pollution & Policy: A Case Essay on California and Federal Experience on Motor Vehicle Air Pollution, 1940–1975* (Berkeley, 1977), 11–12. Railroad substitution of diesel-electric for steam locomotives was quite rapid after World War II.

45. Marx notes that World War II had a great impact on the development and rate of diffusion of the diesel freight locomotive because it caused high utilization and generated public policy that favored General Motors as a producer over steam locomotive manufacturers. See Marx, "Technological change and the theory of the firm," 6–7.

46. Eugene Seskin observes that present policy in regard to automobile emissions has generated a series of modifications and additions to the standard internal combustion engine rather than experiments with different and riskier substitute technologies. See Seskin, "Automobile air pollution policy," 86–87; Lester B. Lave, "Conflicting objectives in regulating the automobile," *Science* 212 (May 22, 1981), 893–899.

PART IV Land, Transport, and Environment

A long-standing tension exists in the United States between the supposed evil city and the moral countryside. The interactions between city and country were not always negative, however. Immigrants to the United States, coming from land-poor European countries, often brought with them a conservation ethic that emphasized the utilization of urban wastes on the land. These wastes included both human excrement removed from cesspools and privy vaults and horse manure that farmers or scavengers collected from the street. What was pollution in the urban context became life-giving fertilizer in a rural setting. The construction of sewerage systems and the adoption of flush toilets, however, limited these practices, eliminating some nuisances but creating problems of their own. A competitor to the water closet, the earth closet—which claimed as one of its virtues the creation of fertilizer for gardens from human wastes—appeared in the late 1860s but failed to attract many adherents. This technology, however, has a direct descendant in today's ecologically sound composting toilet.

Although the construction of sewerage systems as well as heightened sanitary concerns largely eliminated the practice of having scavengers or farmers collect human wastes for use on the land, this conservation tradition did not entirely disappear. In fact, the ideal of using urban sewage to make the land bloom as well as a means to get rid of sewage remained appealing. Actually mandated by Parliament as a sewage disposal method in Great Britain, sewage farming was also used in France and other European countries. Between twenty-five and thirty-five small American cities and towns adopted sewage farming in the late nineteenth and early twentieth centuries, but the numbers never expanded. Today's environmentalists, however, have found the approach attractive because of its ecological soundness, and the Environmental Protection Agency has sponsored several experiments with it over the past two decades, involving fish as well as vegetable farms.

The technological changes of the late nineteenth and early twentieth centuries increasingly eliminated urban processes that linked cities to nature and to the land, however tenuous those connections might be. The rapid disappearance of the horse as a mode of transportation and power in the American city and its replacement by the automobile and the electric streetcar provide an example. Cities with large horse populations that hauled freight and pulled streetcars, as well as furnishing personal transportation, had to dispose of huge amounts of horse manure. In many cities, farmers and scavengers collected this manure for fertilizer and paid for the privilege. The manure, of course, was also a nuisance and a health hazard as well as a resource, and for city dwellers its negative features outweighed the positive. The automobile rapidly replaced the horse in the twentieth century, thereby greatly improving urban sanitary conditions but also producing both technological hazard and environmental pollution—emissions from internal combustion engines—of a new and damaging type. At the same time, the link that horses and horse manure provided between city and country was severed.

While the automobile may have eliminated horses and manure from city streets, it also enabled urbanites who could afford it (as its proponents argued) to move their residences from the crowded city to the bucolic suburb while continuing to benefit from the city's commercial and industrial advantages. The concept, however, that technology could link city and suburb actually predated the automobile and was frequently applied to streetcar systems. As cities became denser and slums expanded, reformers argued that streetcar technology could provide city dwellers with the means to escape to the suburbs, living between city and country but enjoying the benefits of each. In addition, suburban living, by exposing families to clean air and green trees, would result in more moral citizens. Thus, both the automobile and the streetcar became technologies endowed with a moral purpose.[1]

The articles in this section, all written between 1971 and 1975, reflect the concerns of those years with the effects of technology on society, as well as a growing awareness of the strength of the environmental movement. The United States, more than any other society in the world, has suburbanized its population, constantly expanding the metropolitan periphery. A major factor driving suburbanization has been a desire to enjoy environmental goods and benefits not viewed as readily available in the city. Ironically, however, in seeking environmental benefits through suburbanization, Americans have consumed land and resources in prodigious and wasteful amounts, imposing a weighty and destructive burden on the environment.

Photo IV.1. Pumping sewage on crops for fertilizer. Source: *Harper's Weekly*, 1890.

Photo IV.2. Removing dead horses from the city streets. Source: Rudolph Hering and Samuel A. Greeley, *Collection and Disposal of Municipal Refuse* (New York: McGraw-Hill, 1921).

Photo IV.3. A "White Wing" cleaning New York City streets of horse manure. Source: George E. Waring, Jr., *Street-Cleaning and the Disposal of a City's Wastes* (New York, 1898).

Photo IV.4. Horse manure piled up on a New York City Street, 1893. Source: George E. Waring, Jr., *Street-Cleaning and the Disposal of a City's Wastes* (New York, 1898).

Photo IV.5. A sketch drawing of the nation's first scientifically designed sanitary landfill, Fresno, California, 1940. Source: American Public Works Association, *Public Works Engineers' Yearbook* (Chicago: American Public Works Association, 1940).

Photo IV.6. "Throwing Dead Horses into the Harbor of New York at Night." Source: *Frank Leslie's Illustrated Newspaper*, 20 August 1870.

CHAPTER X

From City to Farm

Urban Wastes and the American Farmer

A great city is the most powerful of stercovaries [toilets or places to store manure]. To employ the city to enrich the plains would be a sure success. But the filth is swept into the abyss. All the human and animal manure which the world loses, restored to the land instead of being thrown into the water would suffice to nourish the world. These heaps of garbage at the corners of the stone blocks, these tumbrils of mire jolting the streets at night, these horrid scavengers' carts, these fetid streams of subterranean slime which the pavement hides, what is all this? It is the flourishing meadow, the green grass, the thyme and sage; it is game, it is cattle, hay, corn, bread upon the table, warm blood in the veins.

—Dr. Henry J. Barnes

An important concern of the contemporary ecology movement has been the disposal of sewage in a manner that will not pollute the environment. To deal with this problem, ecologists have adopted an approach that dates back for centuries—the use of human wastes to fertilize the soil. They argue that, under this method, pollution of the surface waters is avoided and the wastes of the urban population are reincorporated into the soil's ecological cycle.[1]

While the use of sewage (human wastes in water carriage) in agriculture dates back to about 1800, the application of human wastes directly to the land has a much longer history. The Romans used human wastes as fertilizers, while Flemish edicts dating back to the early seventeenth century required settlers in peat-marsh colonies to manure their lands with urban refuse. In the 1820s, travelers in the Netherlands often observed on the Schelde barges filled with excrement from Dutch towns destined for the fields. Throughout the nineteenth century, the excrement of English and Scottish towns was collected, often by a method called "the pail system," and used on neighboring farms. As recently as 1923, an Italian scientist noted the agricultural use of human wastes ("Flemish Manure") in Italian and French provinces. The most wide-spread use of this fertilizing method, however, has been in China, Korea, and Japan, where it has been practiced for many centuries. Writing in 1911, agricultural researcher F. H. King calculated that these nations annually applied 182 million tons of human wastes to the soil; the figures for Japan alone were 23,850,295 tons in 1908.[2] These countries still follow this practice.

The utilization of human wastes from urban populations in fertilizing the land in Asia and in Europe suggests that such a practice was probably followed in America before the building of sewerage systems. The existing state of medical knowledge in the nineteenth century posed no barrier to the use of human wastes on the soil. For most of the century, doctors believed that infectious diseases were caused by either the corrupted state of the atmosphere (miasmatic theory) or by specific contagia stemming from decayed animal or vegetable matter. By the Civil War, sanitary reformers were insisting on the removal of filth from towns and cities because they believed that these wastes either generated epidemic disease or threw off "exhalations" that promoted disease. They failed to perceive, however, that human excrement might be the vector for disease and raised no objection to the use of these wastes on the soil; they argued that "in the open country [decomposing matter] is diluted, scattered by the winds, oxydized in the sun; vegetation incorporates its elements."[3]

City wastes were used on the land in America as early as the mid–eighteenth century, if not before. One historian notes that in 1765 much of the "dung and ordure of Manhatten" was used to fertilize farms along the East River, while another observes that farmers utilized the "night soil" of Boston as well as New York in the 1830s and 1860s. The famous Massachusetts Sanitary Commission of 1850 recommended "that, whenever practicable, the refuse and sewage of cities and towns be collected, and applied to the purposes of agriculture."[4] These scattered references, however, give little indication of the extensiveness of the agricultural use of urban wastes that is revealed in the Tenth Census (1880) volumes of *Social Statistics of Cities*, published in 1887.

Social Statistics of Cities appeared as a special two-volume report. It contained profiles of 222 cities and provided both historical and contemporary data. The editor was George E. Waring, Jr., at that time a noted specialist in sanitation. Waring had originally been a student of scientific agriculture; he had lectured to farmers on scientific agriculture, managed several large farms (including those of Horace Greeley and Frederick Law Olmsted), and published several books on husbandry. In 1857 Waring was appointed drainage engineer of New York's Central Park. After serving in the Civil War, he embarked on a career as a sanitary engineer. He became the nation's leading sanitation specialist in the late nineteenth century, published widely on the subject, and installed sewage systems in a number of cities.[5]

In *Social Statistics of Cities* Waring divided the nation into four regions: New England, the Middle States (today the Mid-Atlantic), the Southern States, and the Western States (today the Midwest and the Far West). Much of the material on sanitary conditions and sewerage systems and disposal is fragmentary and is not presented in a uniform manner. It appears, however, that out of the total of 222 cities, only 102 had any sort of sewerage system, with most sewers concentrated in the cities of New England and the Middle States. The great majority of urban households were without water closets connected to sewers and depended on privy vaults or cesspools for human-waste disposal. A rough estimate would be that in 1880 about two-thirds of the households in the cities listed by Waring depended on privy vaults or cesspools and were not connected to sewer systems.[6]

In 103 of the 222 cities in *Social Statistics* farmers or "scavengers" collected the human wastes and either deposited them directly on the land, composed them with earth and other materials and then applied the mixture to the land, or sold them to processing plants to be manufactured into fertilizer. The extent to which these practices were followed appears to be independent of the existence of a sewer system in a city. The regions with the most sewers, New England and the Middle States, were also those where farmers made most extensive use of urban wastes. Farmers utilized the wastes of 43 of 55 New England cities and 31 of 49 Middle cities. In contrast, only 14 of 38 Southern cities and 15 of 80 Western cities practiced this method of waste disposal.[7]

In most of the cities where human wastes were utilized on the land, the law stipulated that cesspools or privy vaults be emptied at night and prohibited the use of night soil on farms within the gathering ground of the city's water supply. Usually the privies were emptied by hand receptacles and buckets, although eleven cities reported using "odorless-evacuators." It is very difficult to estimate the amount of human wastes removed and how much of this was used on the land. Brooklyn reported that 20,000 cubic feet of night soil was taken each year from the city's 25,000 privy vaults and applied to "farms and gardens outside the city." Philadelphia estimated that the city's twenty "odorless" vault-emptying companies removed about 22,000 tons of fluid matter per year, and that the "matters removed are largely used by farmers and market gardeners of the vicinity." And the Boston authorities noted that only one night soil load in ten could be sold for manure, suggesting either an absence of demand on the part of nearby farmers or an oversupply of wastes. The night soil of eight cities, including New York, Baltimore, Cleveland, and Washington, D.C., was manufactured into fertilizer, usually under the trade name of "poudrette." Advertisements boasting of the use of poudrette on lawns, garden vegetables, corn, potatoes, and tobacco appeared in farm journals as early as 1839.[8]

Other urban wastes, such as the manure dropped by horses on the streets or collected in stables, or "offal," refuse food collected from restaurants, markets, and houses, were also utilized by farmers. Offal was used principally for feeding pigs. In Boston, in the year 1880–81, city teams collected about 26,000 loads of offal, averaging between 400–500 cubic feet per load. The city sold the offal for $25,169.74, while the cost of collection was $57,091.17. In both Boston and New York in the late eighteenth and early nineteenth centuries, farmers paid for the privilege of removing manure from the streets and stables. In 1803, New York actually made a profit by the excess of return from selling manure over the costs of street cleaning. By 1880, however, relatively few farmers bothered with street sweepings, probably because the volume of urban traffic so contaminated the manure. Eighteen New England cities, eleven Middle cities, seven Southern cities, and eight Western cities reported that farmers used their street sweepings for fertilizer. These data may be misleading, however, concerning the extent to which farmers utilized urban horse manure. The Waring questionnaire inquired about the disposal of street sweepings but not of stable manure. Long before the Civil War, farmers collected stable manure from nearby towns and cities, while in both 1890 and in 1921 reliable sources reported that farmers located near urban places used "nearly the entire supply of stable manure."[9]

The lack of specific sources makes it difficult to establish how widely urban wastes were put to agricultural uses before 1880. Other types of indirect information, however, suggest that the practice was not widespread until the 1830s and 1840s. One limiting factor was the widespread antipathy of American farmers to using fertilizers of any kind well into the nineteenth century. The failure to use even readily available fertilizers such as farm manures seemed to be a characteristic of an early stage of farming regardless of section. Manures, for instance, were disposed of in the easiest rather than the most effective manner. These practices changed with a decline in land productivity, and farmers made more effective use of fertilizers. By 1850, animal manures were used in the fields in most areas east of the Alleghenies and north of Virginia. Commercial fertilizers also appeared on the market in the 1840s, with Chilean and Peruvian guano having wide popularity in the states of the Upper South and poudrette finding users in both the South and New York and Connecticut. Farmers in the plains states, however, largely ignored manure and other fertilizers until the 1870s.[10]

Even if American farmers had been accustomed to using fertilizers in, for example, 1810 or 1820, the small number of cities would have prevented them from making much use of urban wastes. In 1820, for instance, only 7.2 percent of the population lived in urban areas; only ten cities had populations over 10,000 and only three had over 50,000. The following decades, however, were ones of large urban growth, and by 1860 the nation was 19.8 percent urban with eighty-four cities having over 10,000 population and nine with over 100,000. Cities were also more evenly distributed, with a number located in the interior of the country. By 1880, the census year for which the data for the *Social Statistics of Cities* volumes were gathered, 28.9 percent (or 14,129,000) of the American people lived in urban places of 2,500 or more. Sixteen cities had over 100,000 population, three had over 500,000, and New York City had over a million people. Great cities, of course, created great quantities of waste and raised the question of what to do with it. "It is no exaggeration," noted the *Scientific American* in 1873, "that the problem of the conversion of the excremental waste of towns and people and the refuse of factories into useful materials is now engaging as much of the attention of intelligent minds throughout the world as any social question."[11]

The extent of the usage of urban wastes varied greatly within and between sections. Both New England and the Middle States had long histories of intensive agriculture, problems with land exhaustion, and a high level of urbanization, and farmers in these regions used urban wastes more extensively than in other regions. *Social Statistics* listed eight cities in the Upper South, and farmers utilized the wastes of seven of these cities. The Upper South, of course, had long suffered from problems of land exhaustion, and farmers there also made wide use of commercial fertilizers. Few farmers in the Lower South or in the border states made use of urban wastes, probably because of low levels of urbanization and unfavorable climatic and soil conditions. As for the West, farmers only utilized the wastes of fifteen of eighty cities, with eleven of the fifteen concentrated in the older states of the Middle West.[12]

The costs of transporting urban wastes and the tendency of this material to decompose quickly and become odoriferous meant that only farmers within a close distance of cities could utilize it. In this regard, there appears to be a correlation between the development of truck and garden farming and the use of urban night soil. The modern fruit and vegetable industry rose before the Civil War in response to the emergence of an urban market and improved transportation facilities; with the growth of urban population in the postwar period, truck and garden farming greatly expanded. Most truck farms were located close to the cities because of the perishable nature of the vegetables. In addition, vegetables grew most favorably on highly cultivated and fertilized land, and farmers found a large supply of cheap fertilizer available in the form of urban night soil and manure.[13]

In the 1880 *Social Statistics of Cities*, garden and truck farmers as well as growers of orchards and vineyards are specifically mentioned as using urban night soil as fertilizer. In Baltimore, this practice continued into the beginning of the twentieth century. The city was without a system of municipal sewers until 1912, and human wastes were deposited in over 70,000 cesspools and privy vaults. The vaults were emptied by "night soil" men using either odorless excavators or dippers and buckets and the contents then sold to a contractor for 25¢ per load of 200 gallons. The wastes were then carried by barge eight to ten miles below the city and sold to farmers for $1.67 per 1,000 gallons. Over 12 million gallons a year were sold to farmers and used to grow crops such as cabbage, kale, spinach, potatoes, and tomatoes. According to one reference, "little smell" arose from either the pits where the fertilizing material was stored or the lands to which it was applied.[14]

Baltimore's direct use of night soil on the land, however, was unusual in that it extended into the twentieth century. By the last decades of the nineteenth century, several factors were at work that altered the relationship between urban wastes and agriculture. One important consideration was the widespread availability of cheap commercial fertilizers that could easily be applied to the land. Equally important, however, were the development of public health concerns and the extensive building of sewerage systems in American cities. Even before the formulation of the germ theory of disease, most doctors and sanitarians believed in the "filth theory," which propounded a relationship between the lack of adequate urban sanitary facilities and a high death rate from epidemic and "zymotic" or infectious diseases. In the decades after the Civil War, sanitarians and public health officials launched a campaign for the building of sewerage systems and the connecting of households with these systems. The affirmation of the germ theory of disease and the belief of sanitarians that "the excreta of man and other animals are the principle original vehicles of infection and contagion" stimulated efforts toward sewering towns and cities.[15]

During the last decades of the nineteenth century and the first decades of the twentieth, American cities made vast expenditures on sewerage systems. In 1890 cities with an aggregate population of 14,721,217 were served by 8,199 miles of sewers or 1,795 persons per mile (data for cities with over 10,000 population). By 1909, 20,593,303 people lived in cities with sewers, but the miles of sewer had increased to 24,972 or 825 persons per mile (data for cities with over 30,000 population). Most cities dumped their sewage into available waterways in line with the theory that "running water purifies itself." Many engineers and sanitarians argued, however, that this wasted the valuable fertilizing material in the sewage. Sanitarians in 1873, for example, calculated that the nitrogen, phosphorus, and other chemicals in human excreta were worth about $1.64 to $2.01 per year per individual. But sewage is actually a highly diluted mixture of water and other materials; the leading American textbook on sewage in the late nineteenth century calculated that sewage consisted of ninety-eight parts water, one part mineral, and one part organic. A problem concerning sanitary engineers was how to dispose of billions of gallons of wastewater in the most efficient and healthful manner without wasting its valuable components.[16]

For an indication of the various options available, American engineers and sanitarians looked to Europe and especially to England. From about 1800, several towns in Devonshire, as well as Edinburgh in Scotland, had irrigated neighboring agricultural land with their sewage. The leading advocate of sewage farming in England in the middle of the nineteenth century was the great sanitary reformer Edwin Chadwick, who in 1842 advocated the use of untreated sewage as field manure. Chadwick believed that the sale of urban sewage to farmers would pay for the cost of maintaining urban sewerage systems. In 1865 the British Sewage of Towns Commission, appointed in 1857 to inquire into the most beneficial and profitable method of sewage disposal, reported in favor of the land-disposal method; similar recommendations were made by other English commissions throughout the remainder of the century. By 1880, nineteen English cities with a total population of 738,191 disposed of their sewage on agricultural land. On the continent, Antwerp, Berlin, Brussels, Paris, and Milan all had sewage farms, while Amsterdam converted its sewage into a dry fertilizer.[17]

Beginning in the middle of the nineteenth century, American sanitarians took note of the relevancy of the English and European experience. As early as 1850, for example, the Massachusetts Sanitary Commission, after citing the English and Scottish examples, had recommended the agricultural use of sewage. In 1867 George E. Waring, Jr., in *Draining for Profit and Draining for Health,* also cited the British experience and suggested the beneficial application of New York sewage to the sandy soil of Long Island.

Beginning in the 1870s and continuing throughout the remainder of the century, American engineering and scientific journals, as well as more popular periodicals, carried numerous articles on sewage irrigation (also called "broad irrigation") and other sewage disposal methods overseas.[18]

During the 1870s, several New England institutions, starting with the Augusta (Maine) State Insane Asylum and the Concord (New Hampshire) Asylum, began using their sewage to grow crops. The first municipality to use a sewage farm as a means to dispose of its sewage was Pullman, Illinois, in 1881, followed by Pasadena, California (1888), Colorado Springs, Colorado (1889), and Salt Lake City, Utah (1895), as well as several smaller towns. In 1899 George W. Rafter, in his work *Sewage Irrigation,* listed twenty-four municipal sewage farms or "broad irrigation" projects serving 280,000 people.[19] Among the crops grown on sewage farms were potatoes, wheat, oats, barley, carrots and other garden truck, and fruits. Also, Italian rye grass was very common.

The wide popular interest in sewage farming as a means of disposal that converted urban wastes into useful products was reflected by the technology's prominence in *Young West,* a sequel to Edward Bellamy's utopian novel *Looking Backward.* In the novel, the hero devises a method to make fertilizing bricks from urban wastes; his nation's president lauds him as the man who "has caused *three* blades to grow in place of one." Subsequently, Young West himself is elected president and transforms his country's sewerage system so that "what was taken from the land was returned to it . . . [and] the country bloomed like a garden."[20]

Actually, considering the number of studies made of sewage-farming technology and considering the extent of its use in England and Europe, the number of American projects was small. In 1904, for instance, sanitary engineer George W. Fuller found only fourteen municipal projects serving approximately 200,000 persons. The largest cities utilizing sewage farms were all located in the Far West. Fuller concluded that "the general outlook is clearly toward a decrease rather than an increase in the systematic practice of broad irrigation."[21] The restraints on the more widespread agricultural use of sewage in the United States derived primarily from two areas: those involving technological and economic factors and those reflecting public health concerns.

In 1887 the Massachusetts State Board of Health established the Lawrence Experiment Station to conduct tests for water purification and the most effective methods of sewage disposal. Under the leadership of sanitary engineers Hiram Mills, Allen Hazen, and George W. Fuller, the Lawrence Laboratories demonstrated the efficiency of the intermittent filtration method of sewage disposal. In this process, sewage water was purified by intermittent application to a land filter. While intermittent filtration and sewage farming were occasionally carried on simultaneously, intermittent filtration was usually considered the superior method. It required less land than did the sewage farms and could be used throughout the year rather than only during the crop-growing season.[22] In addition to intermittent filtration, by the turn of the century several other methods of sewage disposal had been developed, such as septic tanks, chemical precipitation, and sprinkling filters, which were replacing sewage farms in the United States and Europe.

Advocates of sewage farming complained that these new technologies wasted the valuable fertilizing components in sewage, but sanitary engineers and chemists argued that the manure value of sewage had been overrated and that the costs of reclaiming these materials on sewage farms were prohibitive. They used evidence from England, France, and Germany—where sewage farming had been conducted on a wide scale for generations—to confirm their unprofitability. Furthermore, they maintained that if sewage farming did not show a profit in these countries, where sewage was less diluted than in America and where farm management was more efficient, then clearly it was inapplicable to the United States, where land and labor costs were higher. In one American case, in Pasadena, California, where the managers of a city sewage farm claimed to have made money, the editors of *Engineering News* charged that the profit had actually been obtained by the use of improper accounting methods.[23]

Equally as important as the economic arguments were those derived from public health considerations. Public health officials and sanitary engineers argued that while sewage farming might be useful as a method of irrigation, it was inefficient as an approach to sewage disposal. They noted that sewage farms, especially in the East, often diverted the flow of sewage to neighboring streams when the crops had absorbed their limit or when there were heavy rainfalls; pollution of the streams frequently resulted. In addition, they maintained that the raw sewage exposed farm employees to possible infection and that the vegetables grown on the farms could be the carriers of "dangerous microbes or other parasites," even though there was no clear evidence to this. Finally, they held that sewage farms produced offensive smells and provided a breeding ground for disease-carrying flies. By 1912, while some engineers continued to maintain that sewage irrigation could serve both the needs of cities for safe sewage disposal and the farmer's need for fertilizer and water, most sanitary authorities disagreed. They argued that while sewage farming might work efficiently in Europe, it was unsuited for America.[24]

In spite of these warnings, some regions, particularly in the West, continued to utilize sewage on the land; in these cases, however, irrigation rather than disposal was the chief motivation. In 1918, the California State Board of Health issued regulations specifying what crops could be irrigated with sewage. Basically, the only prohibition involved the use of sewage on garden vegetables intended to be eaten raw; garden truck that was intended to be cooked, dry vegetables, melons, and fruit trees could be irrigated with sewage. By 1935, one authority found 113 localities in fifteen Western states using sewage irrigation, while another study located 53 California cities and 34 Texas cities growing crops with sewage. In addition, during these decades, a good deal of interest developed in the agricultural use of sewage sludge (sewage that had undergone treatment separating the solid and liquid materials) as a fertilizer. Like sewage irrigation, this was a method that dated back to the middle of the nineteenth century, although it was never widely used. In the 1920s, Milwaukee and Pasadena began marketing their sewage sludge as a fertilizer and the cities of Canton, Ohio, and Rochester and Schenectady, New York, disposed of their sludge to local farmers.[25]

During the last forty years, the land-disposal method of dealing with urban sewage has expanded at a fairly even rate. By 1957 there were 461 "systems" serving 2 million people applying wastewater to the land; by 1972 the number of systems had increased to 571, serving a population of 6.6 million people. The number of land-disposal systems, however, promises to be much larger in the future. The concern over ecology and wastewater disposal that arose during the 1960s has caused a renewed interest in the land-disposal method. An important factor here is that ecologists believe that wastes from municipal sewage-treatment plants are speeding up the eutrophication process in rivers and lakes. Spokesmen for the movement, such as Barry Commoner, have argued that "clearly the ecologically appropriate technological means of removing sewage from the city is to return it to the soil." One project that has been proposed would eliminate the potential health hazards of the earlier sewage farms by creating a "closed system that recycles nutrients, reclaims water to meet drinking water quality standards, and confines and contains wastewater constituents not suitable for recycling."[26]

The last several years have seen an intensification of the debate between those ecologists, and engineers who have espoused the land-disposal method of sewage disposal and those who believe that this method would destroy the soil mantle, permanently contaminate groundwater supplies, and result in immense land-acquisition programs. The entrance into the debate of the Environmental Protection Agency and the Army Corps of Engineers on the side of land disposal suggests that serious attempts will be made to make land disposal one of the foremost methods of the disposal of urban wastes. Ecologists argue that both city and farm will benefit. Cities would dispose of their wastes in an environmentally sound manner and farmers would be freed of their need for inorganic nitrogen fertilizers that distort the aquatic ecosystem.[27]

In a sense, then, the cycle has completed itself. Ideas concerning the profitable disposal of urban wastes on the land that were promoted during the middle of the nineteenth century by sanitarians such as Edwin Chadwick in England and George E. Waring, Jr., in the United States have been adopted by both ecologists and powerful government agencies. The technology and the rhetoric have been modernized, but the essential concepts remain the same: city and country are one rather than separate, and the land provides the essential medium on which they can be united.

Notes

I thank Norma B. Chaty of Envirotech Systems, Inc., for suggesting that I investigate this subject, and the Carnegie-Mellon University Environmental Studies Institute for help in covering the costs of research. My colleague, Professor David E. Wojick, and the students in their joint course "Society and Industrialism" were sources of penetrating ideas about this subject.

1. Chapter epigraph from a speech given at the Suffolk (Mass.) District Medical Society, and reported in *American Contract Journal* 11 (14 June 1884): 297.

2. G. E. Fussell, *The Classical Tradition in Western European Farming* (Madison, N.J.: Fairleigh Dickinson University Press, 1972), 27, 68; B. H. Slicher van Bath, *The Agrarian History of Western Europe A.D. 500–1850* (London: Edward Arnold, 1963), 241, 254–57; Lemuel Shattuck, *Report of the Sanitary Commission of Massachusetts 1850* (Cambridge, Mass.: Harvard University Press, 1948, facs. ed.), 215–16; Henry Robinson, "The Pail System," *Sanitary Engineer* 4 (15 March 1881): 179; Arturo Bruttini, *Uses of Waste Materials* (London: A. S. King and Sons, 1923), 282–89; F. H. King, *Farmers of Forty Centuries: Permanent Agriculture in China, Korea and Japan* (Madison, Wis.: F. H. King, 1911), 193–96; and T. H. Shen, *Agricultural Resources of China* (Ithaca, N.Y.: Cornell University Press, 1951), 32–33.

3. Charles V. Chapin, "The End of the Filth Theory of Disease," *Popular Science Monthly* 60 (January 1902): 234–39; Barbara G. Rosenkrantz, *Public Health and the State: Changing Views in Massachusetts, 1842–1936* (Cambridge, Mass.: Harvard University Press, 1972), 8–41; Charles S. Rosenberg, *The Cholera Years* (Chicago, Ill.: University of Chicago Press, 1962), 65–81; and Shattuck, *Report of the Sanitary Commission of Massachusetts*, 154.

4. Carl Bridenbaugh, *Cities in Revolt: Urban Life in America, 1743–1776* (New York: Alfred A. Knopf, 1955), 241; Percy Wells Bidwell and John I. Falconer, *History of Agriculture in the Northern United States 1620–1860* (New York: Peter Smith, 1941), 234. Several sources mention the utilization of "poudrette," a commercial fertilizer manufactured from urban sewage and night soil, before the Civil War. See Clarence H. Danhof, *Change in Agriculture: The Northern United States, 1820–1870* (Cambridge, Mass.: Harvard University Press, 1969), 267; Lewis C. Gray, *History of Agriculture in the Southern United States to 1860*, 2 vols. (Gloucester, Mass.: Peter Smith, 1958), 2:805; Ulysses Hedrick, *A History of Agriculture in the State of New York* (New York: Hill and Wang, 1966), 347–48; and Shattuck, *Report of the Sanitary Commission of Massachusetts*, 212–18.

5. U.S. Department of the Interior, Census Office, *Tenth Census of the United States, 1880, Report of the Social Statistics of Cities* [George E. Waring, Jr., comp.], 2 vols. (Washington: GPO, 1887); George A. Soper, "George Edwin Waring," *Dictionary of American Biography*, ed. Dumas Malone (New York: Scribners, 1933), 19:456–57; and James H. Cassedy, "The Flamboyant Colonel Waring," *Bulletin of the History of Medicine* 36 (March–April 1962): 1963–76. Waring was an anticontagionist who argued that sewer gas was the source or disease.

6. "Sewerage" is the term used to signify the system used to remove sewage from a town. An earlier survey of sewerage systems in America can be found in Henry I. Bowditch, *Public Hygiene in America* (Boston, Mass.: Little, Brown, 1877), 103–4. Water closets did not gain wide acceptance in the United States until the 1840s and 1850s. Their adoption depended on the introduction of running water and the construction of sewerage systems, although some homes had water closets connected to privy vaults or cesspools. See Nelson Manfred Blake, *Water for the Cities* (Syracuse, N.Y.: Syracuse University Press, 1956), 270–71.

7. These figures are probably understatements, since some cities furnished no data while others made comments such as, "taken out of the city." The figures refer only to those cities that specifically noted that farmers utilized their wastes. The figures by state for New England and the Middle states are as follows (the first number refers to the number of cities where wastes were used on the land; the second number refers to the total number of cities in the state): *New England*: Maine, 5 of 5; New Hampshire, 3 of 5; Massachusetts, 23 of 27; Rhode Island, 4 of 4; Connecticut, 8 of 10; and Vermont, 0 of 2. *Middle states*: New York, 12 of 21; New Jersey, 7 of 10; Pennsylvania, 11 of 17; and Delaware, 1 of 1. The figures for the Southern and Western states are as follows: *Southern states*: Maryland, 1 of 1; Washington, D.C., 1 of 1; Virginia, 5 of 6; West Virginia, 0 of 1; North Carolina, 1 of 2; Kentucky 0 of 4; Tennessee, 0 of 3; Georgia, 2 of 4; Florida, 1 of 2; Alabama, 1 of 3; Mississippi, 0 of 1; Arkansas, 0 of 1; Louisiana, 0 of 2; Texas, 1 of 5; *Middle states*: Ohio, 2 of 15; Indiana, 2 of 8; Illinois, 1 of 12; Missouri, 0 of 4; Michigan, 3 of 7; Wisconsin, 3 of 8; Minnesota, 0 of 3; Iowa, 0 of 7; Nebraska, 0 of 2; Kansas, 1 of 4; Colorado, 0 of 2; California, 1 of 6; Oregon, 1 of 1; and Utah, 1 of 1.

8. See *Social Statistics of Cities*, 1: 484 (Brooklyn), 830 (Philadelphia), and 129 (Boston). The

night soil of Albany, Indiana, was manufactured into a fertilizer called "Bromophyte." For the use of "poudrette" on farms, see Bidwell and Falconer, *History of Agriculture in the Northern United States*, 234; Gray, *History of Agriculture in the Southern United States*, 2:805; Danhof, *Change in Agriculture*, 267; and Hedrick, *A History of Agriculture*, 348.

9. *Social Statistics of Cities*, 1: 126 27; John D. Blake, *Public Health in the Town of Boston 1630–1822* (Cambridge, Mass.: Harvard University Press, 1959), 103, 209, 231; and John Duffy, *A History of Public Health in New York City 1625–1866* (New York: Russell Sage, 1968), 185–88, 357. See *Social Statistics of Cities*, 1: 129. The 1900 Census of Agriculture noted that street sweepings were "sold to farmers and truck gardeners at a small figure, or in some cases given to anyone who will haul them away." U.S., Bureau of the Census, *Twelfth Census of the United States, 1900*, vol. 5, *Agriculture* [Washington: GPO, 1902], Pt. 1, cxl, "Fertilizers"; ibid. *Eleventh Census of the United States, 1890*, vol. 5, *Agriculture* (Washington: GPO, 1895), 598, "Truck Farming"; and Rudolph Hering and Samuel A. Greeley, *Collection and Disposal of Municipal Refuse* (New York: McGraw-Hill, 1921), 578–84.

10. Bidwell and Falconer, *History of Agriculture in the Northern United States*, 233–34; Danhof, *Change in Agriculture*, 251–77; Gray, *History of Agriculture in the Southern United States*, 2: 800–810; and Allan G. Bogue, *From Prairie to Corn Belt* (Chicago, Ill.: University of Chicago Press, 1963), 144–46.

11. U.S. Bureau of the Census, *Historical Statistics of the United States* (Washington: GPO, 1960), 14; "Sanitary Notes—Sewerage and Sewage," *Scientific American*, n.s. 28 (28 June 1873): 405.

12. For general discussion of agricultural conditions in these areas, see Danhof, *Change in Agriculture;* Paul W. Gates, *The Farmer's Age: Agriculture 1815–1860, The Economic History of the United States,* vol. 3 (New York: Holt, Rinehart and Winston, 1960); Fred A. Shannon, *The Farmer's Last Frontier: Agriculture, 1860–1897, The Economic History of the United States,* vol. 5 (New York: Rinehart and Company, 1945); and Rosser H. Taylor, "The Sale and Application of Commercial Fertilizers in the South Atlantic States to 1900," *Agricultural History* 44 (January 1947): 46–52. This pattern is the same as that followed in the use of other fertilizers. See Shannon, *Farmer's Last Frontier*, 115, 170–71; *Twelfth Census, 1900*, vol. 5, *Agriculture*, Pt. 1, cxxxvi–cxlii, "Fertilizers."

13. *Eleventh Census, 1890*, vol. 5, *Agriculture*, 592–99, "Truck Farming"; Gates, *The Farmer's Age*, 255–70; Shannon, *Farmer's Last Frontier*, 259–67; and *Twelfth Census, 1900*, vol. 5, *Agriculture*, Pt. 1, cxxxvi, cxl, "Fertilizers." The 1900 Census noted that most truck farming was "carried on by foreigners in the North and in the far West it is in many places entirely given over to Chinamen or Italians." See ibid., vol. 6, *Agriculture*, Pt. 2, 322. "Vegetables."

14. The cities mentioned in *Social Statistics of Cities* as providing night soil to truck or garden farmers were Chester, Pennsylvania; Dayton, Ohio; Kalamazoo, Michigan; Lynchburg, Virginia; Paterson, New Jersey; Philadelphia, Pennsylvania; Savannah, Georgia; and Wilmington, North Carolina. In Jacksonville, Florida, and Los Angeles, California, urban wastes were used on fruit trees. For information on Baltimore, see Kenneth Allen, "The Sewerage of Baltimore," *Municipal Engineering* 16 (January 1899): 20–26; J. W. Magruder, "The Housing Awakening XIII: Exchanging 70,000 Earth Closets for a Sewer System—Baltimore," *Survey* 26 (2 September 1911): 809–14.

15. Taylor, "The Sale and Application of Commercial Fertilizers," 52; Mazyck Ravenel, ed., *A Half Century of Public Health* (New York: American Public Health Association, 1921), 161–208; and William T. Sedgwick, *Principles of Sanitary Science and the Public Health* (New York: Macmillan, 1918), 89–250.

16. Sedgwick, *Sanitary Science*, 128–39, 231–37; U.S., Bureau of the Census, "Sewers," *Report on the Social Statistics of Cities, Eleventh Census* (Washington: GPO, 1895), 29–32; and U.S., Bureau of the Census, "Sewers and Sewer Service," *General Statistics of Cities: 1909* (Washington: GPO, 1913), 20–23. Ibid., 114–18, has tables listing the waterways into which cities discharged their sewage. "Sanitary Notes—Sewerage and Sewage," 405; George W. Rafter and M. N. Baker, *Sewage Disposal in the United States* (New York: D. Van Nostrand Co., 1894), 153. British sewage was said to be less diluted than American sewage.

17. Shattuck, *Report of the Sanitary Commission of Massachusetts*, 216–17; "A Scotch Sewage

Farm," *Plumber and Sanitary Engineer* 3 (1 September 1880): 378; Edwin Chadwick, *Report on the Sanitary Condition of the Labouring Population of Great Britain 1842*, ed. M. W. Flinn (Edinburgh, 1865), 59–60, 118–19. George W. Rafter, *Sewage Irrigation, Pt. 2, Water-Supply and Irrigation Papers of the U.S. Geological Survey*, no. 3 (Washington, 1897), 43–49; "The Sewerage of Foreign Cities," *Engineering News* 6 (26 July 1879): 237–38; and Henry Robinson, "Sewage Farming," *Sanitary Engineer* 4 (15 September 1881): 482–83.

18. Shattuck, *Report of the Sanitary Commission of Massachusetts*, 212–18; George E. Waring, Jr., *Draining for Profit, and Draining for Health* (New York: Orange Judd & Co., 1867), 227. See, especially, issues of *Scientific American* and *Sanitary Engineer* for articles on European sewage practices. The most comprehensive survey of sewage disposal methods in the late nineteenth century is in Rafter and Baker, *Sewage Disposal in the United States*; see also Rafter, *Sewage Irrigation, Pts. 1 and 2.*

19. Rafter, *Sewage Irrigation Pt. 2*, 41–89; Rafter & Baker, *Sewage Disposal in the United States*, 225–53.

20. Solomon Schindler, *Young West: A Sequel to Edward Bellamy's Celebrated Novel "Looking Backward"* (New York: Arno Press, 1971), 203–8, 225–31, 266–67. Schindler was a Reform rabbi in Boston active in the Bellamy Nationalist movement. For information on Schindler, see Arthur Mann, *Yankee Reformers in the Urban Age: Social Reform in Boston, 1880–1900* (New York: Harper Torchbooks, 1966), 52–72.

21. George W. Fuller, *Sewage Disposal* (New York: McGraw-Hill, 1912), 596–99. One authority reported that in 1902 there were 95 cities and towns of over 3,000 population which had sewage "purification" systems. Or these, 27 used intermittent filtration, 21 used "sewage irrigation," 22 used septic tanks, and 10 used chemical precipitation. See Mansfield Merriman, *Elements of Sanitary Engineering* (New York: John Wiley, 1906), 213–14.

22. Sedgwick, *Sanitary Science*, 142–45; Rosencrantz, *Public Health and the State*; William T. Sedgwick, "Sewage and the Farmer: A Problem in the Conservation of Waste," *Scientific American* 107 (13 July 1912): 38; and Fuller, *Sewage Disposal*, 600–603.

23. Fuller, *Sewage Disposal*, 595–614; "Reflections on Sewage Farming Suggested by Experiences at Reading, England," *Engineering News* 54 (16 November 1905): 515–17; J. A. Voelcker, "The Agricultural Use and Value of Sewage," ibid. 64 (24 November 1910): 566, 571; and "Irrigation with Sewage," *Engineering Record* 65 (20 January 1912): 82–83. See editorial, *Engineering News* 64 (24 November 1910): 571; "Irrigation with Sewage," *Engineering Record* 65 (24 February 1912): 224; and *National Municipal Review* 11 (January 1913): 129.

24. An early statement of warning by a public health physician is G. W. Hosmer, "Sewage, and What Shall Be Done With It," *Harper's Weekly* 34 (1890): 567–68. See also Fuller, *Sewage Disposal*, 596–99; Sedgwick, *Sanitary Science*, 152–54; Sedgwick, "Sewage and the Farmer," 38; Voelcker, "The Agricultural Use and Value of Sewage," 566; and W. C. McNown, "Popular Objections to the Use of Sewage on an Irrigation Project," *Engineering News* 68 (8 August 1912): 270. Sedgwick warned about possible contamination of fruit and vegetables irrigated with sewage but acknowledged that the evidence provided "little or no" grounds for concern. More modern studies concur with this conclusion, although recommending that irrigation with sewage cease a month before harvest. See, for instance, W. Rudolfs, L. L. Falk, and R. A. Ragotzskie, "Contamination of Vegetables Grown in Polluted Soil: VI. Application of Results," *Sewage and Industrial Wastes* 23 (1951): 992–1000. Waring argued that there were "no sanitary objections . . . to the system of sewage disposal by agricultural irrigation." See George E. Waring, Jr, *Modern Methods of Sewage Disposal*, 2d ed. (New York: Van Nostrand, 1896), 114, 130. But see the exchange, "Irrigation with Sewage," *Engineering Record* 65 (20 January 1912): 82–83, and ibid. (24 February 1912): 224, between proponents and opponents of sewage irrigation. In 1904 M. N. Baker argued that differences in soil conditions, personal habits, municipal customs, and professional standards made sewage irrigation acceptable in Great Britain but not in the United States. See M. N. Baker, *British Sewage Works and Notes on the Sewage Farms of Paris and on Two German Works* (New York: Engineering News Publishing Co., 1904), 7. In the early twentieth century, however, the British also turned away from sewage irrigation. See "Reflections on Sewage Farming Suggested by Experi-

ences at Reading, England," 515–17, and Voelcker, "The Agricultural Use and Value of Sewage," 571.

25. Rafter and Baker, in their 1893 work on *Sewage Disposal*, devoted a chapter to "The Use of Sewage for Irrigation in the West." For the California regulations, see Leonard Metcalf and Harrison Eddy, *American Sewerage Practice, Vol. III, Disposal of Sewage*, 3d ed. (New York: McGraw-Hill, 1935), 249–50; Wells A. Hutchins, *Sewage Irrigation as Practiced in the Western States*, U.S. Dept. of Agriculture, Technical Bulletin no. 675 (Washington, March 1939), 2; "Experiences with Sewage Farming in Southwest United States," *American Journal Public Health* 25 (1935): 119–27; and Metcalf and Eddy, *American Sewerage Practice*, 3: 687–94.

26. Richard E. Thomas, "Land Disposal II: An Overview of Treatment Methods," *Journal Water Pollution Control Federation* 45 (July 1973): 1477. This issue of the *Journal* contains several articles on the land-disposal question. For a useful bibliography on this question covering from about 1930 to 1965, see James Law, Jr., *Agricultural Utilization of Sewage Effluent and Sludge: An Annotated Bibliography*, U.S. Dept. of the Interior, Federal Water Pollution Control Administration (Washington, January 1968); Barry Commoner, *The Closing Circle: Nature, Man and Technology* (New York: Bantam, 1971), 186–87; John R. Sheaffer "Pollution Control: Wastewater Irrigation," *DePaul Law Review* 21 (Summer 1972): 992–1007; Duane R. Egeland, "Land Disposal: A Giant Step Backward," *Journal Water Pollution Control Federation* 45 (July 1973): 1468–75; R. C. Toutson and R. E. Wildung, "Ultimate Disposal of Wastes to Soil," *Chemical Engineering Progress Symposium Series* 65: 97 (1969): 19–25; J. R. Peterson, T. M. McCalla, and George E. Smith, "Human and Animal Wastes as Fertilizers," *Fertilizer Technology and Use* (Madison, Wisconsin: Soil Science Society of America, 1971), 557–96; and Emil T. Chanlett, *Environmental Protection* (New York: McGraw-Hill, 1973), 158–60.

27. Commoner, *The Closing Circle*, 186–87; Sheaffer, "Pollution Control," 1002–7; and Egeland, "Land Disposal," 1470–75.

CHAPTER XI

From City to Suburb

The "Moral" Influence of Transportation Technology

Humanity demands that men should have sunlight, fresh air, the sight of grass and trees. It demands these things for the man himself and it demands them still more urgently for his wife and children. No child has a fair chance in the world who is condemned to grow up in the dirt and confinement, the dreariness, ugliness and vice of the poorer quarters of a great city. . . . There is, then, a permanent conflict between the needs of industry and the needs of humanity. Industry says men must aggregate. Humanity says they must not, or if they must, let it be only during working hours and let the necessity not extend to their wives and children. *It is the office of the city railways to reconcile these conflicting requirements.*

—Charles Horton Cooley, 1891

The theme of the evil city and the virtuous countryside has persisted throughout American history. Warnings of the unhealthiness of the urban environment and the threat that the urban masses posed to American ideals emanated from public spokesmen and intellectuals throughout the nineteenth century. Articulators of this point of view, such as Thomas Jefferson, Ralph Waldo Emerson, and Josiah Strong, usually regarded rural life as morally superior and generative of the virtues of health, strength of character, and individualism. Recently, however, Peter J. Schmitt and Scott Donaldson have expanded the classic urban-rural dichotomy. As Schmitt observes, few of those in the late nineteenth and early twentieth centuries who attacked the city and praised their rural childhoods ever returned to farming. While they looked with nostalgia to the rural virtues, they were unwilling to sacrifice the opportunities for the acquisition of wealth found in the city. Rather than being imbued with the philosophy of "agrarianism," which demanded that man draw his livelihood from the soil, they settled for the "spiritual values" of nature. And these values, which Schmitt calls "arcadian," thrived not only in the distant country, but also on the urban periphery in an area easily reached from the crowded city. In short, the suburb, *rus in urbe,* would enable Americans to pursue wealth and yet retain the amenities and values of rural life.[1]

The suburb, in the view of those who praised it, had certain desirable characteristics that distinguished it from the city. While the city was crowded, dirty, and smoky, the suburb had an abundance of "fresh air and clear sunlight, green foliage and God's blue sky." In the suburbs, which were primarily residential, most people lived in single-family dwellings or "cottages" far from the smoke and noise of business and industry. Serenity and calm rather than hustle and bustle were the hallmarks of the suburbs. Natural surroundings, "cottage" living, and peace and quiet provided ideal conditions for family life and for the raising of children, all within easy commuting distance of offices and factories in the core of the city.[2]

Without transportation technology, however, the suburb as the halfway-house between city and country and as the embodiment of the best of these diverse worlds would have been impossible. A number of writers have commented at length on the significance of the concept of technology in American thought, but often they have viewed the machine as opposed to the values of the pastoral or rural ideal.[3] Many influential spokesmen on the urban scene in the late nineteenth and early twentieth centuries, however, viewed technology as putting arcadia within the reach of city-dwellers who would otherwise have been denied its moral benefits. If suburbia was the garden that all urbanites should strive to reach, then it was the machine that made it possible to work in the city but live in that garden.

The "machines" in this instance were various forms of urban transit ranging from the omnibus to the electric street railway, subway, and elevated railroad, as well as the commuter steam railroad. All of these, theoretically, enabled the busy urbanite who labored in the city and resided in congested and unhealthy districts to continue to work in the city but to live in the suburbs in a superior environment. Statements about the role of transportation technology in permitting people to escape crowded and dirty cities occur throughout the literature on the city, but they appear with special frequency at the time of transit innovation. For it was then that Americans, concerned with the social dangers posed by urban growth, reaffirmed their faith in technology and saw in transit innovation the means to escape successfully from urban problems.[4]

Public transit developments had their largest impact on urban and suburban patterns roughly from 1840 to 1910. During these years urban population grew at a rapid rate, rising from 1,845,000, or 10.8 percent of the total population, to 44,639,989, or 45.7 percent. European immigrants, mainly from the rural areas of Ireland, Germany, Austria-Hungary, Italy, Poland, and Russia, accounted for a large part of the new urbanites, a fact that increased the fears of nativists and their spokesmen that the cities threatened American values. At this time, cities also greatly increased their size by annexing contiguous incorporated and unincorporated territory.[5] Within these burgeoning urban areas, public transit systems facilitated the dispersal of population, concentrated commercial activities, and reversed the spatial distribution of socioeconomic classes as compared with the patterns of the pretransit city, thus giving rise to the modern core-oriented metropolitan area.

This transformation of the city's ecological and demographic patterns had its beginnings in the 1840s and 1850s, when many large American cities faced what one historian calls an "urban transportation crisis." Cities in the 1840s and 50s were still primarily walking or pedestrian cities characterized by "crowded compactness." By 1850 New York had a population density of 135.6 persons per acre in its "fully settled area"; Boston, 82.7; Philadelphia, 80.0; and Pittsburgh, 68.4. Most urban residences were two- or three-story wooden or brick buildings. Those occupied by the working class and the poor were often packed with many families and their lodgers. Land uses were not clearly specialized, and middle- and upper-class residences were interspersed with those of lower income groups and located comparatively close to manufacturing and commercial structures. In contrast to contemporary living patterns, the elite often lived close to the business and governmental center of the city, while the majority of working men distributed themselves in the outlying wards. Those who lived in the urban core packed into narrow alley dwellings, tenements, and cellars. As population pressure increased, already crowded lower-class living areas disintegrated into slums.[6]

The growth of manufacturing and business activities accompanied the increased population congestion within cities. Most of the industrial and commercial development was clustered in the older sections of cities, especially near the waterfront in seaports and river towns. These central locations offered savings in transportation costs as well as the benefits of agglomeration economies. As the spatial needs of industries and businesses increased, residential population was pushed out of the central business areas. Many workers who were displaced moved into adjacent sections, creating increased problems and overcrowding and housing deterioration. These declining neighborhoods ringing the city center often attracted a large immigrant population that sought low-cost housing near their places of employment.[7]

The absence of a system of public transit exacerbated the problems of urban congestion. Without public transportation, persons whose work place was separate from their residence were forced to walk to work unless they could afford the expense of a private horse or carriage. The development of factories, banks, stock exchanges, and other such urban enterprises by the mid–nineteenth century greatly increased the number of people confronted with a journey to work, often of some considerable distance. While the peripheral areas of cities as well as towns grew rapidly in the 1830s, 1840s, and 1850s, they were not necessarily bedroom communities. Towns close to the central city, such as Lawrenceville and Birmingham near Pittsburgh and the Northern Liberties outside of Philadelphia, had their own separate economic focus and did not serve as residential sections for a large number of persons employed in the central city.[8]

During the decades before the Civil War, however, several key transportation innovations occurred that eventually made possible a suburban life for many people employed in the central cities. The earliest of these innovations were the omnibus and the commuter railroad, both of which appeared in Boston, New York, and Philadelphia in the 1830s and in Baltimore and Pittsburgh by the early 1850s. The omnibus was usually drawn by two horses and carried about twelve to fifteen passengers over an established route of city streets for a fixed fare. A coachman, mounted on an elevated seat at the front of the vehicle, collected the fares and drove the horses, while the passengers, who entered through a door at the rear of the vehicle, sat on long seats along the side of the omnibus. The steam railway, of course, was not originally intended for city-suburban travel but was rapidly adapted to that use. Boston led the way in the development of commuter railroad traffic, followed by New York and Philadelphia. Regulations that prevented the running of steam locomotives on streets, however, restricted the use of commuter trains in the latter cities, as they did in Pittsburgh, for some years.[9]

The transit innovation that had the most significant impact on urban patterns was the streetcar. First introduced in New York City in the early 1850s, it had spread to Boston, Philadelphia, Baltimore, Chicago, Cincinnati, and Pittsburgh by the end of the decade. The running of cars on rails through city streets was a major technological breakthrough. Alexander Easton, the author of *A Practical Treatise on Street or Horse-Power Railways,* published in 1859, called streetcars the "improvement of the age," and in terms of the increased facility of the intracity transportation of passengers, his enthusiasm was justified. Powered initially by horses and mules, then by cable, and ultimately by electricity, the streetcar dominated urban transit throughout the nation from the Civil War until the 1920s. Because of the much lower average fares of the streetcar as compared to the omnibus or commuter railroad, it had the greatest potential for enabling working people to move to residential areas with suburban characteristics. Although the electric streetcar, developed during the late 1880s, was often referred to as "rapid transit," this title properly belongs to the elevated train and to the subway, both of which operated on paths separate from street traffic and with trains of cars rather than single cars. An elevated system was first constructed in New York in 1871, with the cars pulled by steam locomotives. Elevated systems with electricity as the motive power were developed in Chicago in 1892, Boston in 1894, and Philadelphia in 1905. Boston built the first subway in 1897, followed by New York in 1904 and Philadelphia in 1909. Elevated and subway trains traveled at a much faster rate of speed than did surface cars, but their high construction costs made them feasible only for larger cities with a high volume of traffic. Public transit, therefore, for most cities, involved streetcar systems.[10]

Without the implementation of these transportation innovations, American cities would have developed in a different spatial pattern. Public transit produced an urbanized area roughly characterized by a central business district (CBD) or downtown surrounded by concentric circles or zones with specialized residential and industrial functions. Traction lines radiated from the core into the residential areas. The CBD became, over time, a section devoted almost entirely to business and commercial uses with a concentration of office buildings, banks, specialized retail outlets, and department stores. The residential areas usually had sharply distinguished socioeconomic patterns with the poorer sections near the core and the wealthier neighborhoods toward the fringe. Those areas with suburban characteristics (single-family detached homes and an absence of industry) developed both within and outside the city's boundaries. Many of the residents of those districts worked in the CBD and commuted by public transit; their daily journey-to-work significantly affected the tenor of life in the twentieth-century city.[11]

The two-part city, divided between residential and commercial-industrial sections, developed over the last half of the nineteenth century and first decades of the twentieth century in response to public transit expansion. Heavily congested working-class and immigrant living areas with problems of poor sanitation, high disease rates, and deteriorating housing, however, still persisted in large cities such as New York, Chicago, Boston, Philadelphia, and Pittsburgh. Manhattan's 10th Ward, for instance, had a density of 523.8 persons per acre in 1890 and the 13th Ward 428.8, with the population jammed into five- and six-story tenement houses. In Pittsburgh, in the same year, density in the wards surrounding the CBD had advanced to 121.9 persons per acre in a section with few dwellings over two or three stories. Members of the middle and upper classes worried that these congested living conditions, especially among the poor and the alien, posed a danger to "the moral integrity and the unity of the community."[12]

Many commentators on the evils of urban life in the latter half of the nineteenth and the beginning of the twentieth century argued that the solution of the city's problems lay in the further extension of mass transit. They explained that, through public transportation systems, men who labored in the city would be able to live and raise their families in the superior suburban environment. This theme was first articulated before the Civil War, when omnibus and horsecar lines were begun, and was heard with greater frequency toward the end of the nineteenth century as urban population growth and transportation development continued apace. As more and more centrally employed middle-class citizens moved to suburbs, urban spokesmen advocated the improvement and cheapening of mass transit to make possible suburban life for the working-class people remaining in the city."[13]

In the 1870s, for instance, Congregationalist minister Charles Loring Brace, in his book *The Dangerous Classes of New York,* warned of the deleterious effect of overcrowding on public morals and advocated the dispersal of population from city slums.[14] Specifically, Brace recommended the building of a subway or an elevated railway with cheap fares as the means to enable workers to settle in "pleasant and healthy little suburban villages." In the suburbs each family would have its "own small house and garden," while children would grow up "under far better influences, moral and physical, than they could possibly enjoy in tenement-houses."

Other writers in the 1870s and 80s repeated Brace's arguments about the moral influence of improved transit but also went beyond the social context. William R. Martin in *The North American Review* and L. M. Haupt in *Proceeding of the Engineer's Club of Philadelphia* held that the building of rapid transit would have economic as well as social benefits. Rapid transit, they maintained, would encourage the development of manufacturing within the city, increase real estate values, and stimulate building in open areas. The solution of the city's congestion problems would therefore be accompanied by financial gain for the metropolis's businessmen and builders.[15]

Many cities, however, could not afford the expense of either an elevated rapid transit system or a subway, and their expansion appeared limited by the speed and capacities of the horsecar. The application of cable and electric power to street railways in the 1880s appeared to resolve this problem. During the ten years from 1880 to 1890, the length of street railway track in the United States jumped from 2,050 to 5,783 miles, an advance of 182 percent at a time when urban population increased 56.7 percent. Urbanites in 1890 averaged 111 rides per year, with totals of over 270 rides per year per inhabitant in Kansas City, New York, and San Francisco. In that year horses and mules still supplied the motive force on 71 percent of the streetcar trackage, but this total plummeted during the decade. Most significant about the change in motive power was the increased speed of traction service. Streetcars now traveled at approximately ten miles per hour, just about double the speed of the horsecar, thus greatly expanding the areas within commuting distance of the downtown core. From 1890 to 1902, track mileage increased from 5,783 to 22,577—almost all of it operated by electricity—and rides per urban inhabitant from 111 to 181. Five years later, in 1907, track mileage had jumped 53.5 percent to 34,404 and rides per inhabitant to 250, as the use of public transit continued to far outdistance population increase.[16]

These transit developments gave further encouragement to those concerned with urban congestion that technology could solve the city's congestion problems. City boosters held this view as well as "urbanologists." In Pittsburgh, which began electrifying its streetcar system in the late 1880s, publications intended for visitors boasted of how the city's traction system permitted working people to live in "cosy residences" in the suburbs away from the "noise, smoke and dust of a great city." Suburban life, in turn, said one Pittsburgh guidebook, prevented the "breeding of vice and disease" and "elevated in equal proportion the moral tone of the laboring classes."[17]

The 1890 Census included, for the first time, a volume on transportation, with a special section on the "Statistics of Street Railway Transportation" prepared by sociologist Charles H. Cooley. A number of writers on urban trends in the 1890s used this material to demonstrate that the concentration of people in cities, with its deleterius effects, could be mitigated by improved urban transit. Carroll D. Wright, writing in *Popular Science Monthly,* Thomas C. Clarke in *Scribner's Magazine,* Henry C. Fletcher in *The Forum,* and Cooley himself in his work on *The Theory of Transportation* all agreed that while the conditions of modern industrial and commercial life necessitated concentration, "humanity" required that men and their families live among "sunlight, fresh air, grass and trees." In the words of the U.S. Commissioner of Labor, Carroll D. Wright, adequate urban transportation was "something more than a question of economics or of business convenience; it is a social and an ethical question as well." For, concluded Wright, only suburbs could supply the "sanitary localities, [the] moral and well-regulated communities, where children can have all the advantages of church and school, of light and air . . ." so necessary to "the improvement of the condition of the masses."[18]

By the turn of the century, however, many commentators on urban congestion had come to question whether improved transit alone would make possible a suburban life for the families that filled the city's tenements. In his seminal work of 1899, *The Growth of Cities in the Nineteenth Century,* Adna Weber argued that the development of suburbs rather than other "palliatives" such as model tenements, building laws, and housing inspection offered the best hope of escaping the evils of city life stemming from overcrowding. But while cheap and rapid transit was essential for suburban deveopment, said Weber, it had to be accompanied by a shorter working day and inexpensive subruban homes if working men were to take advantage of an environment that combined "the advantages of both city and country life."[19]

Such sentiments were repeated by other urban reformers during the beginning of the twentieth century. Some, such as Benjamin C. Marsh, secretary of the Committee on Congestion of Population in New York, saw city planning, tax reform, and even municipal land ownership as indispensable accompaniments to improved transit if urban decentralization was to become a reality. Others, such as Frederic C. Howe, believed that a single tax on land values should accompany rapid transit development. Many of the most heated local political battles of the Progressive Era were fought over the issue of municipal regulation or ownership of streetcar lines. Reform mayors such as Hazen Pingree of Detroit, Tom Johnson of Cleveland, and Samuel "Golden Rule" Jones and Brand Whitlock of Toledo advocated better and cheaper traction service as a means to better the lot of urban working men; undoubtedly they believed that such improvements would permit residents of crowded inner-city districts to move to neighborhoods where they could realize the suburban ideal.[20]

In its *Special Report on Street and Electric Railways* (1902), the Bureau of the Census of the Department of Commerce and Labor presented the most comprehensive statement yet on the street railway as a "social factor." The report observed that street railway development had come "in response to an imperative social need" and that urban transit had facilitated the dispersal of population from the city while encouraging the concentration of commercial and manufacturing establishments. Traction companies had also performed an important social service by transporting people from crowded cities to "places of outdoor recreation." But, added the report, because city transit service was often inadequate, it had actually hampered potential suburban growth. The report recommended increased speed, additional cars, and lower fares as one means to deal with transit deficiencies. Satisfactory suburban and interurban transit service in coordination with surface lines was also recommended, but the expense of such systems perhaps precluded a "wholly satisfactory solution of the problem of transportation in great and rapidly growing cities."[21]

By the end of the first decade of the twentieth century, therefore, there existed a large body of literature that viewed urban transit as the means by which men could escape the evils of the crowded city and live in suburbs with the benefits of cottage living, clean air and sunshine, and close communication with nature. Some reformers, however, believed that changes such as tax reform, municipal ownership, and cheap housing had to accompany transit improvements if suburban living for the working class was to become a reality. In reviewing this material, it is difficult not to conclude that many who advocated urban decentralization were as motivated by considerations of social control as by a desire to enable men to live more comfortable or healthy lives. City slums crowded with immigrants usually had high crime rates and poor sanitary conditions, and middle- and upper-class citizens worried about the threat of violence and disease posed by congested working-class areas. Transit systems, by making it possible for working people to leave the unhealthy city for arcadian suburbs, seemingly offered a relatively inexpensive method of curtailing the threat of the slum.[22]

Were the advocates of a technological solution to the problems of the city stemming from congestion—whatever their motivation—misguided? As early as 1866, *The Nation* editorialized that the time and money required for commuting to suburbs eliminated it as an alternative for many members of the working class, a conclusion also reached by the leading contemporary student of the beginnings of mass transit. But what of later traction developments? Writing in 1890, journalist and social critic Jacob Riis pessimistically noted that rapid transit in New York had failed to resolve the problems of the tenement-house slum. Technology had proved ineffective, he held, when faced by the "system" resulting from a combination of "public neglect and private greed." Even the 1902 Bureau of the Census traction report, which had praised the impact of the streetcar, noted that most suburbanites were well-to-do and that poverty and long work hours prevented many workers from utilizing public transit to escape the city. And, in his study *Streetcar Suburbs,* historian Sam Bass Warner, Jr., observes that urban transit provided a "safe, sanitary environment" for only half of the Boston metropolitan population; the remainder were condemned to the crowded city. In addition, says Warner, the emphasis on a system of individualistic capitalism that promised a suburban life for those who could pay the price meant that the society neglected the immense social and housing needs of the working class and immigrant poor.[23]

Some of this criticism is justified. Many members of the working class could afford neither the time nor the money required to commute to suburbs. Moreover, the stress on facilitating the movement of people from city to suburb via urban transit undoubtedly did divert attention from the housing and recreational needs of the poor who remained in the city. But this emphasis suited the American value system—one that combined a strong belief in the capacity of technology to solve social ills, a belief in the moral superiority of a suburban to a city existence, and a commitment to a system of private capitalism and decision making that theoretically allowed each person to make his own choice of residence according to his income and preference.[24]

Given this set of values, the streetcar did not "fail" in its promise. Urban transit systems did enable many citizens to leave the congested areas of central cities for living areas with more amenities. Housing vacated by these groups in turn provided more housing choice for those remaining.[25] That areas of some cities like New York and Philadelphia grew more rather than less congested during the streetcar era resulted as much from the constant influx of new urban residents as from deficiencies in the transit system. Given this rush into the cities, the ability of many of these newcomers to leave congested city areas as quickly as they did is perhaps more astonishing than the fact that congestion remained high in some wards.

Today the private automobile has replaced public transit as the chief means by which Americans commute between suburban residences and city jobs. Not surprisingly, many early auto boosters predicted that it would serve the same function that gave mass transit such an appeal generations earlier—the motor car would open a suburban existence for those who wanted to escape the crowded and unpleasant city. The automobile has been largely successful in this role, and the flow of people leaving central cities for suburban amenities continues year after year. As the 1970 census revealed, more people live today in the suburban rings around central cities than in the cities themselves. The poor and the minorities, however, as during the strectcar era, still seem condemned to the central cities.[26] The "garden" continues to beckon, but obviously it will take more than transportation improvements alone to make possible a suburban life for all those who desire it.

Notes

1. For discussions of this literature, see David R. Weimer (ed.), *City and Country in America* (New York: Appleton-Century-Crofts, 1962); Morton and Lucia White, *The Intellectual versus the City* (Cambridge: Harvard University Press, 1962); Anselm Strauss, *Images of the American City* (Glencoe: Free Press, 1961); Scott Donaldson, *The Suburban Myth* (New York: Columbia University Press, 1969), 24–25; and Peter J. Schmitt, *Back to Nature: The Arcadian Myth in Urban America* (New York Oxford University Press, 1969). xvii–xviii.

2. See for example, Edward E. Hale, "The Congestion of Cities," *Forum* (1888) IV, 532–33; Adna F. Weber, "Suburban Annexations," *North American Review* 166 (May 1898): 616; Donaldson, 24–27; and Sam Bass Warner, Jr., *Streetcar Suburbs: The Process of Growth in Boston, 1870–1900* (Cambridge Harvard University Press and the M.I.T. Press, 1962), 11–14.

3. Leo Marx, *The Machine in the Garden* (New York: Oxford University Press, 1964); Marvin Fisher, *Workshops in the Wilderness* (New York: Oxford University Press, 1967); and Hugo A. Meier, "American Technology and the Nineteenth-Century World," *American Quarterly* 10 (1958): 116–130.

4. Charles N. Glaab and A. Theodore Brown, *A History of Urban America* (New York: Macmillan Co., 1967), 25–27, 93–95, 107–111, 138–142.

5. Kenneth T. Jackson, "Metropolitan Government Versus Suburban Autonomy: Politics on the Crabgrass Frontier," in Kenneth T. Jackson and Stanley K. Schultz (eds.), *Cities in American History* (New York: Alfred A. Knopf, 1972), 442–452.

6. George Rogers Taylor, "The Beginnings of Mass Transportation in Urban America: Part I," *The Smithsonian Journal of History* 1 (Summer 1966): 137–38; Sam Bass Warner, Jr., *The Private City: Philadelphia in Three Periods of Its Growth* (Philadelphia: University of Pennsylvania Press, 1968), 49–62; Warner, *Streetcar Suburbs*, 15–21; Peter G. Goheen, *Victorian Toronto 1850–1900* (Chicago: University of Chicago Department of Geography Research Paper No. 127, 1970), 8–9; and Joel A. Tarr, "Transportation Innovation and Changing Spatial Patterns in Pittsburgh, 1850–1910," (Pittsburgh: Carnegie-Mellon University Transportation Research Institute, 1971), 2–5.

7. Ibid., 3–5; Goheen, 8–9; Allan R. Pred, *The Spatial Dynamics of U.S. Urban-Industrial Growth, 1800–1914* (Cambridge: M.I.T. Press, 1966), 196–197; David Ward, *Cities and Immigrants* (New York: Oxford University Press, 1971), 85–109; Taylor, 39; and Pred, 148–152, 167–177.

8. Taylor, 37; Kenneth T. Jackson, "The Suburban Trend Before the Civil War" (unpublished paper delivered at American Historical Convention, Washington D.C. Dec. 30, 1969), 10–11; and Bernard J. Sauers, "A Political Process of Urban Growth: Consolidation of the South Side with the

City of Pittsburgh, 1872," (unpublished seminar paper, Department of History, Carnegie-Mellon University, 1972), 3–10.

9. Taylor, 40–44; Taylor, "The Beginnings of Mass Transportation in Urban America, Part II," *The Smithsonian Journal of History* 1 (Autumn 1966): 31–38; and *Pittsburgh Post*, Nov. 3, 1860. For railroad commeter traffic in 1890, see John S. Billings (comp.), *Report on the Social Statistics of Cities, Eleventh Census* (Washington: Government Printing Office, 1895), 50; Taylor, "Beginnings of Mass Transportation: Part II," 39–50.

10. Ward, 125–143; George M. Smerk, "The Streetcar: Shaper of American Cities," *Traffic Quarterly* 21 (Oct. 1967): 569–584; and James Blaine Walker, *Fifty Years of Rapid Transit 1864–1917* (New York: Arno Press, 1970, reprint of 1918 edition), i.

11. Blake McKelvey, *The Urbanization of America 1860–1915* (New Brunswick: Rutgers University Press, 1963), 76–85; Smerk, 569–584; Warner, *The Private City*, 177–200; Warner, *Streetcar Suburbs, passim;* and Ward, 125–143. Cities obviously diverged in their patterns of development according to topographical factors and the individual locational decisions of businessmen and householders. I have not meant to endorse any particular theory of uban development but rather to point out that traction increased the tendency towards specialized residential and commercial districts. Some businesses and industries remained scattered throughout the city at the height of the streetcar period. See Raymond L. Fales and Leon N. Moses, "Thünen, Weber and the Spatial Structure of the Nineteenth Century City," to be published in Charles Leven (ed.), *Essays in Honor of Edgar M. Hoover.*

12. Roy Lubove, *The Progressives and the Slums* (Pittsburgh: University of Pittsburgh Press, 1962), 10, 82, 94; Adna F. Weber, *The Growth of Cities in the Nineteenth Century* (Ithaca: Cornell University Press, 1967, reprint of 1899 edition), 460–464; and Pittsburgh data from U.S. Department of Commerce and Labor, *Bureau of the Census, Vital and Social Statistics, Part II: Cities of 100,000 Inhabitants, Eleventh Census* (Washington: Government Printing Office, 1895) XXI, 293–311.

13. For articulation of this theme in regard to the omnibus, see Glen E. Holt, "The Changing Perception of Urban Pathology: An Essay on the Development of Mass Transit in the United States," in Jackson and Schultz, 325–326. On the horsecar, see the Pittsburgh *Post*, Dec. 17, 1859. The *Post* editorialized: "The convenience of these cars to the citizens cannot be over estimated. The effect upon the health and morals of the city by scattering the population is certain to be beneficial to all classes. Those in moderate circumstances will no longer be compelled to rent houses in narrow courts and alleys. . . . People can keep themselves and their children out of the temptation and proximity of vice. In the country, people must hunt after occasions for wickedness, in the city they are thrust upon them." See also Rev. S.C. Aiken, "Moral View of Railroads," *Hunts' Merchants Magazine and Commercial Review* 32 (Nov. 1852): 557–584; Warner, *Streetcar Suburbs*, 26–27. Legislation providing for "workingmen's" fares at a lower than usual cost were required for commuter railroads in Massachusetts in 1872 and for streetcars in Detroit in 1893. Parliament required commuter railroads in the London area to run "workmen's trains" with low fares as early as 1860. For American developments see Charles J. Kennedy, "Commuter Services in the Boston Area, 1835–1860," *Business History Review* 35 (Summer 1962): 169–170, and Melvin G. Holli, *Reform in Detroit: Hazen S. Pingree and Urban Politics* (New York: Oxford University Press, 1969), 47–48; and for London, T.C. Barker and Michael Robbins, *A History of London Transport* (London: George Allen & Unwin LTD, 1963) I, 173–174.

14. Charles Loring Brace, *The Dangerous Classes of New York*, 3rd ed. (New York: Wynkoop and Hallenbeck, 1880), 57–60. (The author is indebted for this reference, as well as those cited in note 25, to Clay McShane of the Smithsonian Institution.) For a study of Brace and his attitude toward the city, see R. Richard Wohl, "The 'Country Boy' Myth and Its Place in American Urban Culture: The Nineteenth-Century Contribution," ed. by Moses Rischin, *Perspectives in American History* (Cambridge: Harvard University Press, 1969) III, 107–121.

15. William R. Martin, "The Financial Resources of New York," *The North American Review* 127 (Nov.–Dec. 1878): 442–443; L.M. Haupt, "Rapid Transit," *Proceedings of the Engineer's Club of Philadelphia* 4 (Aug. 1884): 135–148.

16. Charles H. Cooley, "Statistics of Street Railway Transportation," in Henry C. Adams

(comp.) *Report on the Transportation Business in the United States, Part I, Transportation by Land, Eleventh Census* (Washington: Government Printing Office, 1895) XVIII, 681–684; U.S. Department of Commerce and Labor, Bureau of the Census, *Street and Electric Railways 1907* (Washington: Government Printing Office, 1910), 33.

17. *Illustrated Guide and Handbook of Pittsburgh and Allegheny* (Pittsburgh, 1887), 50; Consolidating Illustrating Company (comp.), *Pittsburgh of Today* (Pittsburgh: Consolidating Illustrating Co., 1896), 69; and Knights Templar, *Official Souvenir 27th Triennial Conclave Knights Templar* (Pittsburgh, 1898), n.p.

18. Charles H. Cooley, "Statistics of Street Railway Transportation," 681–791; Carroll D. Wright, "Rapid Transit. Lessons from the Census. VI," *Popular Science Monthly* 40 (April 1892): 785–792; Thomas Curtis Clarke, "Rapid Transit in Cities," *Scribners Magazine* 11 (May–June 1892): 568–578, 743–758; Charles H. Cooley, "The Social Significance of Street Railways," *Publications of the American Economic Association* 6 (May 1891): 71–73; Cooley, *The Theory of Transportation, Publications of the American Economic Association* 9.3 (May 1894); Henry J. Fletcher, "The Drift of Population to Cities: Remedies," *Forum* 19 (August 1895): 737–745; and Wright, 790.

19. Weber, *The Growth of Cities*, 467–475. See also, Weber, "Suburban Annexations", 617, and "Rapid Transit and the Housing Problem," *Municipal Affairs* 6 (Fall 1902): 409–417.

20. Frederic C. Howe, *The City: The Hope of Democracy* (Seattle: University of Washington Press, 1967, reprint of 1905 edition), 202–206; Lubove, 231–238. See also Charles M. Robinson, "Improvement in City Life," *Atlantic* 83 (April 1899): 531; Delos F. Wilcox, *Municipal Franchises* (New York: Engineering News Publishing Co., 1911, 2 vols.) II, 6–11; Henry C. Wright, "The Interrelation of Housing and Transit," *American City* 10 (Jan. 1914): 51–53; and the summary of the report of the New York City Commission on Congestion of Population in *The Survey* 25 (March 25, 1911): 1064–1067. See, for example, Holli, 33–35; Hoyt Landon Warner, *Progressivism in Ohio 1897–1917* (Columbus: Ohio State University Press for the Ohio Historical Society, 1964), *passim*; and Tom L. Johnson, *My Story*, ed. by Elizabeth J. Hauser (Seattle: University of Washington Press, 1970), passim.

21. U.S. Department of Commerce and Labor, Bureau of the Census, *Street and Electric Railways 1902 Street and Electric Railways 1902* (Washington: Government Printing Office, 1905), 26–43.

22. On the theme of social control and urban decentralization see Lubove, 131, 250–251.

23. "The Future of Great Cities," *The Nation* 2 (Feb. 22, 1866): 232; Taylor, "The Beginnings of Mass Transportation: Part II," 51–52; Jacob A. Riis, *How the Other Half Lives* (New York: Sagamore Press, Inc., 1957, reprint of 1890 edition), 2; and *Electric Railways 1902*, 31–33. See also Edward C. Pratt, *Industrial Causes of Congestion of Population in New York City* (New York: AMS Press, 1968, reprinting of 1911 edition), 191–96; Warner, 160–161. Warner also feels that the suburbs failed to supply the reinforced community life that supposedly was a goal of those who left the city.

24. See *ibid.*, 153–166.

25. Ward, 120–121.

26. See, for example, Harlan Paul Douglass, *The Suburban Trend* (New York: Amo Press, 1970, reprint of 1925 edition); James J. Flink, *America Adopts the Automobile, 1895–1910* (Cambridge: MIT Press, 1970), 108–110; Blaine A. Brownell, "A Symbol of Modernity: Attitudes Toward the Automobile in Southern Cities in the 1920's," *American Quarterly* 24 (March, 1972): 25–26; and U.S. Department of Commerce, Bureau of the Census, *Trends in Social and Economic Conditions in Metropolitan and Nonmetropolitan Areas* (Washington, DC: Government Printing Office, September 3, 1970, Series P-23, No. 33). The current revival of interest in mass transit to connect suburbs with downtown business districts has different roots than the programs and proposals of a century or so ago. Urban transit proposals today are intended to deal with the problems of automobile congestion rather than with people and housing congestion. See John W Dyckman, "Transportation in Cities," in Scientific American, *Cities* (New York: Alfred A. Knopf, 1968), 133–155.

CHAPTER XII

The Horse—Polluter of the City

The immense amount of public concern about automobile pollution has caused Americans to forget that the predecessor of the auto was also a major source of pollutants. Associated with the horse were most of the problems today attributed to the automobile: air contaminates harmful to health, noxious odors, and noise. By the turn of the twentieth century, horse pollution had become so bad in city streets that writers in popular and scientific periodicals were demanding "the banishment of the horse from American cities." The presence of 120,000 horses in New York City, wrote one authority in 1908, is "an economic burden, an affront to cleanliness, and a terrible tax upon human life."[1] The solution to the problems of the horse, agreed the critics, was the adoption of the "horseless carriage"—the automobile.

While a concern with clean streets, and with the horse as the principal cause of dirty streets, had been present in European cities as early as the fourteenth century and in American cities from their beginnings, it required a more statistically minded age to measure the actual amount of manure produced by the horse. Sanitary experts in the early part of the twentieth century agreed that the normal city horse produced between fifteen and thirty pounds of manure a day, with the average being about twenty-two pounds. In a city like Milwaukee in 1907, for instance, with a human population of 350,000 and a horse population of 12,500, this meant

133 tons of manure a day, or an average of nearly three-quarters of a pound of manure per person per day. Or, as the health officials in Rochester calculated in 1900, the 15,000 horses in that city produced enough manure in a year to make a pile covering an acre of ground 175 feet high and breeding sixteen billion flies.[2]

The horse population of some other American cities in the early 1900s, after the automobile and the electric streetcar had caused a decline in the number of urban horses, was: Chicago, 83,330; Detroit, 12,000; and Columbus, 5,000. In total, there were probably 3–3.5 million horses in American cities in 1900, as compared with about 17 million living in more bucolic environments. Today, at a time when horseback riding for pleasure is on the rise, the total number of horses in the United States is approximately 5 million. The ratio between horses and people in the nineteenth-century city was much higher in the days before traction lines were electrified. In 1890, after electrification had already begun, 22,000 horses and mules were still pulling streetcars in New York City and Brooklyn, with another 10,000 performing similar work in Philadelphia and Chicago.[3]

To a great extent, urban life before the development of electric and cable-powered traction and the automobile moved at the pace of horse-drawn transportation. The evidence of the horse was everywhere—in the piles of manure that littered the streets, attracting swarms of flies and creating an offensive stench; in the iron rings and hitching posts sunk into the pavements for fastening horses' reins; and in the numerous livery stables that gave off a mingled smell of horse urine and manure, harness oil and hay. In 1880, when the cities of New York and Brooklyn had a combined human population of 1,764,168 and a horse population somewhere between 150,000 and 175,000, the needs of the quadrupeds were served by 427 blacksmith shops, 249 carriage and wagon enterprises, 262 wheelwright shops, and 290 establishments dealing in saddles and harnesses. On a typical day in 1885, 7,811 horse-drawn vehicles, many with teams of two or more horses, passed by the busy corner of Broadway and Pine streets in New York City.[4]

While some of these conveyances were fine carriages drawn by spirited teams, the most common city horses were commercial or work animals. City streets were crowded with large drays, pulled by teams, that hauled freight and did heavy moving; single-horse spring wagons making deliveries to residential areas and adorned with business advertisements; peddling carts and ice and milk wagons; and omnibuses and hacks that carried passengers back and forth to their destinations. Even after the development of the steam engine, urban civilization still depended on the horse. As *The Nation* noted in 1872, while great improvements had been made in the development of "agents of progress" such as the railroad, the steamboat, and the telegraph, the society's dependence on the horse had "grown almost *pari passu* with our dependence on steam."[5] For, it was the horse that fed the railroads and steamboats with passengers and freight and that provided transportation within the cities.

Vital as it was to the functioning of urban society, city-dwellers recognized early on the problems posed by the horse. The question of clean streets was most obvious. In eighteenth century Boston and New York, money was allocated by the city fathers for street cleaning, and householders were required to sweep the road in front of their doorways. Cities made sporadic attempts during the mid–nineteenth century to improve the quality of street cleaning. In 1855, New York introduced street-sweeping machines and self-loading carts, and in 1865 urban entrepreneurs formed the New York Sanitary and Chemical Compost Manufacturing Company for the purpose of "cleansing cities, towns and villages in the United States" with several varieties of street-sweeping machines. By 1880, almost all cities over 30,000 population employed street-cleaning crews.[6]

American cities made their most sustained efforts to clean the streets during periods of cholera, smallpox, yellow fever, or typhoid epidemics. Many eighteenth- and nineteenth-century medical authorities believed that such diseases were caused by "a combination of certain atmospheric conditions and putrefying filth," among which horse manure was the chief offender. In 1752, Boston selectmen allocated extra funds to clean the streets because of the fear that street dirt might contain smallpox infection; and in 1795, during the yellow fever season, town officials invited neighboring farmers to collect the manure from the streets free of charge. The city fathers of New York, faced by the threat of cholera in 1832, made special efforts to cleanse the cobblestones, thereby divesting the city "of that foul aliment on which the pestilence delights to feed."[7]

But unless jolted by the fear of epidemic, city authorities and citizens tolerated a great deal of filth in their streets. Some cities tried to cover the cost of street cleaning by selling the manure for fertilizer. In 1803 the New York superintendent of scavengers expended about $26,000 for street cleaning and realized over $29,000 from the sale of the manure collected. In those cases, however, where private contractors were responsible for cleaning the streets, citizens often complained that they neglected other forms of rubbish and only collected the salable manure. Sanitation officials in the years after the Civil War often reported that street dirt was too mixed with other forms of litter to be sold as fertilizer. But whatever the salable quality of the street refuse, urban sanitary departments during the nineteenth century were notoriously inefficient. Vexed by graft and corruption, they were staffed by "old and indigent men," "prisoners who don't like to work," and "persons on relief."[8]

Given the state of street cleaning in the nineteenth-century city, it is not surprising to discover that newspapers, diaries, and governmental reports abound with complaints about the problems created in the city by horse manure. Piles of manure collected by street cleaners bred huge numbers of flies and created "pestilential vapours." Litter from wealthy residential neighborhoods was often dumped in poor neighborhoods and left to rot. Streets turned into virtual cesspools when it rained, causing women to accumulate filth on their dresses. In New York, Paris, and other great cities, ladies and gentlemen were aided in their navigation through a sea of horse droppings by "crossing-sweepers." Other complaints derived from the pulverized horse dung that blew into peoples' faces and houses and that covered the wares of merchants with outside displays. The paving of streets accelerated this problem, as wheels and hooves ground the manure against the hard surfaces and amplified the amount of dust.[9]

In many American cities, what paving there was consisted largely of cobblestones, and the noise of the horses' iron shoes and the iron-tired wheels of carts and wagons on the stones created an immense din. Benjamin Franklin complained in the late eighteenth century of the "thundering of coaches, chariots, chaises, wagons, drays and the whole fraternity of noise" that assailed the ears of Philadelphians, and similar comments about urban noise were made by travelers in other cities. Attempts were made quite early to quiet the clamor. In 1747, the Boston town council banned traffic from King Street so that the noise would not interrupt the debates of the Great and General Court; and in 1785 New York City passed an ordinance forbidding teams of more than one horse and wagons with iron-shed wheels from the streets. In London the custom existed of putting straw on the pavement outside sick people's houses to muffle the sounds of traffic, a practice undoubtedly followed in America. As late as the 1890s, a writer in the *Scientific American* noted that the sounds of traffic on busy New York streets made conversation nearly impossible, while the author William Dean Howells complained that "the sharp clatter of the horses' iron shoes" on the pavement tormented his ear.[10]

While the horse created many problems for the city, urban conditions made life for the city quadruped far more difficult than that of its country relations. City horses were notoriously overworked. The average streetcar horse had a life expectation of barely two years, and it was a common sight to see drivers and teamsters whip and abuse their horses to spur them to pull heavy loads. The mistreatment of the city horse was a key factor in moving Henry Bergh to found the American Society for the Prevention of Cruelty to Animals in 1866. City working horses were usually housed under crowded and unsanitary conditions without adequate light or air. Only the pleasure horses of the elite had access to the green fields and open areas enjoyed by the rural horse. Many overworked and mistreated urban horses died on the city streets. In addition, streets paved with cobblestones or asphalt were slipperier than dirt roads, and if a horse broke a leg it was destroyed. In order to minimize this danger, some veterinarians recommended that city draft horses be shod with rubber-padded horseshoes, but few owners followed this advice. A description of Broadway appearing in the *Atlantic Monthly* in 1866 spoke of the street as being clogged with "dead horses and vehicular entanglements." In 1880 New York City removed 15,000 dead horses from its streets; and as late as 1912, Chicago carted away nearly 10,000 horse carcasses. (A contemporary book on the collection of municipal refuse advised that, since the average weight of dead horses was 1,300 pounds, "trucks for the removal of dead horses should be hung low, to avoid an excessive lift.") The complaint of one horse lover that "in the city the working horse is treated worse than a steam-engine or sewing machine" was well justified.[11]

By the 1880s and the 1890s, a combination of factors—the immense population growth of American cities, the need for improved transportation to allow urban populations to spread from overcrowded areas, and an increased concern with sanitation—produced a search for alternatives to the horse as the chief form of urban locomotion. The first major breakthrough came with the development of the cable car and the electric trolley car in the late 1880s, and traction companies were quick to substitute mechanical power for animal power on their streetcar lines. Writing in *Popular Science Monthly* in 1892, U.S. Commissioner of Labor Carroll D. Wright maintained that electric power was not only cheaper than horse power but also far more beneficial to the city from the perspective of health and safety. "The presence of so many horses constantly moving through the streets," wrote Wright, "is a very serious matter. The vitiation of the air by the presence of so many animals is alone a sufficient reason for their removal, while the clogged condition of the streets impedes business, and involves the safety of life and limb."[12]

The expansion of electric traction and the improvements made in the "horseless carriage" and the bicycle caused horse lovers to become defensive about the future of the quadruped. Writing in *The Chautauquan* in 1895, Robert L. Seymour maintained that while the "cheap horse" might be doomed, the "costly, good-looking horse, the horse of history, the heroic horse in action, will probably last long." Can you imagine, asked Seymour, "Napoleon crossing the Alps in a blinding snow storm on a bicycle or Alexander riding heroically at the head of his armies in a horseless carriage?" Another writer in *Lippincott's Magazine* insisted that since "Americans are a horse-loving nation . . . , the wide-spread adoption of the motor-driven vehicle in this country is open to serious doubt." Less romantic observers, however, embraced the possibility of the elimination of the horse with enthusiasm. When William Dean Howells's Altrurian Traveller visited Chicago's white city in 1893, he noted with pleasure that this metropolis of the future had "little of the filth resulting in all other American cities from the use of the horse."[13]

During the opening years of the twentieth century, popular journals such as *Harper's Weekly, Lippincott's Magazine,* and *The Forum,* as well as more specialized periodicals such as *American City, Horseless Age, Motor,* and the *Scientific American,* were filled with articles extolling the automobile and the motor-truck and disparaging the horse. The lines of attack took several directions. Extremely common was the economic analysis, which argued, as did one writer in *Munsey's Magazine,* that "the horse has become unprofitable. He is too costly to buy and too costly to keep." Articles such as these computed the expense of the "horse cost of living" and compared it unfavorably to the cost of automobile upkeep. Other articles pointed out the advantages the motor-truck had over the horse in hauling freight and in preventing traffic congestion. One writer in the *American City* noted that the good motor-truck, which was immune to fatigue and to weather, did on the average of two-and-a-half times as much work in the same time as the horse and with a quarter of the amount of street congestion. "It is all a question of dollars and cents, this gasoline or oats proposition. The automobile is no longer classed as a luxury. It is acknowledged to be one of the great time-savers in the world."[14]

Equally as convincing as the economic arguments for the superiority of the motor vehicle over the horse were those from the perspective of health. "The horse in the city is bound to be a menace to a condition of perfect health," warned Dr. Arthur R. Reynolds, superintendent of the Chicago Health Department. Public health officials charged that wind-blown dust from ground-up manure damaged eyes and irritated respiratory organs, while the "noise and clatter" of city traffic aggravated nervous diseases. Since, noted the *Scientific American,* the motor vehicle left no litter and was "always noiseless or nearly so," the exit of the horse would "benefit the public health to an almost incalculable degree." Also blamed on the horse were diseases such as cholera, typhoid fever, dysentery, and infant diarrhea, which were often transmitted by the house fly. The favorite breeding place of the fly was the manure heap, and in the late 1890s insurance company actuaries discovered that employees in livery stables and those living near stables had a higher rate of infectious diseases such as typhoid fever than did the general public. Sanitation specialists pursued the question, and the first decade of the twentieth century saw a large outpouring of material warning of the danger of the infection-carrying "queen of the dung-heap," *Musca domestica.* The obvious way to eradicate the "typhoid fly," as he was called by L. O. Howard, chief of the Bureau of Entomology of the Department of Agriculture and a leader of the campaign against flies, was to eliminate the horse.[15]

Writing in *Appleton's Magazine* in 1908, Harold Bolce, in an article entitled "The Horse vs. Health," blamed most of the sanitary and economic problems of the modern city on the horse and determined the savings if all horses were replaced by automobiles and motor trucks. According to Bolce, 20,000 New Yorkers died each year from "maladies that fly in the dust" created mainly by horse manure. He calculated that the value of these people's lives, plus the cost of maintaining hospitals to treat them, could be attributed to the failure to substitute automobiles for horses. To this sum he added the cost of street cleaning and rubbish disposal, the higher urban cost of living because of the failure to use motor trucks for horses in transporting goods, and the costs of traffic congestion, reaching a total of approximately $100 million as the price that New York City paid for not banning the horse from its streets. The horse, maintained Bolce, represented one of the last stands of animal strength over science and, as such, it had to go—Americans could no longer afford "the absurdities of a horse-infected city."[16]

While no city ever took such drastic action as banning horses completely from its boundaries, many cities did eventually forbid them the use of certain streets and highways. The number of horses in cities dropped drastically as the automobile and the motor-truck rapidly gained popularity, although the number of horses in the nation stayed high until the 1920s (20,091,000 in 1920).[17] In a sense, the benefits promised by motor-vehicle enthusiasts were initially realized. Streets were cleaner, particle pollution resulting from ground-up manure and the diseases thereby produced were reduced, the number of flies was greatly diminished, goods were transported more cheaply and efficiently, traffic traveled at a faster rate, and the movement of people from crowded cities to suburbs was accelerated by the automobile. These events seemingly justified the spokesmen for the advantages of the motor vehicle over the horse.

And yet, as current difficulties resulting from the use of the automobile attest, motor-vehicle proponents were extremely shortsighted in their optimism that their innovation would not only eradicate the health problems created in the city by the horse but also avoid the formation of new ones. As the number of automobiles proliferated, and cities such as New York and Los Angeles experienced smog conditions that were a serious hazard to human comfort and public health, it became apparent that the automobile was a serious polluter.

Altered environmental and demographic conditions in the city today as compared with those of a century or so ago make specific comparisons between the horse and the automobile as polluters difficult at best. Aside from the disagreeable aesthetic effect created by horse manure, its chief impact on public health seemed to come from wind-blown manure particles that irritated respiratory organs and from the reservoir furnished by the manure for disease spores such as tetanus and, most critically, because horse dung provided a breeding ground for the fly, proven by medical science to be the carrier of thirty different diseases, many of them acute. The pollution created by the automobile, on the other hand, which is also aesthetically displeasing, has primarily a chronic effect on health. The pollutants released by the internal-combustion engine irritate people's eyes and lungs, weakening their resistance to disease and worsening already-present health problems. The immense number of automobiles in cities today has produced environmental difficulties that, unless soon dealt with, will generate problems that will dwarf those produced by horses in the cities of the past.

But the narrowness of the vision of the early advocates of the automobile, and their conviction that it would make urban life more tolerable, can be understood not as their failing alone but as that of most Americans when confronted by technological advance that promises to alter their lives without social cost. Witness the apprehensions voiced presently over nuclear power plants, after an initial flush of enthusiasm that this cleaner and more efficient method of power generation would free us from dependence on fossil fuels. Horses may be gone from city streets, but the unforeseen problems created by their successors beset us today.

Notes

1. Harold Bolce, "The Horse vs. Health," *The Review of Reviews* 57 (May 1908): 623–624.

2. Rudolph Hering and Samuel A. Greeley, *Collection and Disposal of Municipal Refuse* (New York, 1921), 568–69; Charles Zueblin, *American Municipal Progress* (New York, 1916, rev. ed.), 26.

3. Hering and Greeley, *Collection and Disposal of Municipal Refuse*, 569; *Census of the State of Michigan, 1894, Vol. II, Agriculture, Manufactories and Mines*, 148; Phil Strong, *Horses and Americans* (New York, 1939), 297; and United States, Bureau of the Census, *Eleventh Census* (1890), *Report on the Transportation Business in the United States*, Pt. I, *Transportation by Land*, 715–719.

4. *Scientific American* 65 (Nov. 28, 1891): 344; Francis V. Greene, "An Account of Some Observations of Street Traffic," *Transactions of the American Society of Civil Engineers* 15 (Feb. 1886): 194.

5. Clifford Richardson, "Street Traffic in New York City, 1885 and 1904," *Transactions of the American Society of Civil Engineers* 57 (Dec. 1906): 180–203; Lewis Atherton, *Main Street on the Middle Border* (Bloomington, 1954), 33–35; and "The Position of the Horse in Modern Society," *The Nation* 5 (Oct. 31, 1872): 383, 277–278.

6. John B. Blake, *Public Health in the town of Boston 1630–1822* (Cambridge, 1959), 11, 31, 103; John Duffy, *A History of Public Health in New York City 1825–1866* (New York, 1968), 179–184. 367; Lawrence H. Larson, "Nineteenth-Century Street Sanitation," *Wisconsin Magazine of History* 52 (Spring 1969): 246; and "The New York Sanitary and Chemical Compost Manufacturing Company (New York, 1865), prospectus located in the Library of the School of Public Health, University of Pittsburgh.

7. John Duffy, *Sword of Pestilence: the New Orleans Yellow Fever Epidemic of 1853* (Baton Rouge, 1966), 33–34, 137; Charles E. Rosenberg, *The Cholera Years* (Chicago, 1962), 5–7, 75–79, 165–172, 213–225; Blake, *Public Health in the Town of Boston,* 86, 156; W.S. Tryon (comp. & ed.), *My Native Land: Life in America, 1790–1870* (Chicago, 1961), 91; and Duffy, *Public Health in New York City,* 359.

8. *Ibid.,* 185; Hering and Greeley, *Collection and Disposal of Municipal Refuse,* 573–579, 602–603; Duffy, *Public Health in New York,* p.191; George E. Waring, Jr. (comp.), *Report on the Social Statistics of Cities, Pt. I: "The New England and the Middle States,* in U.S., Dept. of the Interior, Census Office, *Tenth Census* (1880) (Washington, 1887) XVIII, pt. I, 129; George E. Waring, Jr., "Modern Methods of Street Cleaning," *American City* 7 (1912): 434–435; Larson, "Nineteenth-Century Street Sanitation," 246; and George E. Waring, Jr., *Street-Cleaning and the Disposal of a City's Wastes* (New York, 1898), 1–18.

9. *Ibid.,* 13; Duffy, *Public Health in New York City,* 359; David Montgomery, "The Working Classes of the Pre-Industrial American city, 1780–1830," *Labor History* 9 (Winter 1968): 3–22; Francis M. Jones, "The Aesthetic of the Nineteenth-Century Industrial Town," in H. J. Dyos (ed.), *The Study of Urban History* (New York, 1968), 173, 183; Peter Quennell (ed.), *Mayhew's London* (London, n.d.), 394–410; John S. Billings, "Municipal Sanitation in New York and Brooklyn," *The Forum,* 16 (Nov. 1893): 352; George E. Waring, Jr., "Disposal of a City's Waste," *The North American Review* 5 (July 1895): 161, 52; and "The Prevention of Dust on City and Suburban Roads," *The American City* 7 (1912): 435.

10. Carl and Jessica Bridenbaugh, *Rebels and Gentlemen: Philadelphia in the Age of Franklin* (New York, 1965, Galaxy ed.), 12; Carl Bridenbaugh, *Cities in Revolt* (New York: Capricorn, 1964); Bayed Still, *Mirror for Gotham* (New York, 1956), 45, 82; H. J. Dyos, *The Study of Urban History* (New York, 1968), 183; "The Horseless Carriage and Public Health," *Scientific American* 80 (February 18, 1899): 93; and William Dean Howells, "Letters of an Altrurian Traveller, IV," *Cosmopolitan* 16 (1893–1894): 416.

11. "The Position of the Horse in Modern Society," *The Nation* 383 (October 31, 1872): 277–278; "Use of Steam Carriages on the Street Railways and Common Roads," *Engineering News* 3 (April 22, 1876): 131; "The Horse or the Motor," *Lippincott's Magazine* 57 (March 1896): 381–382; "Treatment of the City's Horses," *New York Times,* July 18, 1906; Robert West Howard, *The Horse in America* (Chicago, 1965), 237–238; "Through Broadway," *Atlantic Monthly* 18 (December 1866): 717; George E. Waring (comp.), *Report on the Social Statistics of Cities,* Pt. II, "The New England and the Middle States," (Washington, 1887), 591; "Clean Streets and Motor Traffic," *The Literary Digest* 49 (Sept. 5, 1914): 413; Hering and Greeley, *Collection and Disposal of Municipal Refuse,* 413; and "The Position of the Horse in Modern Society," *The Nation* 5 (October 31, 1872): 383, 278.

12. Carroll D. Wright, "Rapid Transit. Lessons from the Census. VI," *Popular Science Monthly* 40 (April 1892): 791.

13. Robert Lewis Seymour, "The Horse and his Competitors," *The Chautauquan* 22 (1895): 196–199. "The Horse or the Motor," *Lippincott's Magazine* 57 (March 1896): 382–383; and William Dean Howells, "Letters of an Altrurian Traveller, II" *Cosmopolitan* 16 (1893–1894): 222.

14. Herbert N. Casson, "The Horse Cost of Living," *Munsey's Magazine* 48 (March 1913): 997;

"The Horse as an 'Economic Anachronism,'" *The Literary Digest* 47 (July 26, 1912): 140, 142; R. W. Hutchinson, "Influence of the Motor Truck in Relieving Traffic Congestion," *American City* 8 (1913): 561–563; and Thaddeus S. Dayton, "Gasoline or Oats?" *Harper's Weekly* 56 (Jan., 6, 1912): 24.

15. "Merely a Matter of Time," *The Motor Age* 5 (Dec. 5, 1901): n.p. "The Horseless Carriage and Public Health," *Scientific American* 5 (Feb. 18, 1899): 80, 93–99; M.N. Baker, *Municipal Engineering and Sanitation* (New York, 1906), 159. "Stables Breeders of Disease," *The Horseless Age* 4 (May 3, 1899): 6; George W. Rafter and M. N. Baker, *Sewage Disposal in the United States* (New York, 1900), 25; and L.O. Howard, *The House Fly: Disease Carrier* (New York, 1911), 156–158. *Ibid.*; Edward H. Ross, *The Reduction of Domestic Flies* (Philadelphia, 1913), 24; Luther S. West, *The Housefly* (Ithaca, 1951), 7; and Edward Hatch, Jr., "The House Fly as a Carrier of Disease," *The Annals of the American Academy of Political and Social Science* 37 (Jan.–June, 1911): 412–423.

16. "The Horse Vs. Health," 623–624.

17. U.S., Bureau of the Census, *Historical Statistics of the United States: Colonial Times to 1957* (Washington, 1960), 289.

PART V

Industrial Wastes as Hazards

The term "hazardous wastes"—symbolizing a danger to human health—has become a familiar concept during the last two decades. It has entered into common discourse, is frequently discussed in the media, and merits encyclopedia and dictionary entries. The definition used today essentially reflects that given in the Resource Conservation and Recovery Act of 1976 (RCRA), the principal legislation that deals with hazardous wastes. RCRA defines hazardous wastes as solid wastes that can cause serious illness or pose a hazard to human health and to the environment when improperly stored, transported, or managed. Most of those who discuss the concept today refer to by-products and wastes that result mainly from manufacturing processes, medical and scientific research, and discarded consumer products.

The meaning of "hazard," however, has changed considerably over time. Historically the term had frequently been used for natural rather than man-made dangers, such as those from earthquakes, floods, and volcanic eruptions. The type of substances that public policy today defines as hazardous received their initial detailed treatment in the industrial hygiene movement, a development of the first third of the twentieth century. Industrial hygiene focused mainly on substances that workers encountered within their work places, not outside, and its concern with hazardous industrial substances seldom extended much beyond the factory walls. Although late-nineteenth-century public health authorities, especially in industrial states such as Massachusetts and Connecticut, considered industrial pollution as a major problem, the focus shifted after the acceptance of the germ theory. Public health officers and sanitary engineers—the professionals most concerned with environmental dangers to health outside the factory—focused on bacterial wastes, especially in water, as the primary threat to human health. When they considered industrial wastes, they concentrated on their nonpathological effects, including the damage they inflicted on stream life, their creation of taste and odor problems in drinking water supplies, and their interference with water- and sewage-treatment processes. It was only after World War II that professionals began to pay greater attention to the health and environmental damages of industrial wastes, and it was not until the 1970s that these received fuller policy recognition.

Environmental historians have done relatively little research on industrial pollution, but it has proven to be a controversial question. Technical issues such as the state of knowledge regarding the dangerous qualities of industrial wastes, the adequacy of various treatment technologies, and the rate of migration of industrial wastes in groundwater have generated disputes. Issues related to the policy process, such as the reasons for the absence of legislation to protect exposed populations and environments or the inadequacy of standards, have also been vigorously debated.

The three essays presented in this section take positions on some of these issues. Since its original publication, however, further research by other scholars has caused me to modify some of the interpretations presented in the first of the essays, "Historical Perspectives on Hazardous Wastes." This article should have paid more attention to the origins and nature of opposition to state regulation of industrial wastes, as well as to the state of knowledge of the flow of pollutants through groundwater. The following article, "Industrial Wastes and Public Health, 1876–1962," has been extended from its original closing date for this edition. And, the final essay, "Searching for a Sink for an Industrial Waste," could have been enriched by a further discussion of the effects of by-product coke ovens on nearby residents. Each year, however, greater numbers of young scholars focus their research on environmental issues, including that of pollution, enlarging the scope of our knowledge and understanding.[2] During the next generation, a flood of research will transform and enrich the field, and hopefully will provide insights into those areas that remain murky or contentious today.

Photo V.1. "Workers and their Dwellings at Pittsburgh Coke Ovens, 1888." Source: *Harper's Weekly,* July 7, 1888, from the collection of the Carnegie Library of Pittsburgh.

Photo V.2. Beehive Coke Ovens, Connellsville, Pennsylvania, 1890s. Source: U.S. Steel Company.

Photo V.3. World's largest concentration of beehive ovens—Jones & Laughlin Iron & Steel Company, located in "Hazelwood," Pittsburgh. Source: Guy C. Whidden and Wilfred H. Schoff, *Pennsylvania and Its Manifold Activities* (Philadelphia, 1912).

Photo V.4. Byproduct Coke Ovens, Clairton, Pennsylvania. Source: U.S. Steel Company

Photo V.5. Coke quenching at the LTV byproduct coke ovens in Pittsburgh, Pennsylvania. Source: Author's photograph.

CHAPTER XIII

Historical Perspectives on Hazardous Wastes in the United States

Introduction

Attitudes in the United States toward wastes and methods of waste disposal have changed markedly over historical time. For our purposes, waste is defined as something "left over or superfluous." Our attitudes toward wastes and our methods of disposing of or reusing them are affected by a host of cultural, economic, and technical factors. Normally we object to wastes for either health or aesthetic reasons but perspectives concerning what is objectionable have shifted markedly over time. What is believed to be a health danger in one time period or in one culture may be viewed from a more benign perspective in another; what one society sees as a nuisance to be eliminated, another may perceive with indifference.[1] What we are primarily dealing with, therefore, are attitudes toward risk in regard to wastes from either a health or socially related perspective.

Normally we conceive of two primary kinds of waste streams. One waste stream is the product of everyday living activities: food wastes or garbage, refuse such as paper or ashes (nonorganic materials), and human body wastes (feces and urine). The second waste stream derives largely from the various productive activities a society engages in, including agriculture, raw-materials processing, and manufacturing. The larger bulk of waste materials is actually generated by the second set of processes, although individuals tend to be most aware of the first waste stream. These waste products may be in either liquid, solid, or gaseous form and may be disposed of in any or all of the media—air, land, or water.

For the purposes of this paper, waste disposal will be viewed in terms of the perceptions of its health effects. Here, there are clear stages of development and change, relating to new hypotheses of disease etiology and available measurement instruments. Over time, alterations in these areas as well as in the value systems of users resulted in important policy changes and new practices in regard to waste disposal.

The Age of the Miasmas

For a considerable part of the nineteenth century, the two dominant theories of disease etiology were the contagionist and anticontagionist theories. Briefly, contagionists argued that epidemics were caused by specific contagia that were transmitted from individual to individual, usually originating from foreign sources. Hence, the proper policy to pursue, once diseases such as cholera, typhoid, or yellow fever were identified, was to declare a quarantine and close off the nation, region, city, or state to suspected carriers. During the years from the 1850s through the 1880s, a rival anticontagionist hypothesis, known as the "filth," "pythogenic," or "miasmic" theory, was most widely accepted. Essentially, this hypothesis held that infectious or "zymotic" disease evolved *de nova* from putrefying organic matter and resulted in epidemics of cholera, yellow fever, or typhoid. The logical policy outcome of this hypothesis was to remove organic matter from the cities before decomposition.[2]

The filth theory of disease was embodied in the nineteenth-century sanitary movement. This movement began in the United Kingdom with the work of Sir Edwin Chadwick and his followers to promote a healthful urban environment by cleansing the cities. Chadwick's ideas greatly influenced the pioneer group of American sanitarians and public health reformers. Municipalities and states passed laws regulating a range of activities relating to health and sanitation, such as cesspool and privy construction and emptying, street cleaning, garbage collection, sewerage development, and water supply.

Pythogenic or miasmic theory held that all organic waste matter was suspect, and this attitude resulted in a concern with organic trade wastes. The first state legislation to control stream pollution was an 1878 Massachusetts law giving the State Board of Health the power to control the river pollution caused by manufacturing wastes. In addition, by the 1880s many municipalities had passed statutes restricting so-called noxious manufacturers to the fringes of the cities and regulating the construction and cleaning of cesspools and privy vaults and garbage disposal. By 1880, most cities with a population above 30,000 had a board of health, a health commission, or a health officer, about half of whom had direct control over the collection and disposal of refuse. Of ninety-nine cities with populations of over 30,000 surveyed in the 1880 U.S. Census, the most common method utilized to dispose of garbage, street sweepings, and ashes was dumping on the ground. Garbage and refuse were also often placed in landfills. Some cities disposed of their street manure and garbage for animal feed or fertilizer (although this required pre-sorting), while others dumped their refuse into adjacent waterways.[3] Most cities had no specific regulations regarding the disposal of manufacturing waste but dealt with it under nuisance provisions of the law.

Few cities in 1880 had systems of sanitary sewers, and human body wastes were deposited in privy vaults and cesspools. Here the soil absorbed some of the wastes, but often privy vaults required periodic emptying to prevent problems of overflow. These wastes were usually dumped in an adjacent waterway or in a land dump, although portions of the privy wastes of over one hundred cities in 1880 were used on the land of neighboring farms as fertilizer.[4]

The Bacterial Revolution and Waste Disposal

Leading sanitarians believed that the city's health problems would be solved by installing water-carriage technology that would remove human wastes from the immediate locale of the household to a remote place for disposal. This location was usually a nearby waterway, and thus the wastes were shifted from a land-disposal sink to a water sink. Underlying the use of waterways as a place of ultimate disposal was the concept that running water purified itself, a hypothesis often confirmed by existing means of chemical analysis. Invariably, the result of placing raw sewage into streams from which downstream cities drew their water supplies was a large increase in morbidity and mortality rates from infectious disease such as typhoid fever for the downstream communities. The irony was clear: cities had adopted water-carriage technology because of an expectation of local health benefits resulting from more rapid and complete collection and removal of wastes, but disposal practices produced serious externalities for downstream or neighboring users.[5]

In the 1890s, bacterial researchers following the seminal work of Pasteur and Koch in establishing the germ theory, identified the processes involved in waterborne disease. The work of William T. Sedgwick and other bacterial researchers at the Massachusetts Board of Health Lawrence Experiment Station was especially critical in clarifying the etiology of typhoid fever and confirming its relationship to sewage-polluted waterways. The challenge of waterborne infectious disease to public health was met by the development of two further technologies—water filtration and chlorination in the pre–World War I period. By 1940, almost all urbanites were drinking treated water, and morbidity and mortality from waterborne disease had ceased to be a serious public health problem.[6]

Sewage treatment, however, lagged far behind. Until the 1930s sanitary engineers and municipal officials believed that the risks involved in using streams for sewage disposal were not sufficient to justify the costs of construction of sewage-treatment plants unless there was a severe nuisance. They argued for full utilization of the natural dilution power of waterways and the adoption of water-filtration technology to protect drinking-water quality.[7]

Compared to sanitary wastes from human populations, industrial wastes were relatively neglected during this period. Because they did not normally contain disease germs, public health authorities argued that "from a purely pathogenic standpoint, their relation to sanitation is remote." Thus, the shift from miasmic theory to bacterial theory caused a focus on human wastes and a reduction of concern with the health effects of organic industrial wastes. From the early part of the century through the 1930s, sanitary engineers identified the following as the main problems caused by industrial wastes: interference with water- and sewage-treatment technologies, consumption of oxygen that reduced the dilution power of streams, the creation of taste and odor problems in drinking water (especially phenols), and devastating effects on fish life. "Few wastes," however, noted one authority in 1938, "are present in most streams in sufficient quantities to become poisonous."[8] In some cases, engineers argued that mine-acid drainage and steel-mill pickling liquor discharge had a germicidal effect on sewage pollution and therefore should not be excluded from streams. However, such acidic pollution also limited the use of the water in these streams for industrial purposes.

The replacement of the filth theory with the germ theory also caused public health authorities to reduce their interest in solid-waste collection and disposal. Leaders of the so-called "New Public Health" argued that public health officials should focus on the control of the diseased or carrier individual rather than environmental sanitation. This reorientation caused many municipalities to remove control over refuse collection and disposal from health departments to sanitation or public works departments. Solid-waste disposal was now viewed as an engineering problem involving nuisance and cost considerations rather than an issue with public health implications.[9]

Questions concerning the means of disposal largely revolved around the cost-effectiveness of different forms of technologies. Incineration and reduction were used by many cities with mixed success. Each involved relatively large capital expenditures. Less capital-intensive methods, such as dumping on the land and hog farms for garbage, remained popular until after World War II. Ocean and waterway dumping of refuse and garbage, however, was largely abandoned because of nuisance, cost, and legal considerations. (Ocean dumping of sewage sludge, however, continued.) All of the above-mentioned techniques were to be seriously challenged after the war.

Waste Disposal in the Post–World War II Period

In the postwar decades, through the landmark environmental legislation of the 1970s, the prime focus was on water and air pollution, with an increasing concern with land pollution after 1965 and especially since the late 1970s. In regard to water pollution control, there has been a pattern of a steadily increasing federal authority, dating from congressional passage in 1948 of the Federal Water Pollution Control Act through the enactment of the Clean Water Act. The goals of this legislation—uniform water-quality standards, such as zero effluent, and water suitable for fishing and swimming—had actually appeared in earlier state enactments such as the Pennsylvania Clean Streams Act of 1905, but never with enforceable provisions. The 1972 legislation, in contrast, embodied the strongest sanctions up to that time.

The sanctions enacted in regard to water pollution, as well as those applied to air pollution by the Clean Air Act, however, had an unpredicted effect. Because surface waters and the air were no longer acceptable sinks for the disposal of wastes, industries turned increasingly to the land. The 1979 report of the Council on Environmental Quality (CEQ) noted, for instance, that "the increasing tempo of the cleanup of lakes and streams is literally driving pollution underground." This effect of the legislation could have been predicted from past experience. For example, in the 1940s, when the Pennsylvania Sanitary Water Board began enforcing the state's Clean Streams Act, numbers of small industrial plants turned to the use of earthen lagoons on plant property as a means of avoiding controls, thus ultimately threatening groundwater supplies, posing nuisances, and creating air pollution.[10] Deep-well injection was another method of land-based industrial waste disposal that expanded because of regulations restricting disposal in surface waters.

Industries have, over the years, actually punched or dug thousands of holes in the ground—usually on their own property—to dispose of wastes. According to T.S. Maugh, "Landfilling has long been the most common method for disposal of hazardous wastes because it has been inexpensive. . . . The costs were low because the technology was simple. Typically, a hole was dug in clay at a selected site, unconsolidated sludge and drums of chemicals placed in it, and the hole was filled and covered with clay to keep out rain and other water."[11]

As the noted sanitary engineer Abel Wolman has observed, the handling of hazardous wastes previous to 1976 was "haphazard, desultory and . . . certainly not carefully reviewed, designed or operated." Such sloppy "housekeeping practices," he says, were a result of a desire to dispose of wastes in the simplest and cheapest manner. Careless "housekeeping" practices had increasingly serious implications for environmental conditions, as the volume and variety of hazardous wastes produced by industry greatly expanded in the postwar period.[12] Also important was the physical expansion of metropolitan areas, causing the encroachment of residences and commercial activities on industrial production and disposal sites.

In the postwar period, municipalities, as well as industries, began using the land more intensely for the purposes of waste disposal. The technique they utilized was the sanitary landfill, a method of garbage and refuse disposal that had originally been developed in the United Kingdom. The sanitary landfill appeared to eliminate the nuisances produced by other methods of disposal at a reasonable cost and apparently caused no public health hazards. The standards required for the proper use of the sanitary landfill were set forth in 1940 by a committee of public health specialists and sanitary engineers headed by the U.S. surgeon general, Thomas Parran. Sanitary landfill, noted the Parran committee report, was a significant health improvement over the open dump; it eliminated undesirable marshes that harbored rats and mosquitoes and provided useful filled-in ground. The commission considered several possible landfill hazards such as fires and low weight-bearing value, but maintained that they could be controlled by proper precautions. No mention was made of possible dangers from leachate runoff or groundwater pollution or the possibility of long-term health hazards. Most of the commission's discussion of risk was in terms of nuisances, not health.[13]

The Parran report, combined with the army's favorable experience with sanitary landfills during World War II, gave the technique wide appeal in the postwar period. In the 1950s, public works officials and public health professionals strongly endorsed the sanitary landfill, especially as a replacement for the open dump. The advantages most commonly cited were those listed by the Parran Commission: the elimination of the nuisances and health hazards associated with open dumps, the filling-in of marshes and swamps with a consequent reduction of rats and mosquitoes, and the creation of land for buildings, parks, and recreational areas. Just a very few articles warned of possible hazards unless proper standards were followed.[14]

In 1961 the Sanitary Engineering Division of the American Society of Civil Engineers (ASCE) published a survey of 250 sanitary landfill sites. Completed landfills were most commonly used for recreational and industrial purposes, although some fills were used for homesites and schools. Of the fills surveyed, 12 percent were less than 250 feet from the nearest dwelling. The article noted that, while groundwater pollution from landfills was a "critical item," it was given minimum concern in site planning, with 79 percent of the sample within 20 feet of groundwater and 27 percent at or near groundwater. Only 9.3 percent of the sites reported that operators had made test ground boring prior to fill operations, and only 14 percent had specially engineered drainage devices.[15]

The 1961 survey also noted that over 70 percent of the landfills examined operated under some sort of municipal or county regulations. During the 1950s, as landfills became more common, cities issued sanitary landfill regulations; states such as California and Illinois suggested operational guidelines; and professional groups, especially the American Public Works Society (APWA), the Sanitary Engineering Division of the ASCE, and the U.S. Public Health Service (USPHS), conducted investigations on standards to avoid undue risk. By the time of the ASCE survey, professional groups involved in solid-waste questions agreed that, while sanitary landfills reduced disposal costs and were superior to the open dump, they presented dangers in regard to leachate seepage, groundwater pollution, poor load bearing, methane, and nuisances such as rats, vermin, and blowing paper. A lack of research on these hazards, however, restricted the availability of technical information that could be used to refine practice.[16]

In 1963, in an attempt to generate interest and research in the solid-waste area, the USPHS and APWA sponsored the first National Conference on Solid Waste Research. In his keynote address, Professor J. E. McKee of the California Institute of Technology, offered four explanations for the lack of research in solid-waste disposal. First, McKee noted that neither cities nor regulatory agencies nor the public demanded such information. Second, solid waste had produced no public health crises equivalent to those in air and water pollution that could have generated such a demand. Third, federal and state government was minimally involved in the area. Fourth, most officials and engineers concerned with solid-waste disposal considered it an economic and political rather than a scientific or engineering problem. All of these factors applied to the sanitary landfill technique, as well as to solid wastes in general, but landfills did not hold an especially prominent place at the conference. Concern for the land as a sink for pollutants did not yet possess the urgency that was beginning to characterize disposal in the air and water mediums in the 1960s.[17]

The Hazards of Land Disposal of Wastes: The Awakening

Conferences such as that sponsored in 1963 by the USPHS and the APWA highlighted the deficiencies in solid-waste research and suggested the need for legislation in the area. More critical for federal involvement, however, was the growing expense of solid-waste collection and disposal. Powerful urban politicians pushed for federal action to lighten the burden on cities. In 1965, Congress passed the Solid Waste Disposal Act. This act created the Office of Solid Wastes and provided the federal government with a more formal role in regard to municipal wastes.[18]

The Solid Waste Disposal Act provided funds for research investigation and demonstration and for technical and financial assistance to state and local governments and interstate agencies in "the planning, development and conduct" of disposal programs. The act's most important impacts were to stimulate research and to inspire state government activity. In 1965, for instance, there was no state-level solid-waste agency in the country; but by 1970 forty-four states had developed programs.[19] During the 1970s, however, the focus of federal legislation moved from research into conventional methods of solid-waste disposal toward the reuse and recycling of resources, as reflected in the passage of the Resource Conservation Act of 1970.

Section 212 of the 1970 Solid Waste Act required that the U.S. Environmental Protection Agency undertake a comprehensive investigation of the storage and disposal of hazardous wastes. This led to a report to Congress in 1974 on the disposal of hazardous wastes and eventually, in 1976, to the passage of the Resource Conservation and Recovery Act. The RCRA attempted to fill the regulatory gaps concerning the disposal of hazardous wastes left by state programs, and early in 1980, acting under the requirements of RCRA, the EPA announced new regulations implementing cradle-to-grave controls for handling hazardous wastes.[20] The use of the land as a sink was now finally to be curtailed.

The various acts passed after 1965 caused a convergence of the different streams of research concerning municipal wastes and industrial hazardous wastes. The point of convergence was landfill operations, with special concern over site construction and groundwater pollution. In 1979, the EPA estimated that there were about 75,000 active industrial landfills, while a 1978 *Waste Age* survey identified more than 14,000 active municipal landfills. The number of abandoned landfills is unknown.[21]

The most serious potential problem involving both municipal and industrial landfills is the threat to groundwater quality. Relatively little attention had been paid to this question before World War II aside from questions involving bacterial wastes or oil field brines. In the postwar years, several state departments of health issued warnings about potential chemical pollution of groundwater from sanitary fills. Studies concerning the effects of industrial effluents such as metal-plating wastes, phenols, oilfield brines, and chemical products on groundwater appeared in the literature. When landfill research accelerated during the 1960s, so did awareness about the hazards of possible groundwater pollution.[22] By 1970, many states had regulations requiring field investigations of groundwater location and into the siting of new municipal and industrial landfills. Problems, however, usually centered around older sites that had been developed without adequate investigation of the risk of possible groundwater contamination.

There are several reasons why the potential for groundwater contamination had been neglected before the 1970s. One is the lack of research in the area of solid-waste disposal in general and the landfills in particular. While considerable study had been done earlier in the century on bacterial contamination of wells because of migration of wastes from privy vaults, little attempt was made to relate this work to industrial wastes. Research and publication regarding the disposal of industrial wastes was also restricted, "owing to the competitive nature of private enterprises, and their reluctance at times to divulge operating problems and techniques." In addition, a lack of analytical instrumentation making possible the tracing of or the detection of extremely low levels of potentially hazardous substances provided a restraint on knowledge. Before federal legislation, there was no incentive system to spur research in analytical chemistry in regard to either groundwater processes or groundwater—leachate—soil exchanges.[23]

Equally important was the absence of a clear hazard or crisis in regard to groundwater pollution from waste disposal. Up to 1970, some incidents involving the pollution of groundwater drinking supplies by industrial wastes had been reported, but municipalities and state governments often ignored the problem. Rather than spend scarce funds on expensive testing and monitoring, governments put their dollars in areas where need appeared more immediate.

Conclusion

The body of federal and state environmental law that has appeared in the last two decades reflects a new set of values. These values include a regard for the quality of the natural environment from an amenity and aesthetic perspective and a concern for health in relationship to the environment.[24] These new values also reflect the findings of the emerging discipline of environmental health, with its focus on the chronic, degenerative diseases. In this regard, we have returned to the original thrust of the nineteenth-century sanitary movement toward cleansing the urban environment in order to ensure freedom from epidemics.

This review of the history of waste disposal has suggested that environmental degradation was not always a result of a willful act on the part of the waste generator. An environmental hypothesis such as "running water purifies itself" provided sanction to cities to dispose of their sewage in nearby streams, while chemical analysis seemingly gave its "scientific" stamp of approval. Sanitary and industrial landfills produced leachates that contaminated drinking-water supplies, but limited monitoring capabilities hindered detection. Research in these areas often only developed after the occurrence of some crisis and as a result of specific public policies, not before. But even after research had pinpointed the mechanisms that produced negative effects, it often proved difficult to persuade the operators of these technologies, be they private or public, to cease using the polluting technology or to stop building new systems that had the same effects. As a result, we must simultaneously deal with both the results of careless past waste-disposal practices and the difficult waste-disposal problems of our own time.

Notes

1. Mary Douglas, *Purity and Danger* (New York: Basic Books, 1966).

2. George Rosen, *A History of Public Health* (New York: MD Publications, 1958).

3. Barbara Rosenkrantz, *Public Health and the State: Changing Views in Massachusetts, 1842–1936* (Cambridge: Harvard University Press, 1972); Charles V. Chapin, "History of State and Municipal Control of Disease," in Mazyck Ravenel (ed.), *A Half Century of Public Health* (Washington, D.C.: American Public Health Association, 1921), 133–60; and Martin V. Melosi, *Garbage in the Cities: Refuse Reform and the Environment* (College Station: Texas A&M University Press, 1981).

4. Joel A. Tarr, "From City to Farm: Urban Wastes and the American Farmer," *Agricultural History*, 49: 598–612.

5. Joel A. Tarr, J. A. McCurley Jr., III, F.C. McMichael, & F.T. Yosie, "Water and Wastes: a Retrospective Assessment of Wastewater Technology in the United States, 1800–1932," *Technology and Culture* 25 (April 1984): 228–239.

6. Ibid., 241–246.

7. Ibid., 244–246.

8. M.O. Leighton, "Industrial Wastes and Their Sanitary Significance," *Reports and Papers of the American Public Health Association* 31 (1905): 29–41; E.B. Besselievre, "The Disposal of Industrial Chemical Waste," *The Chemical Age* 25 (1931): 516–518; and L.F. Warrick, "The Prevalence of the Industrial Waste Problem," in L. Pearse (ed.), *Modern Sewage Disposal* (New York: Federation of Sewage Works Association, New York, 1938), 340–372.

9. J. L. Rice and S. Pincus, "Health Aspects of Land-fills," *American Journal of Public Health* (1940) 30: 1391–1398; Melosi, *Garbage in the Cities*, 81–84.

10. Committee on Environmental Quality (CEQ), *Environmental Quality. 1979* (U.S. Government Printing Office, Washington, D.C., 1979); D.A. Lazarchik, "Pennsylvania's Pollution Incident Prevention Program," in *Proceedings of the 25th Annual Purdue Industrial Waste Conference* 25 (1970): 528–533.

11. T. H. Maugh, "Burial is Last Resort for Hazardous Wastes," *Science* 204 (1979): 1294–1297.

12. Walter Hollander, *Abel Wolman: His Life and Philosophy* (Chapel Hill, NC: Universal Printing and Publishing Co., 1981, 2 vols.) I:480; Michael Greenberg and Robert F. Anderson, *Hazardous Waste Sites: The Credibility Gap* (New Brunswick, NJ: Center for Policy Research, 1984).

13. "Health Experts Endorse Landfills and Recommend Best Practice," *Engineering News Record* (Mar. 28, 1940), 54–55; Rolf Eliassen and Albert J. Lizee, "Sanitary Land Fills in New York City," *Civil Engineering* 12 (September 1942): 483–486; and John L. Rice and Sol Pincus, "Health Aspects of Land-Fills," *American Journal of Public Health (AJPH)* 30 (December 1940): 1991–1998.

14. Joel A. Tarr, "The Search for the Ultimate Sink: Urban Air, Land, and Water in Historical Perspective," in J. Kirkpatrick Flack (ed.), *Records of the Columbia Historical Society of Washington, D.C.* 51 (1984): 21.

15. Committee on Sanitary Engineering Research, American Society of Civil Engineers, "A Survey of Sanitary Landfill Practices," *Journal of the Sanitary Engineering Division, Proceedings of the ASCE* 87 (1961): 65–83.

16. Tarr, "Search for the Ultimate Sink," 22.

17. American Public Works Association (APWA), *Solid Wastes Research Needs* (Chicago: APWA, May, 1962); APWA, *National Conference on Solid Wastes Research* (Chicago: APWA, December 1963).

18. Melosi, *Garbage in the Cities*, 199–200.

19. Ibid., 202–203.

20. CEQ, *Environmental Quality—1979* (Washington: GPO, 1979), 174–188; CEQ, *Environmental Quality—1980* (Washington: 1981), 214–222.

21. Ibid., 88–89

22. David Keith Todd and Daniel E. Orren McNulty, *Polluted Groundwater: A Review of the Significant Literature* (Huntington, NY: Water Information Center, 1976).

23. Craig E. Colton, "A Historical Perspective on Industrial Wastes and Groundwater Contamination," *The Geographical Review* 81 (April 1991): 215–228; G. E. Barnes, "Industrial Waste Disposal," *Mechanical Engineering* 69 (1947): 465–470; and Harvey Brooks, "Science Indicators and Science Priorities," *Technology & Human Values* 38 (Winter 1982): 17.

24. Samuel Hays, "From Conservation to Environment: Environmental Politics in the United States Since World War II," *Environmental Review* 6 (1982) 14–41.

CHAPTER XIV Industrial Wastes, Water Pollution, and Public Health, 1876–1962

As an industrial nation, our manufacturing processes produce a huge stream of wastes of many varieties that enter the environment, some of which are clearly hazardous, others largely inert, and still others whose effects on human health and the environment are not fully understood. Within the last two decades, public health professionals, often representing different disciplines, as well as environmental engineers and scientists have vigorously debated questions about the health effects of the disposal of various industrial wastes in the ambient environment, especially in water supplies. Although such concerns were present over a century ago in the 1870s and 1880s, they diminished in intensity until the post–World War II decades. For most of the period from the 1890s through the 1930s, the public health and sanitary engineering communities focused largely on the health effects of human wastes and paid limited attention to industrial wastes unless their disposal caused extensive nuisance.

This chapter will explore the positions of public health professionals and sanitary engineers on the disposal of industrial wastes in waterways from 1876 through 1962—a period that encompasses both major increases in pollution and the development of critical bacterial and chemical indicators for human and industrial wastes. In 1962, Rachel Carson published

Silent Spring, her dramatic expose of the massive and persistent damages chemical pesticides and insecticides, such as DDT, caused to wildlife. Carson's book was especially significant because it raised public anxiety about the effects of chemical products, as well as industrial wastes, on human health and on wildlife. Engineers and scientists had already begun to raise such questions within their own professional circles, marking a return to the concern with industrial wastes of the 1870s and 1880s. In tracing this circular pattern, I will discuss the rationale for setting priorities about human and industrial pollution, the relationship between industrial wastes and water supplies, and the development of indicators for the presence and measurement of industrial wastes.

Industrial and Human Wastes and the Bacterial Paradigm

During the late nineteenth century, urbanization and industrialization produced water pollution problems in the United States of a scope not previously experienced. Human wastes transported from households, commercial establishments, or work places by sewerage systems and disposed of in waterways produced one type of pollution, while industries that discharged their effluent directly into streams or, in some cases, into sewers, caused the second type. These industrial wastes could be classified in several different ways, such as organic or inorganic, toxic or nontoxic, or as transporting various amounts of sediment, soluble matter, or floating materials.[1] The growing public health profession—composed of an interdisciplinary group of physicians, sanitary engineers, bacteriologists, biologists, and chemists—viewed the great increase in water pollution with particular concern. Before the 1880s and 1890s, however, because of competing medical theories, restricted understanding of the etiology of waterborne disease, and methodological limitations in determining water quality, there were no sharp priorities in dealing with either type of waste.

Municipal and state authorities, for example, in the decades following the Civil War, sponsored investigations of river pollution that often found industrial wastes equal to or exceeding human wastes in producing foul and health-threatening conditions. Since many cities had not yet constructed centralized sewerage systems to deal with household wastes, industrial effluents discharged directly into waterways frequently formed the major pollution burden in rivers used for water supplies. In 1876, the Massachusetts Board of Health commissioned water-quality specialist and civil engineer James P. Kirkwood to examine the state's rivers. After his survey, Kirkwood concluded that the "fluid refuse from . . . factories . . . some of it very poisonous, produced in the processes of cleaning and preparing the manufactured article . . . forms the chief element in the pollution of these streams." He warned that while "the exact influence or even presence" of industrial pollutants and of sewage might be so minute that they could not be analyzed, "they may be sufficient . . . to render the water . . . not merely repulsive or suspicious, but more or less dangerous for family use." In New Jersey in the 1870s and 1880s, both the state and local governments commissioned Albert R. Leeds, a geologist who taught at the Stevens Institute of Technology, to perform analyses of the Passaic River, which supplied drinking water for Newark and Jersey City. Leeds found that factories along the lower stretch of the river, "each of which pours out its own peculiar filth," including various acids, dyes, and chemicals, were responsible for much of the river's pollution.[2] Investigations of river pollution in other manufacturing states, such as Connecticut and Rhode Island, found similar conditions.

By the turn of the century, however, both the relative components of river pollution and the methodology of water analysis had sharply altered. Hundreds of towns and cities had constructed thousands of miles of sewers and were discharging millions of gallons of untreated sewage into adjacent rivers and lakes each day, causing severe pollution problems. While sewerage systems improved public health in upstream cities, their discharges often contaminated the water supplies of downstream cities, causing substantial increases in mortality and morbidity from typhoid fever and other waterborne diseases.[3]

In the early 1890s, a skilled group of bacteriologists, chemists, and sanitary engineers under the direction of biologist William T. Sedgwick at the Massachusetts Board of Health's Lawrence Experiment Station clarified the etiology of typhoid fever and confirmed its relationship to sewage in waterways. Innovations in slow-sand and mechanical filtration of water supplies followed, as did chemical treatment (chlorination in 1908), with the result that death rates from waterborne disease sharply declined in those cities with treatment systems. Other developments in the first decade of the twentieth century included innovations in methods of quantitative water analysis based primarily on the work of sanitary engineers George W. Fuller and George C. Whipple and the members of the Committee on Standard Methods of Water Analysis of the American Public Health Association (APHA). Their contribution involved the use of bacterial analysis to determine the presence of coliform bacteria in water. Since coliform bacteria are present in great numbers in human and animal feces, but are not typical water organisms, their significant presence in waterways served as an indicator of fecal pollution and possibly pathogenic organisms. The United States Public Health Service (USPHS) used bacterial analysis in 1914 for standard setting when it issued the first standards for water served in interstate commerce. Many states copied these standards, and, in conjunction with the various editions of *Standard Methods of Water Analysis* formulated by the APHA, the American Water Works Association, and other groups, they came to furnish the basic measure of water quality throughout the nation.[4]

By placing their highest priority on the dangers posed by sewage pollution to water supplies, public health professionals diminished the attention paid to industrial wastes. As Marshall O. Leighton, hydrographer of the U.S. Geological Survey and a former health officer, commented to the APHA in 1905, "it is accepted as a matter of course that when a body of sanitarians considers polluting wastes, city sewage is the subject of discussion." Since industrial wastes did not normally contain disease germs, Leighton added, "from a purely pathogenic standpoint, their relation to sanitation is remote." Industrial wastes were thought to have potential for direct infection only in regard to the risk of contracting anthrax from tannery and wool-scouring wastes.[5]

Leighton's own interests, as well as those of the USGS, however, were in the wider uses of rivers, not merely the relation of sewage pollution to human health. In 1903, the Geological Survey organized a Division of Hydro-Economics, headed by Leighton, to investigate the economic value of water supplies. The division would be particularly concerned with factors that reduced water quality, such as turbidity, color, hardness, and various chemical and mineral constituents. From Leighton's perspective, industrial wastes were the "great pollution problem of today" because of their damage to the total life of the stream, interference with natural processes, destruction of fish and vegetation, and deposits of solids and silt. Public health officers, he maintained, should be concerned with industrial wastes because they provided food for disease bacteria introduced by sewage, interfered with sewage disposal and water filtration processes, and reduced stream assimilative capacities. In addition, he noted that industrial wastes usually destroyed the potable quality of drinking-water supplies, making them of major concern to the state departments of health that normally had responsibility for their purity.[6]

The professional and public response to Leighton's injunction was relatively limited. The APHA, for example, as reflected in its journal, demonstrated only infrequent concern with the industrial waste problem. In 1902, the APHA appointed a Committee on Trade Waste Disposal, but it presented only one preliminary report before disappearing. The committee, noting that the problem was extremely complicated with hundreds of different type of industries producing objectionable wastes, each requiring "distinct and individual consideration," made no recommendations. In 1916 the APHA Committee on Sanitary Control of Waterways prepared a report and bibliography on sewage and industrial wastes, but no reports followed for some years and the bibliography was not published. In 1927, the association created a Committee on Disposal of Sewage and Industrial Wastes, but it issued no major reports.[7] Several industrial waste papers were delivered at the APHA annual meetings in the 1920s and 1930s (usually before the association's Sanitary Engineering Section, renamed the Public Health Engineering Section after 1926 and the Environment Section in 1969) and published in the *American Journal of Public Health,* but their number was insignificant compared to those dealing with sewage and the bacterial pollution of water supplies.

State departments of health were usually responsible for water pollution because of its relation to infectious disease. Physicians headed these departments, although by the 1920s most also had sanitary engineering divisions. The stringency of water pollution regulations varied from state to state. According to one authority, because most states recognized stream control as a public health function, action was limited in regard to "serious pollution capable of working great economic injury, although of little or no public health significance." In addition, while industrial wastes often severely damaged stream life, causing fish kills and destroying vegetation, state departments of health seldom acted against the polluters because they did not consider aquatic life within their realm of responsibility.[8]

Some states, such as New York and Wisconsin, created state conservation commissions in the 1920s. Their mission was usually to cooperate with the state departments of health to restrict pollution and to supply technical advice to manufacturers in reducing their effluents. Disputes often occurred, however, between these agencies over the relative emphasis on research and on the taking of action against polluters. Many states had commissions, but they possessed little power to force industries to cease discharging harmful wastes. In 1915, the Massachusetts State Legislature failed to enact a bill that would have forbidden the pollution of streams by substances that made them "poisonous or dangerous to fish or animal life" or vegetation. Two years later, the Pennsylvania legislature passed a law that prohibited the discharge into streams of any matter deleterious to fish, but in 1923 new legislation gave all anti–stream pollution authority to the Sanitary Water Board. While the board included representation from the State Fish Commission, it did not insist on the absolute prohibitions of the 1917 law, permitting some streams to be classified as "C"—not for recreational purposes and not capable of supporting fish life. Even states such as Massachusetts, with its relatively stringent antipollution regulations, exempted certain rivers and industries from enforcement. Health departments were hesitant to move against industrial polluters because of political concerns over limiting industrial growth, and polluting industries often lobbied to limit legislation and/or implementation of strong pollution-control regulations.[9]

Between 1917 and 1926, three industrialized states—Connecticut, Ohio, and Pennsylvania—created boards with special responsibilities for industrial wastes. The Connecticut Industrial Waste Board and the Ohio board were originally located within the state departments of health, while Pennsylvania formed a separate Sanitary Water Board chaired by the state's secretary of health and comprised of the heads of other departments relating to water resources, such as the Departments of Forests and Waters, Fish, and the Public Service Commission. Connecticut eliminated its Industrial Waste Board in 1921, but in 1925 it created a separate State Water Commission with the power to order the elimination of pollution and to prescribe the means by which the pollution would be abated. These state boards believed that pollution abatement would only come through industrial cooperation, not legal sanctions, and that they needed to supply technical advice as well as investigate stream conditions. Clean water, it was emphasized, was important for industrial processes as well as other purposes.[10] The ideal situation was one where wastes could be profitably recycled.

The Pennsylvania Sanitary Water Board established three classes of streams: those that were relatively pure; those in which pollution needed to be controlled; and those that were so polluted that "it is not now necessary, economical or advisable to attempt to restore them to a clean condition." For many sanitary engineers, Pennsylvania furnished a realistic model that provided for both environmental quality and industrial advance. In this emphasis on cooperation to achieve socially responsible progress, they shared the ethos of the "associative" state of the 1920s. Not all public health authorities, however, accepted the Pennsylvania model. Earle B. Phelps, an important figure in the development of stream sanitation, argued that the Pennsylvania law reflected "too general application of a strictly water supply point of view" because of its "C" classification. A true "conservationist," he maintained, would not be so willing to abandon a river to pollution.[11] Most sanitary engineers and public health officers did not agree with this position.

Industrial Wastes and Drinking-Water Supplies

Departments of health throughout the period were especially concerned with the effects of industrial pollution on drinking-water supplies. Their emphasis, however, was on the indirect rather than direct health impacts of industrial wastes. Before World War I, numerous cases of damage to water supplies by industrial pollutants were reported in the literature, but it was not until the 1920s that a systematic survey was conducted. In 1922, the Committee on Industrial Wastes in Relation to Water Supply of the American Water Works Association presented a report indicating that industrial pollutants had damaged at least 248 water supplies in the United States and Canada. The report commented that the presence of industrial wastes in water supplies used for potable purposes increased the difficulty and expense of coagulation and filtration and caused problems of color, turbidity, taste, and odor. While the report also warned of the presence of specific toxic chemicals, such as cyanide, they received more limited attention. Wastes from sugar refining, coal mining and washing, gas and by-product coke works, wood distillation, corn products, dye and munitions manufacture, oil-producing wells and refining, metallurgical processes and mining, textiles, tanneries, and paper and pulp mills caused the most problems.[12]

While health officers and sanitary engineers were aware that inorganic industrial wastes, especially heavy metals such as lead, zinc, and arsenic from metallurgical operations, had potential health effects, they had not been extensively studied. As one speaker at the 1921 APHA annual meeting noted, the "ordinary water-works laboratory concerns itself mostly with the biological phases of the supply and certain routine chemical tests useful in controlling operation." Because of a failure of the PHS Advisory Commission on Drinking-Water Standards to agree on specific limits, it was not until 1925 that maximum permissible concentrations were established in the standards for lead, copper, or zinc. Chemists and sanitary engineers in water-supply laboratories commonly analyzed water for substances that would make it unsuited for industrial use, paying more attention to the clearly observable effects of metals and acids in water supplies on industrial boilers than on the possibility of human health effects. The industrial hygiene movement, which began in the early teens, studied the toxicity of industrial materials and their impact on workers' health, but engineers and scientists in industrial hygiene and in water quality seldom communicated. As Christopher Sellers has noted, "the sanitary engineers who developed water pollution as a specialty pursued questions that differed markedly from the toxicological ones of their industrial hygiene colleagues. . . . They gave little or no consideration to the diverse chemicals that comprised industrial pollution, much less to these substances' human toxicity."[13]

During the 1920s, water-quality specialists believed that phenol wastes from gas and by-product coke works and mine acid drainage were the most damaging industrial wastes to water supplies. The difficulties caused by phenols, primarily taste and odor, were susceptible to control because they derived from point sources—by-product coke ovens and manufactured gasworks. Gas house wastes had been causing nuisance problems in cities since the middle of the nineteenth century, but by the 1920s the worst problems were under control. The problems from phenol wastes from by-product coke ovens were a post–World War I phenomenon and were largely centralized in the Ohio River Basin, although they also appeared in Chicago, Cleveland, and Detroit.[14]

Before 1917, beehive ovens produced most of the coke made in the United States. Coke was the only product of these ovens, and no attempt was made to reclaim coal by-products. By the 1910–14 period, coke manufacturers and the steel industry had begun to shift to the by-product coke oven. When the war eliminated the importation of coal tar products from Germany, the transition accelerated: by 1929, 75 percent of the coke manufactured in the United States came from the new technology. While the by-product coke ovens produced less air emissions than the beehive ovens, they caused serious water pollution problems. Liquid effluents containing phenols from the ammonia stills at the coke plants were customarily discharged into adjacent rivers, causing serious taste and odor problems for the water supplies of downstream cities. If municipalities chlorinated the water supply in order to control pathogenic organisms, chemical reactions with the phenols caused the water to become even more distasteful. The most serious difficulties occurred in the drinking water drawn from the Ohio River and its tributaries, where by 1924 twenty-five water supplies had been affected.[15]

Writing in the *American Journal of Public Health*, F. Holman Waring, chief engineer of the Ohio Board of Health, noted that phenolic wastes had several injurious effects from a public health perspective: the phenols could affect the health of people drinking the polluted water; the offensive tastes and odors might discourage individuals from drinking the amount of water required for good health or to drink more aesthetically pleasing but biologically contaminated water; and waterworks managers might be tempted to reduce the amount of chlorination to avoid creating chlorinated phenolic compounds, thereby increasing the risk of exposure to infectious disease.[16]

By 1924, concern with these effects, as well as pressure from state health officials, led U.S. surgeon general Hugh S. Cumming to call a conference of the state health commissioners of Ohio, Pennsylvania, and West Virginia to explore the problem of the pollution of interstate streams in the Ohio River Basin. The conference led to the formation of the Ohio River Interstate Stream Conservation Agreement and to understandings with the by-product coke industry to control phenol contamination of water supplies. Undoubtedly, a factor in their voluntary action was a concern over the possibility of legal action. By 1929, seventeen of the nineteen firms in the basin had installed phenol elimination devices, sharply reducing the most severe tastes and odors. The coke companies followed a procedure that used the phenolic wastes to quench the coke rather than disposing of it in streams. This resulted in air pollution problems, but air pollution had a much lower priority in the 1920s than did taste and odor in water supplies.[17]

Of all industrial wastes, mine acid drainage was most damaging to water supplies. Again, the problem was particularly severe in the bituminous mining areas of western Pennsylvania and West Virginia, affecting the Allegheny, Monongahela, and Ohio rivers and their tributaries, although mine acid also damaged water supplies in the anthracite regions of eastern Pennsylvania. Mine acid drainage increased the costs of water treatment and distribution. The acid also made river waters harder, caused corrosion of iron and steel, and damaged concrete structures. Thus, industries needing good quality process water, as well as municipalities, had an interest in controlling mine acid drainage.[18]

Public health officials and sanitary engineers, as well as mine owners and government officials, often argued that, since mine acid drainage benefited the public health by killing bacteria in streams, it should not be controlled until cities constructed sewage-treatment facilities. In 1909, for example, a Pennsylvania health bulletin maintained that, "so far as the risk of the most serious of all water pollutions is concerned, that by the typhoid bacillus and its companions and index, the colon bacillus . . . the attempt to exclude mine water and spent tan-liquor from streams which may eventually become sources of drinking water would be a mistake." A 1924 PHS report on the Ohio River drew a similar conclusion concerning the germicidal effect of acid pollution.[19]

But while acid wastes might disinfect streams, they were also variable in their strength and effects. As the chief engineer at the Pittsburgh filtration plant warned, overreliance on the sterilizing role of acid wastes could create a false sense of security about bacterial hazard. The Pennsylvania Sanitary Water Board attempted throughout the interwar period to deal with the mine acid drainage problem through cooperative agreements with mining companies, but accomplishments were limited. In contrast to the phenol wastes, there were many sources, both large and small, and inactive as well as active mines caused difficulties.[20] To this day, mine acid drainage is a major threat to Pennsylvania streams.

Petroleum from wells and refineries was another industrial waste impacting water supplies in several states, especially Kansas, Pennsylvania, and Texas. Oil wells discharged oil and brine, giving water supplies a brackish taste and odor, increasing their hardness, and interfering with water- and sewage-treatment processes. Caustic solutions from refinery wash waters also created taste and odor problems. As in the case of phenols from by-product coke ovens, improvements came with industry cooperation. After the passage of the Oil Pollution Act of 1924—which only applied to coastal waters and made no reference to water supplies—the American Petroleum Institute (API) created a Committee on Disposal of Refinery Wastes. In 1932, this committee began investigating methods of controlling refinery pollution and, in cooperation with representatives of state departments of health, conservation groups, and water-supply engineers, devised control methods that could be applied within the refinery. In 1935 the API issued a manual on the *Disposal of Refinery Wastes* that was gradually adopted throughout the industry on a voluntary basis, substantially reducing the damage from refining wastes to water supplies.[21]

The Public Health Service and Indicators of Industrial Pollution

The regulation of water pollution remained almost entirely a state responsibility until after World War II. The one piece of national legislation between the Refuse Act of 1899 and the Water Pollution Control Act of

1948 was the Oil Pollution Control Act of 1924. This legislation resulted from concern over oil pollution damage to commercial fisheries and resorts, as well as the creation of fire hazards in harbors, rather than from threats to the public health. (One sanitary engineer wrote that oil pollution was "detrimental to public health principally because it discourages healthful water and shore recreation.") The U.S. Corps of Engineers was responsible for enforcing the act.[22] While the Corps also surveyed industrial polluters on navigable rivers and estuaries and reported its findings to Congress, no legislative action resulted.

The persistence of outbreaks of waterborne infectious disease in the first decades of the twentieth century indicated the inability of many state health departments to protect drinking-water supplies against pollution. In 1912, concern over these health problems led Congress to assign the PHS the function of investigating "the diseases of man and conditions influencing the propagation and spread thereof, including sanitation and sewage and the pollution, either directly or indirectly, of the navigable streams and lakes of the United States."[23] The PHS investigations were especially important to the understanding of the comparative effects of both sewage and organic industrial wastes on stream quality, the natural assimilative characteristics of streams, and the measurement of stream pollution.

Public health authorities focused on organic industrial wastes which, by consuming dissolved oxygen, reduced the natural assimilative abilities of streams, thus increasing the likelihood that human wastes containing harmful bacteria would remain unoxidized. Organic wastes from industries such as beet sugar, dairies, canneries, tanneries, pulp and paper mills, textile mills, and meat packing caused some of the severest stream pollution. The PHS also investigated aspects of the relationship between drinking-water quality and industrial pollutants such as phenols. Finally, it conducted investigations of pollution conditions in interstate river basins such as the Potomac, the Illinois, and the Ohio, as well as in the Great Lakes.[24]

Beginning in 1913, in response to the congressional mandate, the PHS launched a significant research effort by a team of sanitary engineers, medical officers, chemists, biologists, and bacteriologists at what became its Center for Pollution Studies in Cincinnati. These stream pollution investigations studied the biochemical oxygen demand (BOD) characteristics of wastes, the capacity of streams for natural oxidation, and methods of treatment for industrial wastes. The station was directed by Dr. Wade H. Frost, who, although a physician, noted that the Cincinnati laboratory was "primarily an engineering station . . . [and] the chief engineering station of our service." Frost conceived of the station's stream pollution studies as the beginning of a "systematic program to acquire accurate, practically useful knowledge of the pollution of all streams in the country." In addition, he believed that "abstract studies of the fundamental principles underlying the physics, chemistry and biology of stream pollution are absolutely essential to real scientific progress." For these studies he relied on MIT graduate Earle B. Phelps, chief of the Division of Chemistry at the PHS Hygienic Laboratory, who directed research on the biochemistry of sewage and of industrial wastes as well as the treatment and disposal of industrial wastes.[25]

Phelps was especially important in regard to the industrial pollution studies. Trained at the Massachusetts Institute of Technology in both chemistry and bacteriology, he served as an assistant bacteriologist at the Massachusetts State Board of Health's Lawrence Experiment Station from 1899 to 1903. From 1904 to 1906, also at Lawrence, he directed studies in the treatment of strawboard wastes for the USGS Division of Hydro-Economics as an assistant hydrographer. From these early experiences, Phelps imbibed the wider view of the uses of streams articulated by Marshall O. Leighton and other Progressive Era conservationists. Phelps became an assistant professor of chemical biology at MIT in 1908. In 1910–11, he conducted path-breaking investigations of oxidation processes in New York Harbor with Colonel William M. Black of the U.S. Corps of Engineers for the New York Board of Estimate and Apportionment. The Black and Phelps study was especially important because it espoused for the first time the use of dissolved oxygen measures as a way of determining water quality.[26]

Phelps believed that industrial wastes, as well as human wastes, required attention because they constituted "an indirect menace to the public health in so far as they may draw upon the stream's natural purifying power, thereby delaying or preventing the ultimate disposal of directly infectious matter." In addition, Phelps noted, both organic and inorganic industrial wastes were undesirable because they reduced the efficiency of water purification and sewage-treatment plants. The very presence of large amounts of these objectionable pollutants, he argued, "dulls the esthetic sense of a community, and, by presenting apparently insuperable barriers to any real progress towards a clean stream, may delay or permanently prevent the proper treatment of the more dangerous sewage pollution." The effects of industrial wastes on the assimilative power of streams called for a strategy of reducing or eliminating wastes or finding alternative methods of disposal. Methodologies that made reuse of wastes possible (and profitable) were considered most desirable. Under Phelps's direction, studies were performed of strawboard, tanning, tomato, and creamery wastes, all effluents with large oxygen-consuming characteristics that were not confined to any one particular section of the country.[27]

After these investigations, however, because of a shortage of funds, the PHS avoided further examinations of specific industrial wastes and focused on stream studies. The PHS followed the recommendations in late 1922 of a consulting group who advised that studies of specific sewage problems and of industrial wastes were "secondary in importance" to general studies of water pollution. In taking this position, the PHS resisted pressure from professional organizations such as the American Society of Civil Engineers that it conduct a broad investigation of "the cause, extent and effect of pollution of waters by industries, that methods of mitigating such evils be investigated, and that existing legislation be reviewed in order to determine what if any legislation is required to cope with the situation." PHS leadership, however, was reluctant to act because it believed that it was limited by their charter to study "the causes and effects of stream pollution . . . , either direct or indirect, upon the public health." Only in cases where industrial wastes had a "sanitary aspect," would the PHS intercede.[28]

PHS officers normally took a conservative attitude in regard to their responsibilities in regard to industrial pollution. Dr. Wade H. Frost, director of the Cincinnati station, believed that industrial waste disposal was "a duty of the industry . . . , and that the function of the governmental agency should be to stimulate and perhaps direct the efforts of the industry toward this end rather than to take over the responsibility." Frost warned that "if the governmental agency assumes responsibility for devising a treatment process the industrial interests are apt to hold the government responsible for any shortcomings in the process and to sit back complacently awaiting a perfected process."[29] He recommended to the Surgeon General that the PHS follow three principles in regard to industrial wastes policy: (1) preference be given to the study of wastes that were "definitely objectionable" in their effects, required extensive research, and were "widely distributed" throughout the country; (2) industry itself be responsible for "proper disposal" if its wastes; and, (3) the PHS cooperate with, and through, the states, with industry, especially where special problems existed. Frost advised that the PHS establish a Division of Stream Pollution Investigations, centered at the Cincinnati station.

Frost believed, as did most of the engineering and scientific personnel at the Cincinnati station, that one of its main objectives was to establish "fundamental relationships which may be applied to studying and remedying stream pollution conditions in general, thus greatly simplifying and improving the methods at present applicable." The development of such a general theory of stream purification was primarily the work of Earle Phelps and sanitary engineer H. W. Streeter as part of their examination of the oxidation processes in the Ohio River. A major problem in controlling and regulating industrial pollution as compared to that from human wastes was the existence of a variety of different types of effluents, variations in the types of treatment required, and uncertainty in regard to effects. Without generalizable indicators, these characteristics made the effective control and regulation of industrial wastes extremely difficult. Indicators of the oxygen-consuming characteristics of organic industrial wastes were especially critical in the context of the total assimilative capacity of the stream and its ability to assimilate human wastes.[30]

The Streeter and Phelps studies resulted in the formulation of the "oxygen sag" curve, by which the self-purification characteristics of streams were understood in terms of the measurable phenomenon of dissolved oxygen (DO) and the BOD characteristics of various wastes under certain conditions of time and temperature. One sanitary engineer compared the process to a financial exchange system in which the "dissolved oxygen may be likened to cash in a bank, and the Bio-chemical Oxygen demand to the checks which may be drawn against it. The dissolved oxygen or cash is being constantly replenished by re-aeration, which latter may be likened to continuing deposits." The use of the Streeter-Phelps model provided an approach that circumvented the heterogeneous character of the industrial wastes by agreement on a common characteristic that furnished a basis for pollution control. As Phelps and Streeter noted in a key 1924 Public Health Service study of the Ohio River, this was "the only procedure which permits any quantitative comparison, however imperfect, between industrial wastes and other sources of pollution in such a broad area as the Ohio watershed."[31]

The Streeter-Phelps "oxygen-sag" curve provided the first quantitative model available for the analysis of water-quality changes due to inputs of various types of wastes. Following the development of these analytical procedures for pollution assessment, health departments adopted a standard approach for estimating the comparative effects of industrial and human pollution on streams. These procedures involved the determination of dissolved oxygen content through stream surveys, the calculation of the stream's assimilative capacity, and the requirement that municipalities and polluting industries reduce the BOD load of their wastes. Because those investigating the stream were primarily concerned with measuring its total assimilative capacity, they described industrial wastes in terms of their population equivalents (e.g., some unit of the industrial waste was equivalent in its BOD characteristics to so many persons). In some cases DO and BOD studies would be combined with bacterial (coliform counts) and biological surveys (plankton studies) to form a measure of stream sanitation and the adequacy of dilution. In addition, investigators often performed chemical analyses in regard to factors such as suspended solids, hardness, acidity and alkalinity, and the presence of chlorine and various nitrogen compounds. This set of experimental techniques, according to Phelps, laid the foundation "for a real science of potomology, or river science." It would now be possible, he added, "to determine not only the present condition of a stream but also what . . . change in the pollution at its source is necessary to bring about a specified improvement in the stream." This approach, he argued, promised "ultimate success" in the attack on stream pollution.[32]

Industrial Wastes and the New Deal

While the PHS developed improved stream pollution indicators during the 1920s, and states and municipalities eliminated some severe industrial and sewage nuisances, progress toward clean streams was limited. Aside from the Oil Pollution Control Act of 1924, Congress enacted no environmental legislation, relying entirely on state regulation and voluntary interstate compacts. In the 1930s, however, the political revolution produced by the Great Depression brought to Washington government officials with a conservationist ideology. In addition, the need to put men to work resulted in the federal government spending hundreds of millions of dollars on the construction of urban sewers, sewage-treatment facilities, and water-purification plants. Some federal funds came in direct expenditures and some in grants-in-aid. By 1938, federal financing had helped construct 1,165 of the 1,310 new municipal sewage treatment plants built in the decade, while the population served by sewage-treatment increased from 21.5 million in 1932 to more than 39 million by 1939. On the whole, however, because cities only provided about 40 percent of the urban population with sewage treatment, the focus remained primarily on sewage pollution and its effect on public health.[33]

Concern with environmental planning resulted in several important studies of industrial pollution at the federal level. In 1935, for instance, the Special Advisory Committee on Pollution of the National Resources Committee (NRC) surveyed state health agencies concerning the extent of industrial pollution in their states, and named those industries producing wastes most damaging to water bodies. It identified a range of industries, including textiles, pulp and paper, food canning and milk preparation, coke and gas manufacture, chemicals, coal mining, and petroleum refining as serious polluters. In addition, the report noted that their "incomplete" survey indicated that there were 523 industrial waste-treatment plants in twenty-three states. Some industries, either as a result of cooperation with government agencies or their national associations, had reduced their waste streams, often through reuse and recycling. On the whole, however, the report observed that the diversity and complexity of industrial wastes required that each industry, and perhaps even each plant, be treated as a special problem. Cooperation between industry and government authorities, therefore, was a necessity.[34]

The NRC committee also listed those conditions it believed responsible for limited action against industrial water pollution. Among these it noted an absence of industrial and municipal cooperation; a lack of uniformity in regulations that often led to unfair competition between states for industries; the costs of treatment; industry failure to reuse wastewater or recycle by-products; and, the absence of "practicable and economical" methods for the treatment of certain types of wastes, although it did note that "practicable and economical" methods were available for many. The committee maintained that only with "the active cooperation of the industries involved" and with "flexible and reasonably administered water-pollution legislation" could the problem be solved.[35]

Legislation dealing with water pollution, however, bogged down in questions of federal vs. state control and the extent to which streams should be used for purposes of waste discharge. The NRC report, for instance, called for the use of "limiting standards of pollution for certain classes of waters," but only after "systematic and thorough" studies had been conducted. It noted that state agencies wanted the federal government to assume a "cooperative" role and to act in a "guiding, stimulating and advisory capacity," but not to extend its control. The committee observed that "public apathy" and "the cost of remedial measures," rather than a lack of legislation, explained the failure of action against water pollution. The PHS also opposed federal control, arguing that the states were the proper regulatory bodies and that the service should play a cooperative and coordinating role. Or, as Surgeon General Cumming wrote, "Too stringent legislation would be practically impossible to enforce and would have a tendency to retard rather than accelerate stream pollution abatement." The opposition to federal control by well-known engineering and administrative figures (such as composed the NRC committee), state health agencies, and industrial groups limited the possibilities of obtaining any regulatory legislation in the 1930s. Although various pieces of water pollution control legislation were introduced in the Congress from 1935–40, none became law, either because of congressional inaction or presidential veto.[36]

While federal regulation of industrial wastes was not achieved in the 1930s, sanitary engineers gained considerable knowledge of their composition and their treatment. State agencies and industries conducted a number of studies of treatment methods for wastes with high BOD. In some instances, state regulation, often motivated by the demands of conservationist organizations such as the Izaak Walton League for clean streams, became tougher. The growth of industrial waste discharge into sewers also provoked regulations. Wastes with high BOD and suspended solids considerably increased sewage-treatment costs, and municipalities and states levied charges and restricted the use of the sewers for industrial wastes without proper pre-treatment.[37]

Between 1937–42, the USPHS and the Corps of Engineers conducted a massive investigation of the Ohio River, arguably the nation's most polluted interstate waterway. From a methodological perspective, the study reflected a culmination of advances in water-quality analysis of the past quarter century. The investigation performed bacteriological tests for coliforms and physical and chemical tests for BOD, DO, acidity and alkalinity, temperature, turbidity, and suspended solids at a number of sampling stations on the Ohio and its tributaries. Estimates of the quality of wastes produced by the basin's 1,800 major industries were converted into BOD population equivalents, producing a population equivalent of over 9 million compared to a total of slightly over 6 million for domestic wastes. The investigators found that acid wastes, primarily from mine drainage but also from the metals industries, which were not in the BOD inventory, caused the "most widespread damages." These included taste and odor problems and damages to industrial water supplies, navigation and hydroelectric power structures, and aquatic recreational facilities. The report commented that "Untreated domestic sewage damages mainly public health and water supplies, recreation, plant and animal life, and aesthetic values. Industrial wastes, by their variety, effect damages to all water uses, but, in general, have a lesser effect on public health than have domestic wastes." The largest sanitary survey of a major river and its tributaries made to World War II, the Ohio River Basin Study provided a base and a model for stream surveys in the coming years.[38] Its conclusions concerning the effects of industrial wastes on the public health, based as they were on the limited psychological evidence available at the time, meant that industrial pollution would continue to be lower on the research and regulatory agenda than human wastes.

The approach to industrial wastes formulated by investigators such as Earle Phelps and Harold Streeter in the late teens and the 1920s and utilized in the Ohio River Pollution Investigation formed the basis for stream investigations up through the 1950s. This body of research and theory supplied an important set of indicators about stream health and sanitation, especially in regard to biological conditions and aquatic life. However, these indicators had direct reference to the public health only in regard to the reduction of the dilution power of streams, although they might serve as surrogates or indicators (in the way that coliform counts did for bacterial wastes) of organic wastes with potentially dangerous characteristics. These indicators also supplied no direct information concerning the presence or danger of toxic metals and toxic organic and inorganic complexes, although again they might serve as surrogates. For many of these substances only limited qualitative or quantitative methods of analysis existed, and information was lacking on their effects on both aquatic and human life.[39]

These limitations in analysis extended not only to the stream but also to an understanding of the industrial wastes themselves. Writing in 1931, industrial waste specialist Willem Rudolfs noted that the analytical procedures utilized for industrial wastes had been primarily borrowed from water and sewage treatment and in many cases were "inadequate . . . and faulty." No knowledge existed concerning treatment procedures for some wastes, while information about the necessary degree of treatment was lacking for others. Even in cases where treatment processes had been studied and recommended by the PHS or state boards of health, only limited application had been made by polluting industries because of the limitations of statutory legislation and reliance on a policy of cooperation.[40]

New Products, New Methodologies, and New Concerns

At the time of American entrance into World War II, the status of industrial waste disposal, treatment, and analysis was approximately where it had been a generation earlier, with the Ohio River pollution investigation representing the state of the art. Investigators had made limited advances in analytical and treatment methods, but no major alterations had occurred in the basic approach to water-quality examination and standards. A few methods were available for specific organics in water or wastes, such as oil, grease, organic nitrogen, cyanide, and phenol, and a rudimentary method existed for total organic carbon. When a committee of water-quality experts made recommendations for revisions of the PHS Drinking Water Standards in 1942, for example, the major change that they made involving industrial wastes, as well as naturally occurring substances, was the establishment of maximum permissible concentrations for lead, fluoride, arsenic, and selenium, while salts of barium, hexavalent chromium, heavy-metal glycosides, and other substances with severe physiological effects were forbidden in water systems.[41]

A number of factors combined, however, during the war and immediately after, to cause critical changes in regard to industrial pollution. One was the large expansion by traditional industries (metals, coal, food, petroleum, etc.) without concomitant attention to pollution control because of wartime restrictions and a lack of legislation and enforcement, leaving a heavy pollution burden for the postwar years. The second factor involved development and production of a range of new substances such as chlorinated hydrocarbons and synthetic materials with large polluting potential. In retrospect, the chemical products, such as chlorinated hydrocarbons (DDT, chlordane, benzene hexachloride, endrin, dieldrin, aldrin, etc.) and synthetic detergents (alkyl benezene sulfates) had the most severe long-term effects on water quality and stream life, although other new products such as rayon and artificial rubber had more immediate and observable impacts. These developments, according to one authority, meant that in the field of sanitary engineering, "after many years of indifferent or complete lack of attention, followed by slowly awakening interest in this important problem, it appears that in the postwar period the disposal of industrial waste is to be the leading topic."[42]

Increased attention to industrial wastes was reflected in the reorientation of old organizations and by the formation of new. In 1946, for instance, the American Chemical Society sponsored a symposium on "Industrial Wastes—A Chemical Engineering Approach to a National Problem." Driving the symposium was not only problem-recognition but also a desire to enlighten Congress in its deliberations over pollution-control legislation then under consideration. The meeting featured talks dealing with a range of industries and pollution problems, including coal and coke, fine chemicals, pulp and paper, and textiles and dealt with air as well as water contamination.[43] The papers from the symposium emphasized the failures as well as the successes of industrial control of wastes and were published in the May 1947 issue of *Industrial and Engineering Chemistry.* The editors called it "one of the most important issues" in the thirty-nine years of the journal's existence. Other developments in these immediate postwar years included formation of the Purdue University Conference on Industrial Wastes—the first annual conference devoted entirely to the topic (1945); establishment of a regular monthly column called "Industrial Waste Digest" by the journal *Public Works* (1946); the start of a regular "Industrial Waste Forum" at the annual meetings of the Federation of Sewage Works Association (1948); the publication in the National Safety Council's *Industrial Safety Series* of a manual on "Industrial Waste Disposal" (1948); and the expansion by *Sewage Works Journal* of its name to *Sewage and Industrial Wastes* (1950).

The PHS attempted to respond to the new focus on industrial pollution. In 1936 President Franklin D. Roosevelt had appointed Dr. Thomas Parran the new surgeon general to replace Hugh S. Cumming, who had held the position since 1920. Parran was a more aggressive administrator than his predecessor, and he pushed the PHS into a larger role in regard to pollution control. "Flushing a toilet," he had commented, "does not end the problem of the proper disposal of wastes." Speaking at the 1946 American Chemical Society Symposium, he noted that industrial wastes were the nation's "largest source of pollution" and had a "serious effect on our public health and welfare." He noted that industry had not accepted its full responsibility in regard to industrial pollution and that the American people should not tolerate the "gross pollution of our public waters." He called for more state action and more interstate agreements, as well as a larger federal role in promoting these agreements through funding for programs and research. Because of pending pollution-control legislation in the Congress, however, he made no recommendation on a federal regulatory role.[44]

Scientific developments and application of new and improved methods of water sampling and analysis derived from advances in the fields of analytical chemistry and zoology also reinforced the new attention to industrial pollution. From approximately the turn of the century to 1940, there had been few improvements in the instrumental techniques available to organic chemists, aside from advances in polarography and column chromatography. In the 1930s, however, an "instrumental revolution" began and then accelerated after the war. Innovations in organic sampling and analysis, for instance, transformed organic pollutant analysis. The development in 1950 of the carbon absorption apparatus vastly improved sampling methodology. Mass spectrometry, although developed earlier, was first applied to water pollution in 1953. Water analysts began using infrared spectrometry, ultraviolet spectrometry, chromatography, and gas chromatography in the 1950s and 1960s, facilitating the identification of various pollutants. Important advances (labeled a "revolution" by one authority), occurred in bioassay techniques used to assess toxic affects.[45] These developments stimulated interest in nonbiological substances in water supplies and raised a variety of questions about the relationship between various pollutants, both old and new, and the public health.

Diffusion in the 1950s of this knowledge beyond limited circles, however, was relatively slow. Writing in 1954, Edward J. Cleary, executive director and chief engineer for the Ohio River Valley Water Sanitation Commission, complained that personnel in "the field of sanitation relating to water supply and waste disposal have been peculiarly unresponsive to problems of toxicity" and continued to focus on "bacteriological hazards." He further observed that there had been no organized investigation to determine if trace constituents from industrial and other wastes caused "unsuspected" public health hazards. Only "meager information" existed, Cleary said, on what was "toxic to man and animals," and there was virtually no information of how concentration levels of a continually ingested substance might "affect degenerative diseases, shorten life, or simply disturb one's sense of well-being."[46]

The PHS's organization of the first major conference on the Physiological Aspects of Water Quality in 1960 attempted to assess the state of knowledge in these regards. The mission of the conference was "To investigate the present state of knowledge with respect to the physiological effects of certain chemical constituents of water, to discover if possible what types of research can and should be undertaken in order to determine both the adverse and beneficial effects of such chemical constituents, and to learn how they may be altered or eliminated in the interest of creating a healthier environment." A wide spectrum of engineers and scientists, including biologists, chemists, toxicologists, and sanitary engineers, presented papers at the conference. These dealt with inorganic and organic constituents in water supplies, specific minerals and trace elements, insecticides and organics, and their physiological and toxic effects on nature and on man. The papers were cautious in their estimates of hazards from industrial wastes, stressing the absence of research and evidence. The conference evaluator, the eminent sanitary engineer Abel Wolman, a past president of the American Public Health Association and a member of the 1937 NRC Water Pollution Committee, specifically noted the absence of basic data showing the effect of chemical pollutants in water supplies on the public health. He called for research to identify epidemiological evidence "on the biological effects of the things that worry us and should worry us." Without such evidence, he added, "we are almost permanently restrained in intelligent action." Wolman asked for more investigation, particularly by industry, and warned that human society had to "move toward a rational, conscious equilibrium between the use of technology and its misuse."[47]

During the decade of the 1960s, stimulated by the publication of Rachel Carson's *Silent Spring*, concern over the environmental and human effects of industrial pollutants, and especially of chlorinated hydrocarbon insecticides and pesticides such as DDT and endrin, accelerated, as did research and legislation. While water pollution studies continued to utilize the traditional parameters of DO, BOD, suspended solids, and acidity, investigators added monitoring for various trace elements and toxics that could not be identified in the past or were not present. In the 1940s and 1950s, for instance, colorimetric techniques made it possible to measure ten parts per million (ppm) of DDT; but by the late 1950s and 1960s, paper chromatography made possible identification of one ppm, and in the 1970s gas chromatography permitted identification of a "few" parts per billion. Similar detection improvements occurred for a number of other organic and nonorganic compounds. Throughout this period, water quality in the nation's major rivers improved considerably from the perspective of the traditional parameters. Long-term measures of DO depletion in the Delaware and Potomac rivers and New York Harbor, for instance, peaked in the decade of the 1940s and declined in the 1960s and 1970s as a result of new state and federal water pollution control legislation and more advanced sewage-treatment methodologies.[48]

Simultaneously with the DO and BOD improvements, however, concern over the environmental and health effects of the newly generated and identified chemical and other toxic pollutants resulted in demands by environmental groups for increased monitoring and control of industrial wastes. For many environmentalists and public health professionals, an environmental health paradigm had replaced the older bacterial paradigm, and industrial wastes had supplanted sanitary wastes as the focus of their concern.[49] Older chemical compounds became suspect as carcinogens, at the same time as industry generated thousands of new chemicals each year and disposed of them in waterways used for drinking-water purposes. Largely unmonitored, and with unknown physiological consequences, they raised new concerns over the safety of drinking-water supplies.

Conclusion

This essay has focused on the lower priority accorded the health effects of industrial wastes compared to human wastes by the public health and sanitary engineering communities in the period from 1876 through 1962. The potential for acute health effects from human wastes as compared with the belief that industrial wastes had only indirect effects drove this prioritization. State departments of health nominally responsible for pollution control usually only responded to industrial wastes when they endangered the potable nature of water supplies, interfered with water- and sewage-treatment processes, or caused obvious nuisance. As Harvey Brooks notes in his essay, "Science Indicators and Science Priorities," industrial wastes constituted a very "messy" research problem—one that does "not lend itself to elegant and widely applicable generalizations." These characteristics tended to retard and complicate research concerning control and reduction and to make their regulation difficult.[50]

The formulation by Streeter and Phelps of a theory of stream purification with a set of general quantitative indicators was of particular importance for providing a measure of the wide variety of high-oxygen-consuming organic industrial wastes. The Streeter-Phelps model enhanced the possibility of successful control of organic wastes by circumventing their heterogeneous nature. The DO and BOD measures were restricted, however, in regard to inorganic effluents and toxicity. The immediate postwar generations witnessed an analytical revolution in regard to analyzing and identifying trace elements of various inorganic and organic industrial wastes in water supply that made evident the need for new indicators and for new research concerning the relationship between industrial pollution and the public health. These developments, however, had only limited effects on industrial waste pollution until the environmental movement and the passage of the federal Clean Water Acts of the 1970s.

Historically, the ability to measure wastes and their effects has been critical for regulating and controlling them. The use of any particular set of indicators, whether for environmental or social phenomena, reflects a definition of the problem that often defines the scope of attempted problem control. Effective public policy is dependent on accurate indicators, and the indicators themselves legitimate the need for policy. New analytical capabilities, however, have consistently produced new thresholds of concern. Changing values and cultural attitudes, as well as scientific measurements and epidemiological knowledge, have influenced our attitudes toward wastes. Thus, as a society, we move to higher and higher levels of concern over perceived risks, as values interact with the increasing ability of science to detect potentially hazardous substances.[51]

Notes

This essay is an expanded and revised version of "Industrial Wastes and Public Health: Some Historical Notes, Pt. I, 1876–1932," which originally appeared in the *Journal of the American Public Health Association* 75 (September 1985):1059–67. It is based on research supported by the National Oceanic and Atmospheric Administration Grant No. NA82RAD00010 and the National Science Foundation Grant No. SES-8420478. Any opinions, findings, and conclusions or recommendations expressed in this publication are those of the author and do not necessarily reflect the views of these agencies. The author would like to thank Francis C. McMichael, John Modell, Granger Morgan, M. Allen Pond, Robert F. Rich, Bruce Seeley, and Maurice Shapiro for their comments on the original article.

1. M. D. Leighton, "Industrial Wastes and their Sanitary Significance," *Public Health Reports and Papers of the American Public Health Association* 31 (1905): 29–41; L. Pearse, "Processes Available for the Treatment of Industrial Wastes," *Engineering and Contracting* 43 (1915): 363–364.

2. James Kirkwood, "A Special Report on the Pollution of River Waters," *Annual Report Massachusetts State Board of Health—1876* (New York: Arno Press Reprint, 1970); A. R. Leeds, "The

Monstrous Pollution of the Water Supply of Jersey City and Newark," *Journal of the Chemical Society* 9 (1887): 81–97.

3. Joel A. Tarr, James McCurley, Terry F. Yosie, "The Development and Impact of Urban Wastewater Technology: Changing Concepts of Water Quality Control, 1850–1930," in Martin. V. Melosi (ed.), *Pollution and Reform in American Cities, 1870–1930* (Austin: University of Texas Press, 1980), 59–82.

4. Joel. A. Tarr, James McCurley III, Francis C. McMichael, and Terry Yosie, "Water and Wastes: A Retrospective Assessment of Wastewater Technology in the United States, 1800–1932," *Technology and Culture* 25 (April 1984): 239–246; Samuel C. Prescott and C. E. A. Winslow, *Elements of Water Bacteriology*, 5th ed. (New York: John Wiley & Sons, 1931); and John Borchardt and G. Walton, "Water Quality," in American Water Works Association, *Water Quality and Treatment: A Handbook of Public Water Supplies*, 3rd ed. (New York: McGraw-Hill, 1971).

5. H. W. Clark, "The Pollution of Streams by Manufactural Wastes, and Methods of Prevention," *Journal of the New England Water-Works Association* 15 (1901): 500–501; Leighton, "Industrial Wastes and their Sanitary Significance," 29–41; George A. Johnson, "The Disposal of Trade Wastes," *Engineering News* 63 (1910): 637; and Harold Eddy, "Industrial Waste Disposal," *Journal of Industrial Engineering and Chemistry* 9 (1917): 696–700.

6. Samuel P. Hays, *Conservation and the Gospel of Efficiency: The Progressive Conservation Movement, 1890–1920* (Cambridge: Harvard University Press, 1959); Leighton, "Industrial Wastes and their Sanitary Significance," 29–41; and Mary C. Rabbitt, *Minerals, Lands, and Geology for the Common Defense and General Welfare, Volume 2, 1879–1904: A History of Geology in Relation to the Development of Public-Land, Federal-Science, and Mapping Policies and the Development of Mineral Resources in the U.S. During the First 25 Years of the U.S. Geological Survey* (Washington: GPO, 1980), 353–354.

7. E. B. Phelps, "Stream Pollution by Industrial Wastes and Its Control," in M. Ravenel (ed), *A Half Century of Public Health* (New York: American Public Health Association, 1921); A. L. Fales, "Progress in the Control of Pollution by Industrial Wastes," *American Journal of Public Health* (henceforth, *AJPH*) 18 (1928): 715–727.

8. See "State Laws Governing Pollution by Industrial Waste," *Chemical and Metallurgical Engineering* 38 (1931): 506–507; Phelps, "Stream Pollution by Industrial Wastes and Its Control;" E. F. Murphy, *Water Purity: A Study in Legal Control of Natural Resources* (Madison: University of Wisconsin Press, 1961); and C. M. Baker, "What Should Be the Policy of the State Board of Health in the Control of Stream Pollution?" *Transactions, 6th Annual Conference of State Sanitary Engineers, Public Health Bulletin No. 160* (Washington, DC: Government Printing Office, 1926).

9. Anonymous, "Stream Pollution and Industrial Wastes in Wisconsin," *Engineering News-Record* 99 (1927):664; H. Eddy, "The Cleaning Up and Improvement of a Stream Polluted by Sewage and Trade Wastes," *Engineering and Contracting* 46 (1916): 144; W. L. Stevenson, "The State vs Industry or the State with Industry," *Transactions, American Institute of Chemical Engineers* 16 (1924): 201–216; E. B. Besselievre, "Statutory Regulation of Stream Pollution and the Common Law," *Transactions, American Institute of Chemical Engineers* 16 (1924): 217–230; and Murphy, *Water Purity: A Study in Legal Control of Natural Resources* (Madison: University of Wisconsin Press, 1961).

10. Connecticut Department of Health, *First and Second Biennial Reports of the Industrial Wastes Board* (Hartford, CT: The Department, 1918–21); John. E. Monger, "Administrative Phases of Stream Pollution Control," in Abel Wolman, "Domestic and Industrial Wastes in Relation to Public Water Supply: A Symposium," *AJPH* 16 (1926): 788–794; Terry. F. Yosie, "Changing Concepts of Stream Pollution Control in Pennsylvania," in *Retrospective Analysis of Water Supply and Wastewater Policies in Pittsburgh, 1800–1959* (Pittsburgh: Carnegie-Mellon University, 1981, unpublished dissertation), 266–323; and Monger, "Administrative Phases of Stream Pollution Control," 788–794. In 1906, Massachusetts created legislation with a similar requirement. The Board of Health was required to advise the polluting factory owner regarding "the best practicable and reasonably available means of rendering the wastes or refuse harmless." Fales, Progress in the Control of Pollution by Industrial Wastes," 715–727; Stevenson, "The State vs Industry or the State

with Industry," 201–216; and Monger, "Administrative Phases of Stream Pollution Control," 788–794.

11. Yosie, "Changing Concepts of Stream Pollution Control in Pennsylvania," 266–323; T. Saville, "Administrative Control of Water Pollution," *Transactions, American Institute of Chemical Engineers* 27 (1931): 74–77; and Earle B. Phelps, "Discussion: Domestic and Industrial Wastes in Relation to Public Water Supply: A Symposium," *AJPH* 16 (1926): 795–797.

12. "Progress Report of Committee on Industrial Wastes in Relation to Water Supply," *Journal of the American Water Works Association* (henceforth, *JAWWA*) 10 (1923): 415–430; Walter Donaldson, "Industrial Wastes in Relation to Water Supplies," *American Journal of Public Health* 11 (1921): 193–198.

13. Walter Donaldson, "Industrial Wastes in Relation to Water Supplies," *AJPH* 12 (1922): 420–421; Borchardt, "Water quality;" Harold Babbitt and J. J. Doland, *Water Supply Engineering* (New York: McGraw-Hill, 1931), 521; Christopher Sellers, "Factory as Environment: Industrial Hygiene, Professional Collaboration and the Modern Sciences of Pollution," *Environmental History Review* 18 (Spring 1994): 69–74. The USPHS was also engaged in industrial hygiene studies; see Ralph C. Williams, *The United States Public Health Service 1798–1950* (Washington, DC: Commissioned Officers Association of the USPHS, 1951), 279–286.

14. "Progress Report on Recent Developments in the Field of Industrial Wastes in Relation to Water Supply," *JAWWA* 16 (1926): 302–303.

15. "Progress Report of Committee on Industrial Wastes in Relation to Water Supply," *JAWWA* 10 (1923): 415–430; "Discussion of Report of Committee No. 6 on Industrial Wastes in Relation to Water Supply," *JAWWA* 12 (1924): 411–415.

16. F. H. Waring, "Results Obtained in Phenolic Wastes Disposal Under the Ohio River Basin Interstate Stream Conservation Agreement," *AJPH* 19 (1929): 758–770.

17. Fales, "Progress in the Control of Pollution by Industrial Wastes," 715–727; Robert Dunlap and Francis C. McMichael, "Reducing Coke Plant Effluents," *Environmental Science and Technology* 10 (1976):654–657.

18. "Progress Report of Committee on Industrial Wastes," *JAWA* 10 (1923):415–430.

19. W. L. Stevenson, "The Sanitary Conservation of Streams by Cooperation," *Transactions, American Institute of Chemical Engineers* 27 (1931): 9–30; Pennsylvania State Board of Health, "The Germicidal Effect of Water from Coal Mines and Tannery Wheels upon Bacillus Typhosus, Bacillus Coli and Bacillus Anthracis," *Pennsylvania Health Bulletin* 5 (1909): 1–10; and W. A. Frost, *A Study of the Pollution and Natural Purification of the Ohio River, Public Health Bulletin No. 143* (Washington, DC: GPO, 1924), 159.

20. C. F. Drake, "Water Purification Problems in Mining and Manufacturing Districts," *JAWWA* 23 (1931): 1264–1265; ________, "Effect of Acid Mine Drainage on River Water Supply," *JAWWA* 23 (1931): 1474–1493; and A. B. Crichton, "Disposal of Drainage from Coal Mines," *Proceedings of the American Society of Civil Engineering* 53 (1927): 1656–1666.

21. A. L. Fales, "Progress in Control of Oil Pollution," *JAWWA* 18 (1927): 587–588, 692; W. B. Hart, "Waste Disposal in Retrospect," *American Petroleum Institute Reporter* 36 (1956): 349; and W. B. Hart, "Disposal of Refinery Waste Waters," *Industrial and Chemical Engineering* 26 (1934): 965–967.

22. J. Pratt, "The Corps of Engineers and the Oil Pollution Act of 1924" (unpublished manuscript, College Station, TX: Texas A&M University, 1983); Fales, "Progress in Control of Oil Pollution," 587–588, 692.

23. Williams, *The United States Public Health Service*, 312–313. The USGS had conducted stream surveys that detailed both human and industrial pollution, as well as biological and hydrological factors, since 1897.

24. E. F. Eldridge, *Industrial Waste Treatment Practice* (New York: McGraw-Hill, 1942); Williams, *The United States Public Health Service*, 312–320.

25. Wade H. Frost, "The Natural Purification of Streams Bacteriologically, Biologically, and Chemically," *Transactions, 5th Annual Conference of State Sanitary Engineers, Public Health Bull No. 154* (Washington, DC: GPO, 1925); Wade H. Frost to The Surgeon General (Rupert Blue), July

30, 1913, Records of the PHS, National Archives, Record Group 3450; and Frost to Assistant Surgeon General J.W. Kerr, Sept. 4, 1914, Records of the PHS, National Archives, Record Group 3450.

26. "E. B. Phelps," in *National Cyclopedia of American Biography* (New York: James T. White & Co., 1955) 40:411–412; W. M. Black and E. B. Phelps, *Report Concerning the Location of Sewer Outlets and the Discharge of Sewage into New York Harbor* (New York: NY Board of Estimate and Apportionment, March 23, 1911); and George S. Soper, et al, *Main Drainage and Sewage Disposal Works Proposed for New York City: Reports of Experts and Data Relating to the Harbor* (New York: Report of the Metropolitan Sewerage Commission of New York, 1913).

27. E. B. Phelps, *Studies of the Treatment and Disposal of Industrial Wastes, Public Health Bull No. 97* (Washington. DC: Government Printing Office, 1918), 5–7; Earle B. Phelps, "Memorandum on Sewage and Industrial Waste Studies (Informal)," Nov. 30, 1914, Records of the USPHS, National Archives, Record Group 3450; Phelps to Surgeon General (Homer Cumming), May 1, 1915, ibid.; and, H. W. Streeter, "Discussion on Bearing of Industrial Wastes on Stream-pollution Investigation," *Transactions, 5th Annual Conference of State Sanitary Engineers,* Public Health Bull No. 154 (Washington, DC: Government Printing Office 1925), 111–112.

28. Wade H. Frost to Surgeon General Hugh S. Cumming, Dec. 15, 1922, in Records of the USPHS, National Archives, Record Group 3450. The consulting group was composed of Dr. S.A. Forbes, Edwin O. Jordan, Langdon Pearse, and Earle B. Phelps, Joseph W. Ellms, Dr. Lowell J. Reed. See Williams, *The USPHS,* 317; Charles Haydrock to USPHS, May 15, 1922, in Records of the USPHS, National Archives, Record Group 3450; J.W. Schereschewsky to Haydrock, May 24, 1922, and Haydrock to Surgeon General (Homer S. Cumming), June 2, 1922, in Records of the USPHS, National Archives, Record Group 3450.

29. Wade H. Frost to Assistant Surgeon General A. M. Stimson, May 28, 1926, ibid.

30. Frost to Surgeon General [Homer S. Cumming], Dec. 22, 1919, ibid.; "Stream Pollution Investigations," Annual Report for the Fiscal Year. . . 1920;" J. W. Clark, W. Viessman, Jr., M. J. Hammer, *Water Supply and Pollution Control*, 3rd ed. (New York: Harper & Row, 1977), 287–288; E. J. Theriault, *The Oxygen Demand of Polluted Waters, Public Health Bull No. 173* (Washington, DC: GPO, 1927); and Streeter and Phelps, *Factors Concerned in the Phenomena of Oxidation and Rearation, Public Health Bull No. 146* (Washington, DC: GPO, 1925).

31. R. S. Weston, "The Use, Not the Abuse of Streams," *Transactions, American Institute of Chemical Engineers* 27 (1931): 4–5; Streeter and Phelps, *Factors Concerned in the Phenomena of Oxidation and Rearation*; and W. H. Frost, *A Study of the Pollution and Natural Purification of the Ohio River II. Report on Surveys and Laboratory Studies, Public Health Bull No. 143* (Washington, DC: GPO, 1924), 80–84.

32. Stevenson, "The Sanitary Conservation of Streams by Cooperation," 27:9–30; N. T. Veatch, "Stream Pollution and Its Effects," *JAWWA* 17 (1927): 58–63; E. B. Phelps, *Stream Sanitation* (New York: John Wiley and Sons, 1944); and Phelps, "Discussion: Domestic and Industrial Wastes in Relation to Public Water Supply," 795–797.

33. See "Stream Pollution Control Activities in the U.S., 1928," Records of the PHS, National Archives, Record Group 3450; Joel A. Tarr, "The Evolution of the Urban infrastructure in the Nineteenth and Twentieth Centuries," Royce Hanson (ed.), *Perspectives on Urban Infrastructure* (Washington: National Academy Press, 1984), 40–42; C.R. Koppes, "The Department of the Interior, 1933–1953," in Kenneth E. Bailes (ed.), *Environmental History: Critical Issues in Comparative Perspective* (Lanham, MD: University Press of America, 1985), 440–450; and National Resources Committee, Special Advisory Committee on Water Pollution, *Report on Water Pollution* (Washington, D.C., July, 1935), 59 (henceforth cited as *NRC*). The NRC committee was composed of W.B. Bell, Glen E. Edgerton, A.C. Fieldner, Elmer Higgins, Thorndike Saville, R.E. Tarbett, and Abel Wolman.

34. *NRC,* 45–48.

35. *NRC,* 43–48, 60.

36. *NRC,* 57–58. Homer S. Cumming to Hon. M.H. Thatcher, Mar. 3, 1932; Cumming to the Hon. Augustine Lonergan, Sept. 30, 1933; Longeran to Cumming, Sept. 30, 1933; C. C. Pierce (Acting Surgeon General) to Longeran, Oct. 5, 1933; Cumming to Seth Gordon, Feb. 4, 1935, all in PHS

Records, National Archives, Record Group 2323. See, also, untitled and unsigned ms. in PHS files, dated Dec. 30, 1934, outlining the PHS position; Joel A. Tarr, et. al., "The Development of a Federal Role in Water Pollution Control," in *Retrospective Assessment of Wastewater Technology in the United States, 1800–1972: A Report to the National Science Foundation* (Pittsburgh: CMU, 1978); and Colten, "Creating a Toxic Landscape," 96–98. .

37. L. F. Warrick, "The Prevalence of the Industrial Waste Problem," in Langdon Pearse (ed.), *Modern Sewage Disposal* (New York, 1938); F. W. Mohlman, "The Disposal of Industrial Wastes," *Sewage Works Journal* 11 (1939): 646–656.

38. USPHS and U.S. Corps of Engineers, *Ohio River Pollution Control, Part I. Report of the Ohio River Committee, House Document No. 266*, 78th Congress, 1st Sess. (Washington: GPO, 1944), 31; Williams, *The U.S. Public Health Service*, 222.

39. See, for instance, the methods of analysis of industrial waste suggested in E.F. Eldridge, *Industrial Waste Treatment Practice* (New York: McGraw-Hill, 1942), 36; also Gordon M. Fair, J. C. Geyer, and J. C. Morris, *Water Supply and Waste-Water Disposal* (New York: John Wiley, 1954), 867–868.

40. Willem Rudolfs, "Stream Pollution in New Jersey," *Transactions of the American Institute of Chemical Engineers* 27 (1931): 41–47; E. F. Eldridge, *Industrial Waste Treatment Practice* (New York: McGraw-Hill, 1942), 3–4.

41. J. K. Hoskins, "Some Developments is the Water Pollution Research Program of the Public Health Service," *AJPH* 30 (1940): 527–531; Harvey Brooks, "Science indicators and Science Priorities," *Science, Technology, and Human Values* 7 (1982): 14–31; and Aaron A. Rosen, "The Foundations of Organic Pollutant Analysis," in Lawrence H. Keith (ed.), *Identification and Analysis of Organic Pollutants in Water* (Ann Arbor: Ann Arbor Science, 1976), 3–14.

42. F. W. Mohlman, Director of Laboratories, Chicago Sanitary District, noted that "During the war, many instances of pollution have been condoned by public authorities in order to place no barriers in the path of production of war essentials." See National Safety Council, "Chemical Waste Disposal: A Symnposium," in National Safety Council, *Transactions* (1944), 145–157. In 1951 the Interstate Sanitation Commission (New York, New Jersey, and Connecticut), estimated that the BOD demand of the industrial wastes discharged into New York harbor waters was equivalent to that of the raw sewage discharged by over two million people. See Colten, "Creating a Toxic Landscape," 102–104; William McGucken, *Biodegradable: Detergents and the Environment* (College Station: Texas A&M University Press, 1991); Thomas R. Dunlap, *DDT: Scientists, Citizens, and Public Policy* (Princeton: Princeton University Press, 1981), 3–97; John H. Perkins, *Insects, Experts, and the Insecticide Crisis: The Quest for New Pest Management Strategies* (New York: Plenum Press, 1982), 3–60; and George E. Symons, "Industrial Wastes," *Sewage Works Journal* 17 (1945): 558–571

43. *Industrial and Engineering Chemistry* 39 (May 1947): 558.

44. Ibid., 560–561. Since pollution control legislation was pending in the congress, Parran was careful to take no position on the question of a federal role in regulation.

45. H. Kolnik and K. M. Reese (eds.), *A Century of Chemistry: The Role of Chemists and the American Chemical Society* (Washington: American Chemical Society, 1976), 342–350; Brooks, "Science indicators and Science Priorities,"14–31; Rosen, "The Foundations of Organic Pollutant Analysis," 4–14; H. A. Clarke, "Characterization of Industrial Wastes by Instrumental Analysis," *Proceedings of the Purdue Industrial Waste Conference* (1968), 26–34; and John Cairns Jr., "Don't Be Half-Safe: The Current Revolution in Bioassay Techniques," ibid. (1966), 559–567.

46. In 1954, Edward J. Cleary, the Executive Director and Chief Engineer of the Ohio Valley Water Sanitation Commission, wrote that "workers in the field of sanitation relating to water supply and waste disposal have been peculiarly unresponsive to problems of toxicity." He observed that there had been "no organized program of investigation to determine if unsuspected public health hazards may exist as a result of trace constituents from industrial and other wastes," that there was only "meager information" on what was "toxic to man and animals," and "virtually no information as to the concentration levels at which a substance continually ingested may affect degenerative diseases, shorten life, or simply disturb one's sense of well-being." See Edward J.

Cleary, "Determining Risks of Toxic Substances in Water," *Sewage and Industrial Wastes* 26 (Feb. 1954): 203–209. Craig Colten observes that, during the 1950s, while chemical firms possessed "internal expertise on toxicology, it was seldom applied to chemical wastes except when public agencies pressured individual companies." See Colten, "Creating a Toxic Landscape," 108–109.

47. H. A. Faber and L. J. Bryson (eds.), *Proceedings of the Conference on Physiological Aspects of Water Quality* (Sept. 8–9, 1960) (Washington: USPHS, 1960); Faber and Bryson (eds.), *Physiological Aspects of Water Quality*, 233–238.

48. Samuel P. Hays, *Beauty, Health, and Permanence: Environmental Politics in the United States, 1955–1985* (New York; Cambridge University Press, 1987), 171–177; Dunlap, *DDT*, 98–245; Council on Environmental Quality (CEQ), "Trends in Analytical Chemistry Detection Techniques," in *Environmental Quality, Tenth Annual Report of the CEQ* (Washington: GPO, 1979), 238; and M. Gordon Wolman, "Crisis and Catastrophe in Water-Resources Policy," *JAWWA* (Mar., 1976), 136–142. See also M.Gordon Wolman, "The Nation's Rivers," *Science* 174 (1971): 905–918, for a critique of DO and BOD as adequate measures of stream quality. These were the same grounds on which Phelps had earlier attacked the reliance on public health and drinking water considerations as a measure of stream pollution. Phelps, of course, was advocating the use of DO and BOD as a means of estimating the injurious effects of industrial wastes. See his comments recorded in the discussions following the papers in Abel Wolman, "Domestic and Industrial Wastes in Relation to Public Water Supply: A Symposium, *AJPH* 16 (1926): 795–797

49. See CEQ, "Toxic Substances and Environmental Health," *Environmental Quality, Tenth Annual Report of the CEQ* (1979), 174–255.

50. Brooks, "Science indicators and Science Priorities," 14–31.

51. See Robert W. Kates, Christoph Hohenemser, and Jeanne X. Kasperson, *Perilous Progress: Managing the Hazards of Technology* (Boulder, CO: Westview Press, 1985), for a discussion of these issues.

CHAPTER XV Searching for a Sink for an Industrial Waste

The creation of wastes from any process, be it a natural, consumer, or production process, requires location of a place of deposit, or a "sink." Much of the history of industrial waste disposal, as well as the disposal of wastes from other sources such as an urban population, involves the search for a sink in which wastes could be disposed of in the cheapest and most convenient manner possible.[1] Often, however, such sinks proved only temporary, and substances placed in them created severe pollution, interacted with other substances to produce serious nuisance and health consequences, or leaked out or migrated into other media. Society reacted against such pollution problems in various ways, depending on the problem's severity, the existence of technological means to reduce or control the pollution, the prevailing legal and legislative norms relating to the environment, and the existence and availability of other possible sinks. Often, action or policy intended to remove a polluting waste from one environmental sink resulted in its deposition in another sink.

This essay examines the pollution problems and environmental damages resulting from the production of fuel to be used in the smelting of iron. In the United States iron makers first used charcoal as a metallurgical fuel in their furnaces but began switching to mineral fuels in the 1830s and

1840s. Anthracite coal was used initially, but in the post–Civil War decades iron and steel manufacturers increasingly replaced anthracite with coke—the solid residue of almost pure carbon left when heat is used to drive out the volatile matter from bituminous coal. By 1911, coke fueled the smelting of 98 percent of the nation's pig iron. From about the 1850s through World War I, the technology used to produce coke was the beehive oven, but after that date there was a rapid transition to the by-product coke oven, the technology in use today.

All processes for the making of fuels for iron manufacturing were (and are) polluting to a degree, but changes in the technology and in its location resulted in significant differences in the environments and populations affected. This essay will examine the transition from the beehive to the by-product oven and the resultant shift in location of much of the nation's coke-making facilities from rural to urban sites—a change that moved the pollution burden to different environmental media and which resulted in the harming of drinking-water quality in a number of metropolitan areas. It will also explore the different policy options followed by government and the courts at a time when industrial waste disposal was of limited public concern.

Iron-Making Fuels before Coke

Mineral fuel, in either the form of anthracite or bituminous coal, did not become important in the making of iron and steel in the United States until the mid–nineteenth century. Before this time, the fuel most commonly used was charcoal, an ideal furnace fuel because it was relatively free from sulphur or phosphorus impurities and because its ash furnished part of the flux required to smelt the ore. Charcoal manufacture usually took place on what was called an iron plantation, a large forested tract of land that also contained an iron furnace, a charcoal "pit" or "hearth," casting beds, and workers' housing. On the plantation, bundles of cord wood in 6- to 10-foot lengths were stacked in a cone with a base of about 25 feet in diameter, covered with damp leaves and turf, and burned for between three and ten days. No attempt was made to condense any of the by-product wood chemicals vented from the stack during the charcoaling process.[2]

Charcoal production had major environmental effects both because of the amount of timber consumed and because it involved the slow burning of wood with no emissions control. Geographer Michael Williams estimates that it took 150 acres of woodland to produce 1,000 tons of pig iron, the annual output of an average iron furnace. Plantations ranged in size from the 3,000 acres of woodland usually required for a profitable iron plantation to more than 10,000 acres. In 1862, a year of low production, 25,000 acres were cleared, while in 1890, the peak year of charcoal-iron production, 94,000 acres (147 square miles) were cleared. While timber harvested for iron production constituted only 1.3 percent of the land cleared for agriculture, it was heavily concentrated in certain states such as Ohio, Pennsylvania, and West Virginia.[3]

In addition to forest depletion, the charring process produced severe local air pollution effects. While the timber was being charred, the smoking piles of wood, covered with wet leaves, gave off a dark, heavy smoke with a disagreeable odor that gathered over the stream valleys where the operations of the iron plantations were often located. Many Pennsylvania boroughs passed ordinances prohibiting charcoal manufacture within town limits. The iron furnaces themselves were also heavy air polluters, as the opening at the top of the furnace permitted carbon monoxide, heat, and smoke to escape and created great clouds of smoke and fumes.[4]

As technological innovations that improved productivity were gradually made in both charcoal making and iron furnaces during the course of the nineteenth century, growing demand for iron depleted conveniently located timber supplies. Beginning in the 1830s, however, the pressure on fuel supply was relieved by the increasing substitution of anthracite coal for charcoal. Anthracite possessed almost pure carbon content and burned with a small blue flame that produced intense heat and little smoke. A ton of anthracite was equal in heat energy to 200 bushels of charcoal, giving it a strong cost advantage. The primary deposits of anthracite were in the rugged, mountainous areas of northeastern Pennsylvania, making access difficult, but in the 1830s the construction of three major coal canals sharply reduced the cost of transportation and permitted large-scale production.[5]

Iron-makers had seldom used anthracite in their blast furnances before the 1830s because the fuel was difficult to ignite. In the middle of that decade, however, it was discovered that the "hot blast," an innovation developed in Great Britain in 1828 for use in charcoal furnances, faciliated the use of anthracite as a fuel. Anthracite had greater strength than charcoal and could be burned in larger furnaces with higher thermal efficiency. Furnaces utilizing the mineral fuel sprang up along the established coal trade routes close to urban areas, with the Lehigh Valley of Pennsylvania becoming the center of the anthracite iron industry. By 1853, there were 121 anthracite blast furnaces in eastern Pennsylvania, the nation's center of iron manufacture.[6]

From the 1850s until the end of the 1870s, anthracite served as the dominant fuel in the making of iron. The shift to anthractite from charcoal partially shifted the environmental burden of providing fuel for iron making from the forests to the coal-bearing lands of eastern Pennsylvania.[7] But while anthracite retained its position as a desirable fuel in manufacturing and in domestic heating until the middle of the twentieth century, its importance in iron making was relatively short lived. In 1852, the railroad crossed the Appalachian Mountains, opening the rich bituminous coalfields of western Pennsylvania to extensive exploitation and development. In the process, some of the most fertile valleys in western Pennsylvania were exposed to widespread environmental damages from the mining of coal, the making of coke, and the smelting of iron.

Bituminous Coal and Coke

Bituminous coal located in the Connellsville fields of western Pennsylvania (part of the Pittsburgh seam) was the world's richest coking coal—it was the critical factor in making the Pittsburgh region the nation's iron and steel center in the late nineteenth century. "Coking," noted John Fulton in 1895, "is the art of preparing from bituminous or other coal a fuel adapted for metallurgical and other special purposes." It involves using heat to expel volatile matter such as water, sulphur, and hydrocarbons from the coal, leaving a solid residue consisting primarily of fixed carbon, ash, and the nonvolatilizable sulphur. The best coking coal was a coal with few impurities such as sulphur or phosphorus, with a sufficient proportion of volatile or gaseous matter to supply the necessary heat in coking, and that would produce a coke with a porous structure and strong physical strength.[8]

Coke was low in cost compared to anthracite and ideal in structure for blast furnace use. It possesses a porous structure that permits it to be burned at a rapid rate and also ample strength at high temperature to carry the burden of the blast furnance charge. These advantages of cost and quality led to coke's rapid displacement of anthracite and charcoal in the iron-making process. In 1854, for instance, charcoal iron composed about 47.5 percent of the nation's total pig iron production, anthracite pig iron 45 percent, and bituminous pig iron 7.5 percent. By 1880, however, the percent of pig iron made with bituminous coal and coke had risen to 45 percent, while mixed anthracite and coke had dropped to 42 percent and charcoal to only 13 percent. In 1911 bituminous coal and coke provided the fuel for 98 percent of the pig iron manufactured in the nation, and charcoal had practically disappeared as a source of fuel for the process.[9]

Coke was initially produced following a technique borrowed from that used in charcoal preparation: bituminous coal would be placed in piles or rows about fourteen feet wide on level ground surfaced with coal dust; wood was interspersed to ignite the mass. In the most effective process, the coal was burned slowly in a moist, smoldering heat, driving off the sulphur and the hydrocarbon gases and leaving a silvery white coke high in carbon. When the burning of the gaseous matter had ceased, the heap was smothered with a coating of dust or duff, followed by application of a small quantity of water. No attempt was made to capture any of the escaping gases. The time necessary for coking a heap was usually between five and eight days, with a yield of approximately 59 percent of coke.[10]

Most of the coke produced in the nation from about 1850 to 1920 was manufactured by the so-called beehive coke oven. These ovens were first constructed in the 1850s in western Pennsylvania, and by the late nineteenth century most coke makers had adopted the technology. The oven consisted of a circular, domelike chamber lined with fire brick and with a tiled floor. Although there were improvements in oven design over the years, changes were relatively minor. The ovens were first preheated by a wood and coal fire and then charged with coal from a car (a "Larry") running on tracks above the oven. The coke was leveled by hand with a scraper and the front opening bricked up with clay, with a 2- or 3-inch opening at the top. The coking process proceeded from the top downward, with the burning of the volatile by-products escaping from the coal producing the required heat. The coking process was complete when all the volatile material was burned. The brickwork was removed from the door, and the coke was cooled by a water spray from a hand-held hose and then removed from the oven by hand or by mechanical means. On average, it took about 1.5 tons of high-quality bituminous coal "coked" over a period of forty-eight hours to produce a ton of coke, the composition of which was between 85 and 90 percent carbon.[11]

In 1855 there were 106 beehive ovens in the nation; by 1880 the total had increased to over 12,000 ovens in 186 plants, and by 1909 there were almost 104,000 ovens in 579 plants, the maximum number ever to exist. Initially almost all of the beehive ovens were located in the Connellsville coke region of western Pennsylvania, and as late as 1918, more than half the country's ovens were still in this area. Beehive ovens were normally arranged in banks of single or double rows, often built into a hillside in order to conserve heat. Normally, they were constructed in banks of about 300, considered the optimum scale for efficient operation. Numerous small operations were scattered over the countryside in coking areas. Some installations, however, were much larger, with 600 to 900 ovens. The Jones & Laughlin Iron and Steel Company beehive coking plant in the Hazelwood area of Pittsburgh was the world's largest, with 1,510 ovens. The J&L ovens were also unusual in that they were located close to the blast furnaces of an integrated mill, whereas most beehives were sited in proximity to coal deposits in order to reduce transportation costs.[12]

The uncontrolled emissions vented from the beehive ovens had a devastating impact on the nearby environment. The various hydrocarbons, fumes, and ash released by the coal distillation process killed and stunted trees and crops within the locale of the ovens and often left a layer of coal dust, ash, and particles on the surrounding fields. Writing in 1900, a Pennsylvania state botanist observed that "the most conspicuous feature in coke oven surroundings is the general wretchedness of everything of the nature of shrub or tree, either individual or collective." In those areas where there were large numbers of ovens scattered over the countryside, he wrote, "the district becomes almost continually one of a highly vitiated air, seldom without the overhanging clouds of smoke."[13] In 1905 and 1906, the U.S. Forest Service, interested in the possibility of reclaiming "waste holdings," conducted studies at the request of the H. C. Frick Coke Company concerning possible reforestation of some of their western Pennsylvania holdings. These studies, both performed by Samuel N. Spring, a graduate of the Yale forestry program, emphasized the extreme effect of the "sulphur smoke" from the coke ovens on vegetation around the ovens and on vegetation on hills some distance downwind.

"Cloud by day and fire by night" was the phrase often used to characterize the coke region. But while the damage from the sulfurous fumes did the most environmental damage, solid wastes from the coking process also caused injuries. Coke oven operators often dumped materials such as coal wastes and ashes into nearby creeks, damming their flow, causing flooding, and undoubtedly damaging stream ecology. Local courts heard several cases in the early twentieth century brought by farmers suing for damages because their lands had been damaged by flooding and the silting up of bottomlands caused by the dumping of coking wastes.[14]

In several cases in the late nineteenth and early twentieth centuries, Pennsylvania courts held that beehive coke plants, like other manufacturing firms, were liable for damages caused by their operations. In these cases, the courts made a clear differentiation between responsibility for injuries from coking operations and those resulting (as they held in the Sanderson cases of the 1880s) from the "natural and necessary result of the development of his own land by the owner." Since the manufacture of coke was not the "natural and necessary use" of property, even on coal lands, individuals whose property was injured by such operations were entitled to compensation.[15]

But while judges might award damages, they seldom would issue injunctions closing down a firm's operations. Rather, they usually performed a rough type of cost-benefit analysis that Christine Rosen has called "private" or "social cost" balancing. In such a situation, judges weighed the benefits the victims would receive from an abatement of pollution with the economic injury either the firm (in private balancing) or society (with social cost balancing) would suffer if the court required the polluter to reduce its pollution.[16] Invariably, in late-nineteenth-century Pennsylvania, the courts found the projected costs to society of an injunction to be too high, even though they awarded damages for property damages to individuals. These awards, however, were usually relatively small since most coking operations were in rural areas, and the cost of damages to surrounding farmers were minor compared to the potential cost of pollution-control equipment such as tall stacks.

Because of the heavy pollution produced by beehive ovens, cities were reluctant to allow them to operate within their boundaries. In 1869, for instance, a Pittsburgh ordinance forbade the construction of coke ovens within the city limits and assessed a penalty of $100 per day for operating an oven. The ordinance, however, was probably not enforced, since the 1884 Sanborn maps for Pittsburgh show well over a hundred coke ovens within the city boundaries. In 1892 the city revised its policy and enacted a new ordinance permitting coke oven construction if the ovens were supplied with "smoke control devices" approved by the chief of the Department of Public Safety. The Jones & Laughlin Steel beehive oven plant in Hazelwood (constructed between 1899 and 1907), for instance, was fitted with tall stacks to burn and disperse the oven gases and smoke away from nearby residential areas.[17]

While these stacks undoubtedly reduced the emissions flow compared to the average beehive oven, the pollution was still substantial. In 1914 the Mellon Institute's "Smoke Investigation of Pittsburgh" reported that coke ovens located in the Pittsburgh district (but not specifically naming Jones & Laughlin) smoked "almost constantly, although the smoke . . . is no denser than 60 percent black," a figure probably derived from the smoke measurement device known as the Ringlemann chart. Although by today's standards this would be considered very substantial air pollution, the level was not unusual for industrial emissions in Pittsburgh in the first decade of the twentieth century. The city passed several smoke-control ordinances in this period, but they either exempted iron and steel mills or were unenforceable. Court records show no lawsuits that specifically referred to the urban coke operations.[18]

Another drawback of the beehive process was the waste of the valuable coal by-products vented during coking. For instance, in 1910, the maximum year for beehive production, 53 million tons of coal were processed into coke. One study suggests that this amount of coal had the potential of producing approximately 530 billion cubic feet of gas, 400 million gallons of coal tar, nearly 150 million gallons of light oils, and 600,000 tons of ammonium sulfate. A few attempts were made to construct beehive ovens in which the by-products could be captured, but engineers had little success except in regard to waste heat. Even if the by-products were recovered, aside from the coal gas, few markets existed in the U.S. in the late nineteenth century for other coal distillation by-products.[19]

In addition to wasting valuable by-products, the beehive oven possessed other disadvantages as a production technology. Most important was that it required very high quality bituminous coal in order to produce good coke, and it was a difficult technology with which to achieve scale economies. The beehive's principal advantage over alternatives was its low cost and ease of construction and operation. As long as high-quality bituminous supplies existed and markets for coal distillation by-products were limited, there was little incentive to alter the technology.

The primary alternative coke-making technology to the beehive oven at the turn of the century was the retort or by-product oven, developed by Belgian, French, and German engineers because of the absence of good coking coal in Europe. The by-product or retort oven was a narrow, slot oven constructed in batteries in which coking chambers alternated with heating chambers. Coal was charged through openings in the top of the oven and the coke pushed out by a power-driven ram at the end of the combustion process, to be quenched outside the oven. The gas evolving from the coal supplied the heat required for distillation and was also used for other purposes throughout the mill.[20]

The by-product oven had the advantage of capturing the volatile elements freed by the coking process. The yield of by-products was determined by the quality and quantity of coke desired. The freed gases were collected by a system of pipes at the top of the oven and then cooled, resulting in condensation of tar and water vapor into a liquid of about 70 percent tar and 30 percent of ammonia in water. The ammonia underwent further separation at a still and was eventually transformed into various ammonium compounds. In addition to the ammonia compounds, other by-products from the coking process included benzene, toluene, naphtha, and xylene. Because it salvaged useful materials that might otherwise be wasted, one engineer observed that the by-product oven was "a part of the movement for the conservation of our natural resources," an interesting attempt to place this industrial technology in the conservation movement of the early twentieth century.[21]

The by-product oven had several other advantages over the beehive oven in addition to capturing the coal distillation products: it could produce high-grade coke using blends of various qualities of coal; it produced a higher coke yield per ton of coal (average of 70 percent compared to 64.5 percent); and it provided for more rapid coking. But the technology also had important disadvantages. One involved the concern of pig iron furnacemen that by-product coke, because of its smaller size and dull black color, was actually inferior to the silvery beehive coke. The second, and probably the most critical, was the high capital and operating costs involved in producing by-product coke. Markets for by-products could, in principle, make up some of this differential, but the United States was largely without a well-developed chemical industry capable of utilizing coal distillation products such as tar, ammonia, and benzene. Coal tar products produced in Germany dominated the American markets. These factors, plus the availability of large supplies of superior coking coal, retarded the technology's widespread adoption in this country.[22]

Coal, iron, and steel firms did construct several by-product ovens at the beginning of the century, transferring and adopting different versions of the European technology, but more rapid development did not occur until after 1914. One important factor stimulating change was the discovery by U.S. Steel that coke oven gas and tar could produce important fuel efficiencies in the integrated steel mills. In addition, the cutting-off of organic chemical supplies from Germany by the war provided an especially strong impetus for rapid development. Faced by the loss of synthetic dyes, drugs, and solvents, as well as benzene, toluene, and phenols for the production of explosives, the United States had to rapidly create a chemical industry largely based on aromatic compounds.[23]

Many by-product plants were constructed to take advantage of the new markets. They were most frequently located at integrated steel mills, where

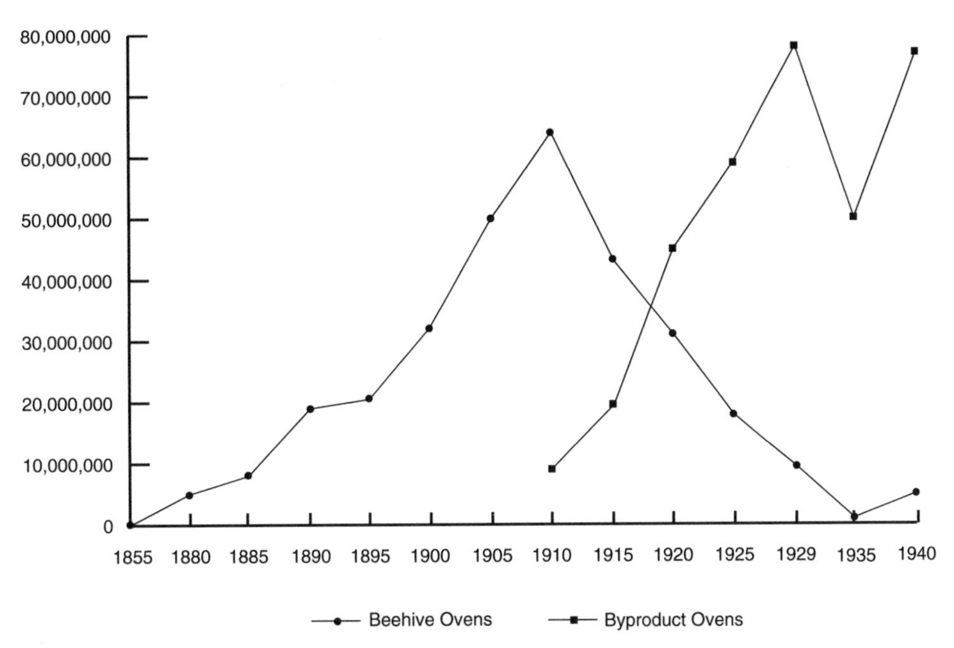

Figure 15.1. Net Tons of Coal Used, Beehive and Byproduct Coke Ovens

the coke could be used in the blast furnaces and the gas and tar could be used as fuel to provide heat in the iron and steel works. In addition, a number of cities, including St. Paul, St. Louis, Baltimore, and Jersey City, began obtaining their municipal gas supplies from by-product ovens. By 1929, in what was a "revolutionary" industrial change, the new technology was supplying 75 percent of the coke manufactured in the United States. Between 1909 and 1940, the number of beehive ovens in the country shrank from nearly 104,000 to about 15,000, located in seventy-five different plants and consuming 4,802,996 tons of coal. By the same year, the number of by-product ovens had increased to 12,734 in eighty-nine plants, consuming 76,582,780 tons of coal.[24]

The By-Product Coke Oven and Drinking Water Supplies

By-product coke ovens were customarily located on the banks of rivers or lakes, usually in close proximity to the blast furnaces of an integrated steel mill. The water bodies provided them with the water necessary for their operations, with cheap barge transportation for the coal and with a place to dispose of their liquid wastes. These locations were usually in or near urban areas, in contrast to the mostly rural locations of the beehive ovens. Although the by-product oven produced fewer air emissions than did the beehive oven, it still had a heavy pollution flow both to the air and to the water. Its wastewater stream, for instance, contained heavy concentrations of ammonia, cyanide, and phenolics, as well as various acid, base, and neutral hydrocarbons. While all of these substances could damage stream life and the quality of water supplies, the phenol-containing effluent stream discharged by the ammonia stills had the most noticeable impact, creating severe taste and odor problems in drinking water. The volumes discharged were extremely large; one chemical engineer estimated in 1923 that by-product coke plants discharged approximately 38,000,000 tons of still wastes in that year.[25]

While phenols would produce obnoxious tastes and odors in any water supply, the effect on drinking water was especially severe if the receiving waters were chlorinated. During the late nineteenth and early twentieth centuries, a number of American cities with centralized water systems attempted to protect their supplies from infectious sewage wastes by adopting filtration technology or chlorinating their supplies. The first successful use of chlorination occurred in 1908, when Jersey City proved the effectiveness of the technique in protecting its water supply from pathogens. A number of other municipalities, also faced by the threat of sewage pollution of their water supplies, adopted the technology. As chlorination spread to various industrialized areas, however, waterworks managers observed that the water supply often developed strong medicinal tastes and odors. This was initially blamed on the chlorine alone, but in 1918 chemists from the Milwaukee Department of Health and the city waterworks determined that the tastes and odors were due to interaction between the chlorine and coal tar wastes.[26]

Cities in the Ohio River basin were particularly subject to these problems because of their heavy use of chlorine to protect their water supplies (drawn from sewage-polluted rivers) and the concentration of by-product coke ovens in the region. In the mid-1920s, the U.S. Public Health Service

(USPHS) identified twenty-five cities in the Ohio River Valley where the interaction of chlorine with phenol wastes made the water almost undrinkable. Other cities in different watersheds, such as Chicago, Cleveland, Detroit, Milwaukee, Rochester, and Troy, also experienced similar problems.[27] In this case, two new technologies had interacted chemically to produce harmful results for drinking-water quality.

Aside from the taste and odor nuisances, public health officials worried about the health effects of the phenols in water supply. They had four principal concerns: they could injure the health of persons drinking the polluted water; the offensive tastes and odors might discourage individuals from drinking sufficient water for good health or perhaps cause them to drink biologically contaminated water; and, because phenols reacted with chlorine, waterworks managers might reduce the amount of chlorine treatment, thereby increasing the risk of exposure to infectious disease. A typhoid outbreak in 1925 in Ironton, Ohio (eighteen cases, three deaths), attributed to the fact that "choloro-phenol" tastes in the municipal water supply were so offensive that local residents were forced to drink unprotected water, highlighted these fears.[28]

The focus on the health risks presented by the phenol wastes, rather than on their injuries to property, represented a shift in concern in regard to pollution from coke manufacture. The emissions from beehive ovens were viewed as producing primarily private property damages that could be compensated for monetarily, but the threat to water supply threatened the public good. Public authorities had several courses of action, the most available of which, given the absence of specifically applicable state statutory industrial waste regulations, was to use the courts to secure relief under the common law.[29] Here state authority in regard to the police power was a well-established basis for court action to protect the public health.

The most important court action occurred in McKeesport, Pennsylvania, an industrial city of approximately 50,000 people located at the junction of the Youghiogheny and Monongahela rivers about fifteen miles above Pittsburgh. From 1881 until 1916, McKeesport obtained its water supply from the Youghiogheny River, but increased pollution by mine acid drainage and other industrial wastes caused the city to shift to the Monongahela River in the latter year. Although the Monongahela was also polluted by industrial wastes, its waters required less treatment than that from the Youghiogheny, significantly lowering the city's treatment costs.[30]

Approximately eighteen months after McKeesport had moved to the Monongahela for its supplies, the Carnegie Steel Corporation (subsidiary of the U.S Steel Corporation) brought on line the world's largest by-product coke operation, located at Clairton, Pennsylvania. This plant was sited on the banks of the Mononghahela River about two miles above the McKeesport water intake and was meant to provide coke for the U.S. Steel integrated steel mills in the Mononghahela Valley. The plant initially had 768 ovens, each of which had a capacity of 13.2 tons of coal per charge and produced about nine tons of coke. In addition, the plant produced large quantities of coal by-products such as gas, tar, benzene, and ammonium sulphate.[31]

The Carnegie by-product works at Clairton discharged the liquid wastes from their plant through a pipe into Peters Creek, a stream that ran through their property and into the Monongahela River. The city of McKeesport charged that in the river the wastes "permeated" the water and caused "foul" odors and "offensive and nauseating taste" in its water supply. In September of 1918, McKeesport sued in the county courts for an injunction to prevent the Carnegie Steel Corporation from discharging these polluting wastes, charging that the plant's discharges rendered its water "unwholesome, unpalatable and unfit for drinking purposes and domestic uses, as well as injurious to health." The courts, following the well-established precedent that the police power could be used to protect the public health, granted the injunction, observing that the city's "duty to supply its inhabitants reasonably pure and palatable water cannot be questioned; nor can the right of its inhabitants to such quality be injuriously affected by foul and noxious fluids." If it wanted to continue operations, Carnegie Steel would have to find a different means to dispose of the wastes from its by-product coke plant.[32]

The Clairton decision reflected the fact that, by 1918, legal precedent existed under the police power to prevent the pollution of water supplies by industrial wastes if they clearly injured the public health. In the case of by-product coke oven wastes, cause and effect were usually clearly definable, and the courts were willing to act. In regard to other industrial wastes, however, cause and effect were less obvious, and therefore the courts and municipal and state health authorities were much more reluctant to use the police power to take action.[33]

Another strategy to deal with industrial pollution was to push for new legislation to enlarge the power of state health departments over industrial wastes. In most states, control over water pollution originally rested in the department of health because of its relation to infectious disease, and, aside from Massachusetts and Rhode Island, little attention was paid to industrial pollution in the first decades of the twentieth century. In the 1920s, however, such pollution increasingly became a matter of concern, and several states enacted legislation to attempt to regulate its discharge. In Pennsylvania, for instance, the legislature established a Sanitary Water Board in 1923 to determine the proper use of streams in the commonwealth and to control pollution. In 1924, the deputy state attorney general ruled that the Sanitary Water Board had the power to prevent the pollution of public water supplies by industrial wastes (such as phenols) that could cause the public to avoid drinking the water because of tastes and odors. In Ohio, in 1925, the legislature approved a bill that gave the State Health Department the authority to zone sources of water supplies and to require new industries, or industries developing new processes, to file plans for satisfactory treatment plants. The state director of health called it "the most advanced piece of public health legislation bearing on the subject of stream pollution prevention in the country."[34]

In both Pennsylvania and Ohio, the directors of the relevant administrative bodies argued that cooperation rather than legal action was the proper strategy to follow in regard to industrial pollution, an approach to regulation that was typical of the 1920s. Underlying this strategy in the case of industrial pollution was a belief that "no existing satisfactory and economically possible means of treatment existed for the wastes of many important industries" and that legal action could drive industry out of the state, rather than produce environmental improvement.[35] State authorities believed that firms themselves wished to find a means to deal with pollution and would be willing to cooperate in seeking means of remedial action.

Pennsylvania pioneered in the area of cooperative agreements and during the 1920s made compacts with industries such as pulp and paper and tanning to achieve voluntary pollution reduction. This strategy was also used in regard to phenol pollution, and by 1928 the Sanitary Water Board had reached agreements with all Pennsylvania by-product coke operators to prevent taste and odor-producing substances from reaching state waters. In Ohio, similar agreements between the state and industry were reached during the decade. In spite of these agreements, state authorities worried about the limitations on their ability to secure action by firms located on interstate streams such as the Ohio River. Without such agreements, it would be extremely difficult to secure voluntary compliance by intrastate firms because of their fear of being at a disadvantage in competitive markets.[36]

In a 1923 letter to U.S. Surgeon General Hugh S. Cumming, Dr. John E. Monger, director of the Ohio Board of Health, argued that the problem of interstate pollution required federal leadership, since "it is apparent that any particular State is powerless to secure a remedy." The federal government, however, had limited authority in regard to water pollution. In 1912, the U.S. Congress had assigned the Public Health Service the task of "investigating" the relationship between health and water pollution. The USPHS was concerned with industrial wastes primarily because of their relation to the quality of drinking-water supplies. Beginning in 1913, it launched a series of stream pollution investigations centering on the Ohio River that were directed from what later became its Center for Pollution Studies in Cincinnati. In February 1923, after having received considerable complaints about phenol pollution of water supplies, the USPHS launched a national survey of the problem.[37]

Dr. Monger urged the USPHS to take a leading role in addressing the interstate problem, but Surgeon General Cumming was extremely cautious about overstepping his authority. It was only after considerable urging that he agreed to take the minimal step of sponsoring a national conference on the phenol question. This conference, held in Washington on May 18, 1923, was attended by representatives from fifteen state health departments, as well as federal agencies such as the USPHS, the Bureau of Mines, and the Bureau of Standards. Although the federal role remained limited, the meeting served important organizational and informational purposes.[38]

A second national conference, also called by the surgeon general, was held in Washington in January 1924 and led to the formal organization of the Ohio River states. Three months later, in April 1924, health department officials from the Ohio River states met in Pittsburgh with representatives of the basin's by-product coke firms to see if they could reach agreement to control phenol wastes in the Ohio River. The tone of the meeting was cooperative rather than confrontational. By this time, many of the firms had already signed agreements with the Ohio and Pennsylvania departments of health to control their phenol wastes, and now the remainder agreed to take similar action.[39]

At the Pittsburgh meeting, the health department representatives from Ohio, Pennsylvania, and West Virginia also signed the Ohio River Interstate Stream Conservation Agreement. In this compact, the signatories agreed on uniform policy in regard to the protection of water supplies from phenols and other tarry acid wastes. The chief engineers of the three departments of health were constituted a "Board of Public Health Engineers of the Ohio River Basin," with the responsibility for protecting the watershed from industrial pollution. The prominent role of the engineers rather than the public health physicians (directors of state public health departments were required to be physicians in almost all states) in this case reflects both the rise to prominence of engineers in public health activities and the extent to which industrial pollution was viewed as a technical problem.[40]

By 1929, seventeen of the nineteen basin firms had installed phenol elimination devices, sharply reducing the most severe taste and odor problems. The procedure utilized by fourteen of the plants having 88 percent of the nation's coke-making capacity was to use the phenolic wastewater to quench the glowing coke, vaporizing the wastewater into the atmosphere rather than disposing of it in waterways. The Carnegie Steel Corporation had adopted quenching at its Clairton Works in 1918 after the McKeesport injunction, and it was to become the model for most of the other by-product plants, actively pushed by Judge E. H. Gary, chairman of the board of directors of the U.S. Steel Corporation. Having themselves installed a quenching system, U.S. Steel was anxious to persuade its competitors to adopt a technology at least as costly as the one they were utilizing. After a visit to the Clairton works to examine their quenching operation, Monger wrote to coke plant managers urging them to adopt the technique. "We trust," he said in a thinly veiled warning, "that voluntary action will be taken by various coke producing companies in Ohio without formal order by this department."[41]

Quenching with the phenolic wastewater was an expensive environmental control technology, especially for companies not used to expenditures on pollution control. The capital costs of installing such a system at one plant of the Youngstown Sheet and Tube Company with a 5,000-ton daily capacity in the mid-1920s was $250,000, with much higher costs at larger plants. The companies disliked quenching, not only because of its capital costs but also because it greatly increased the rate of equipment and structural corrosion. U.S. Steel's maintenance costs at its Clairton works (30,000-ton capacity) in the mid-1920s were over $60,000 per year. The phenolic wastewater used in the quenching system also imparted an odor to the coke, limiting its use for home heating. The coke manufacturers, however, had little choice, since alternative approaches such as phenol recovery or chemical treatment were either uncertain or prohibitively expensive.[42] In the absence of other more cost-effective recovery methods, and under pressure from the threat of lawsuits, the companies voluntarily (and reluctantly) adopted the quenching technique.

The importance of the threat of legal action in achieving "cooperation" from industry was illustrated in the case of phenol pollution in the Chicago area. Chicago, as well as other smaller cities in Illinois and Indiana, drew their water supply from Lake Michigan. In 1927, phenolic wastes from coke ovens in the Calumet region of Indiana, located on the lake, so badly polluted the water supply that it was undrinkable for a week. As in the Ohio River basin, conferences were held between industry representatives and the state health authorities. On June 13, 1928, representatives of the health departments of states bordering the Great Lakes (Illinois, Michigan, Wisconsin, Indiana, Minnesota, Ohio, Pennsylvania, and New York) signed the Great Lakes Water Covenant, an agreement similar to that in the Ohio River basin in regard to reducing phenol pollution.

At the time of the signing of the agreement, the involved industries gave the health authorities "verbal assurances" that they would control their phenol wastes within a two-year period. In spite of the "assurances," however, no action was taken, and during the following year a number of episodes of tastes and odors in the water supply related to phenol wastes occurred. The Chicago health authorities "realized that unless pressure was exerted by the Department of Health in an unmistakable way, the citizens of Chicago might be forced to drink unpalatable water for many years to come," and the department threatened to obtain a court injunction shutting down the firms. Faced by the threat of legal action, the Indiana coke companies capitulated and installed quenching technology.[43]

Coke quenching with the phenol wastewater did not actually eliminate the waste stream but shifted the burden of the pollution from the water to the air. While engineers originally thought that quenching destroyed the pollutants in the wastewater, it soon became apparent that the volatile components, such as phenol, cyanide, and ammonia, were actually being steam-distilled and discharged to the atmosphere. Some engineers believed this an undesirable solution. At the 1927 convention of the American Water Works Association, for instance, an engineer and chemist for the Rochester Gas & Electric Corporation, a firm itself having problems with its phenol wastes, commented that quenching was "one of the most objectionable things that could be inflicted on any community because we simply delay the time of getting them (the phenols) into the water supply." But this was a minority view, and most engineers and public health figures accepted quenching as the only option that came close to meeting both environmental and economic constraints.[44] In the absence of air pollution statutes, and with only a very limited understanding of the effects of airborne industrial pollutants, the air became a preferable sink to the water.

Conclusion

This study of the connection between the production of metallurgical fuels for the iron-making process, especially that of coke, and the creation of environmental damage and pollution, illuminates several aspects of the relationship between industrial technology, the environment, and public policy in the period before stringent local, state, or federal environmental legislation.

As technologies, both the beehive and the by-product coke ovens were major polluters, but their effects on the environment were substantially different. Beehive coke oven operators, as has been noted, made only very limited attempts to capture emissions, most of which went into the atmosphere, fouling the air and destroying nearby vegetation. Since these ovens were largely located in rural rather than urban areas (as were charcoal plantations), a somewhat smaller population was affected, but because the ovens were usually scattered over the landscape, they impacted a larger geographical area. Air quality, for instance, in the Connellsville coke area, with the largest concentration of beehive coke ovens in the world, was notoriously bad. A city such as Pittsburgh could either ban the ovens outright or attempt to impose pollution-control requirements, although enforcement was uncertain at best. The only recourse urban and rural dwellers often had was the common law—an option that was both expensive and cumbersome and brought limited damages.

By-product coke ovens were most frequently located near heavily populated metropolitan areas where the potential existed for pollution impacts on a larger population. In contrast to the beehive ovens, the by-product ovens were viewed as an improvement in regard to air emissions. The Mellon Smoke Investigation of Pittsburgh, the first comprehensive study of the impact of smoke pollution on a major American city, for instance, found it encouraging in 1914 that by-product coke ovens, "which give no [visible] smoke," would soon replace the city's "antiquated" beehive ovens.[45] Unanticipated, however, was the phenol waste stream's severe impact on chlorinated drinking-water supplies. Thus, two new technologies, one of which produced significant public health improvements and the other that improved productivity and reduced air emissions, had interacted to produce a health and nuisance-related environmental problem.

The quenching technology utilized to keep the phenol wastewater out of waterways, and thus to protect drinking-water quality itself, ultimately proved problematical from an environmental and health perspective. Quenching shifted the pollution burden from the water to the air, but it was not until the 1960s, at a time of growing environmental consciousness, that the technique received a serious challenge. Until this decade, air pollution (because of uncertainties about health effects) clearly held a lower position on the environmental policy agenda than did water pollution.[46]

Policy-making by government also reflected a tolerance for industrial pollution unless it had obvious health effects. The Ohio River basin agreement between coke producers and public health authorities that removed the phenolic wastes from water supplies was the result of a cooperative strategy (backed by a threat of court action) rather than a policy of command and control. Such an approach became the norm in Pennsylvania and many other industrial states in regard to industrial pollution for most of the period from the 1920s through the 1960s. State government officials, as well as industrial managers, were concerned over possible federal inroads into their authority and uniform standard setting from a distance. In 1926, for instance, when Congress was considering national legislation controlling stream pollution, the Ohio River Basin Association passed a resolution opposing federal control. The leadership of the USPHS, which had strong ties with the state health organizations, also held this position. Agreements such as the compact over phenols, said Dr. Wade H. Frost, director of the Cincinnati Research Station of the Public Health Service, gave "infinitely more promise of success than the enactment of federal laws to regulate the pollution of interstate streams." This skepticism about federal command and control regulation and a preference for interstate agencies and state standard setting remained the preferred line of attack on industrial pollution for Ohio River states through the 1970s.[47]

The case of waste disposal from coke manufacturing highlights the extent to which, in the past, because of economic and technological factors, we have often shifted our pollution burdens from one medium and one sink to another, rather than adopting a more holistic approach to the environment. Lack of knowledge and expensive control costs clearly limited choices at the time of the innovations, but progress in pollution control was often slowed by company resistance to large expenditures for control technology. As the case illustrates, society attempted a range of options in attempting to cope with industrial pollution, including, in this example, local and county ordinances, state legislation, regional compacts, and eventual federal standards. The movement from one governmental level to another reflects the difficulties of regulating interstate pollution, public frustration over the slowness of state regulation and industry response, and a shift from concern over property damages to the environmental and health effects of pollution.

Today our society confronts the necessity of paying for past waste-disposal practices that both affect environmental quality and pose potential health dangers. Many current environmental problems involve cross-media effects, such as groundwater pollution by runoff from hazardous waste dumps or sanitary landfills, that reflect past disposal choices. History beckons us, therefore, to sharpen our sensitivity to the interconnected nature of the environment and the implications of these connections for environmental policy.

Notes

I would like to thank Nicholas Casner, Edward W. Constant, II, Francis C. McMichael, Richard G. Luthy, Harry W. Paxton, Jeffrey Stine, Sam Bass Warner, Jr., and two anonymous reviewers for helpful comments on this essay. I would also like to thank Fredric L. Quivik for permitting me to quote from his paper, "Industrial Pollution on the Southwest Pennsylvania Countryside: Beehive Coking in the Connellsville Coke Region, 1860–1920," delivered at the 1993 Biennial Conference of the American Society for Environmental History, Pittsburgh. The research on which my article is based was funded by National Science Foundation Grant No. SES-8420478.

1. For elaboration of this concept, see Joel A. Tarr, "The Search for the Ultimate Sink: Urban Air, Land, and Water Pollution in Historical Perspective," J. Kirkpatrick Flack (ed.), *Records of the Columbia Historical Society of Washington, D.C.* 41 (Charlottesvile, Va., 1984): 1–29.

2. Arthur C. Bining, *Pennsylvania Iron Manufacture in the Eighteenth Century* (Harrisburg, Pa., 1973), 19–38; Michael Williams, *Americans and Their Forests: A Historical Geography* (New York, 1989), 104–10; and Paul F. Paskoff, *Industrial Evolution: Organization, Structure, and Growth of the Pennsylvania Iron Industry, 1750–1860* (Baltimore, Md., 1983), 1–38, 91–105. Some charcoal kilns, developed shortly before the Civil War, attempted to capture wood chemicals; charcoal retorts, developed in the 1870s and 1880s, were specifically designed to capture volatile wood chemicals. See R. H. Schallenberg and D. A. Ault, "Raw Materials Supply and Technological Change in the American Charcoal Iron Industry," *Technology and Culture* 18 (July 1977): 453–54.

3. Peter Temin, *Iron and Steel in Nineteenth-Century America: An Economic Inquiry* (Cambridge, Mass., 1964), 82–3; Paskoff (n. 2 above), 18; and Williams (n. 2 above), 147–50, 347–44. Charcoal production, however, was much less important as a cause of deforestation than clearing for agricultural purposes.

4. Bining (n.2 above), 59.

5. Schallenberg and Ault (n. 2 above), 436–66; Alfred D. Chandler, Jr., "Anthracite Coal and the Beginnings of the Industrial Revolution in the United States," *Business History Review* 46 (Summer 1972): 151–64. Chandler maintains that this substitution occurred not because of demand for new energy sources, but because anthracite operators acquired the means to develop and make available the supply of anthracite.

6. Paskoff (n. 2 above), 101–02; Temin (n. 3 above), 58–62.

7. The total production of charcoal iron rose until 1890, but the proportion made with the fuel diminished sharply as a fraction of total production. See Temin (n. 3 above), 266–269; and

Williams (n. 2 above), 337–44. For a discussion of the anthracite region, see Donald L. Miller and Richard E. Sharpless, *The Kingdom of Coal: Work, Enterprise, and Ethnic Communities in the Mine Fields* (Philadelphia, Pa., 1985).

8. The Connellsville coke region was 42 miles long, an average of 3.54 miles wide, and covered 147 square miles. John Fulton, *Coke: A Treatise on the Manufacture of Coke and Other Prepared Fuels and the Savings of By-Products* (Scranton, Pa., 1906), 145; John Fulton, "A Report on Methods of Coking," in *Second Geological Survey of Pennsylvania: 1875* (Harrisburg, Pa., 1876), 122–23; Raymond Foss Bacon and William A. Hamor, *American Fuels* (New York, 1922, 2 vols.) 1: 6–9; and Frederick Moore Binder, *Coal Age Empire: Pennsylvania Coal and Its Utilization to 1860* (Harrisburg, Pa., 1974), 118–19.

9. Temin (n. 3 above), 268–69; Kenneth Warren, *The American Steel Industry, 1850–1970: A Geographical Interpretation* (Oxford, England, 1973), 77–80, 110, 200–06; and Chandler, "Anthracite Coal," 141–81.

10. Ibid., 122–23.

11. Fulton (n. 8, above), 146–57. The process of drawing coke by hand was immensely difficult. Not only was the work heavy but the laborers were also exposed to the heat of the ovens and dust and fumes. Workers willing to perform these difficult tasks were difficult to find and attempts were made to develop mechanical coke drawers as well as mechanical levelers and watering devices. See Walter W. Macfarren, "Coke Drawing Machines," in *Proceedings of the Engineers Society of Western Pennsylvania* 23 (November 1907): 451–509; and F. C. Keighly, "The Connellsville Coke Region," *Engineering Magazine* 20 (October 1900): 39. The yield of high class Connellsville coal coked in beehive ovens in 1875 was 63 percent. Franklin Platt, *Special Report on the Coke Manufacture of the Youghiogheny River Valley, Second Geological Survey of Pennsylvania: 1875*, 63; in 1911 the U.S. Geological Survey reported the average yield nationally for beehive ovens was 64.7 percent. See F. H. Wagner, *Coal and Coke* (New York, 1916), 296–99, and Fulton (n. 8 above), 147–48, Higher quality coke could be produced by increasing the length of the coking process to 72 hours.

12. H. N. Eavenson, *The First Century and a Quarter of the American Coal Industry* (Baltimore, Md., 1942), 380–84, 579–83; David Demarest and Eugene D. Levy, "A Relict Industrial Landscape," *Landscape* 29 (1986): 30–32; Muriel E. Sheppard, *Cloud By Day: The Story of Coal and Coke and People* (Uniontown, Pa., 1947), 22–24; and William L. Affelder, "Jones & Laughlin's Coke Plant," *Mines and Minerals* 29 (December, 1908): 195–99.

13. W. A. Buckhout, "The Effect of Smoke and Gas Upon Vegetation," *Annual Report of the Pennsylvania State Department of Agriculture* (Harrisburg, Pa., 1900), 180–82. One observer of the beehive coke region wrote that the ammonia "thrown into the atmosphere . . . by coke ovens . . . no doubt add[s] materially to the fertility of our land; for nature never wastes anything, and it is in all probability precipitated." See John W. Boileau, *Coal Fields of Southwestern Pennsylvania: Washington & Greene Counties* (Pittsburgh: Privately printed, 1907), 56.

14. See Shephard (n. 12 above). Some accounts describe the whole Connesllville region as frequently blanketed with fumes emitted from beehive coke ovens. See, for instance, William A. Metcalf, "On Smoke," *Proceedings of the Engineer's Society of Western Pennsylvania* 8 (1892): 29–30. Miners and coke oven workers often planted gardens close to the ovens but were careful to place them down-wind of the ovens. The fumes from the ovens trapped heat in the locality, leading to a longer growing season. For pictures of "prize winning" gardens, as well as of beehive coke oven plants, see John K. Gates, *The Beehive Coke Years: A Pictorial History of Those Times* (Uniontown, Pa., 1990). Joseph Lentz v. Carnegie Bros. & Co, 145 *Pa.* 613, (1891); and Fredric L. Quivik, "Industrial Pollution on the Southwest Pennsylvania Countryside: Beehive Coking in the Connellsville Coke Region, 1860–1920," unpublished paper presented at the 1993 Biennial Conference of the American Society for Environmental History, Pittsburgh, PA, March 6, 1993. For a discussion of cases resulting from damages from smelter fumes in the Columbia Basin, see Mark Cioc, "The Trail Smelter Dispute: International Arbitration and Transboundary Air Pollution in the Columbia Basin, 1927–1941," unpublished paper delivered at The ASEH Conference, 1993. Fumes and other emissions from the beehive ovens, as well as contact with the tarry coal wastes, almost certainly affected the health of the workers at the ovens and possibly the health of nearby resi-

dents, but the author has been unable to find a study documenting this. Industrial hygiene, which would have dealt with worker's health, was in its infancy at the time and attention focused on accidents rather than long-term chronic effects of industrial work. See, for instance, Sara J. Davenport, *Bibliography of Bureau of Mines Publications Dealing with Health and Safety in the Mineral and Allied Industries 1910–46, U.S. Bureau of Mines Technical Paper 705* (Washington, D.C., 1948), 23–24.

15. For an interesting interpretation of common law rulings on pollution cases in the nineteenth and early twentieth centuries, see Christine Rosen, "A Litigious Approach to Pollution Regulation: 1840–1906," (Unpublished paper, School of Business, University of California, Berkeley, 1990). In the Sanderson Cases, the Pennsylvania Supreme Court held that injured parties could not collect damages because of harm done to their property from mine acid drainage, because mine acid was a result of the "natural and necessary use" of property. See Sanderson v. Pennsylvania Coal Co., 86 *Pa.* 401 (1878) and 102 *Pa.* 370 (1883); Pennsylvania Coal Co. vs. Sanderson 94 *Pa.* 302 (1880) and 113 *Pa.* 126, 146 (1886); Brown et. al. v. Torrence 7 *Pa. State Reports* 186 (1880); Adam Robb v. Carnegie Bros. & Co., 145 *Pa.* 324 (1891); Joseph Lentz v. Carnegie Bros. & Co., 145 *Pa.* 612 (1891); and Campbell v. Bessemer Coke Company, 23 *Pa. Superior Ct.* 374 (1903).

16. Christine Rosen, "Cost Benefit Analysis in Pollution Nuisance Law 1840–1904," (unpublished paper, University of California at Berkeley, 1990). See, for instance, Adam Robb v. Carnegie Bros. & Co., 145 *Pa.* 324 (1891).

17. J. F. Slagle, *A Digest of the Acts of Assembly and a Code of the Ordinances of the City of Pittsburgh* (Pittsburgh, Pa., 1869), 269; and Hiram Schock (comp. and ed.), *Digest of the General Ordinances and Laws of the City of Pittsburgh to March 1, 1938* (Pittsburgh, Pa., 1938), 384–85. Affelder (n. 12 above),197. No mention has been found of other attempts to control pollution from beehive ovens, although at the Continental No. 1 Plant of the H.C. Frick Coke Company, located near Uniontown, Pennsylvania, a system was developed that used a tall stack to burn by-products and generate steam. See, also, Charles Catlett, "Increasing the Coke Yield from Bee-Hive Ovens," *Cassier's Magazine* 24 (October 1903): 534–42.

18. Mellon Institute, *Some Engineering Phases of the Smoke Problem*, Bulletin no. 6, Mellon Institute Smoke Investigation (Pittsburgh, Pa., 1914), 68–69. The report notes that in these coke ovens, the gases were "conveyed by long passages to the stacks where they are discharged." Since most beehive coke ovens vented through the roof, the use of the stacks suggests that the J & L ovens were included in the observation. In 1904, however, in the case of Sullivan v. Jones & Laughlin Steel Company, the Pennsylvania Supreme Court sustained an injunction issued by a lower court that enjoined J & L Steel from operating its blast furnaces in a manner that caused substantial damage to nearby residential areas. The suit had been engendered by J & L's use of fine Mesabi ore that produced severe dustfall problems. See Sullivan, Appellant, v. Jones & Laughlin Steel Company 208 *Penn. State Reports* 540 (1904). See, also, "Decision of the Supreme Court of Pennsylvania on the Use of Mesabi Ores," *The Bulletin of the American Iron and Steel Association* (July 10, 1904), 100–01. Full compliance with the injunction, however, was difficult to achieve given the nature of the technology, and the courts held in 1908 that the modifications installed by J & L constituted adequate compliance with the decree, in spite of evidence that the plaintiffs were still suffering injury. See, Sullivan v. Jones & Laughlin Steel Co., 222 *Pa* 72 (1908).

19. Sam H. Schurr and Bruce C. Netschert, *Energy in the American Economy, 1850–1975* (Baltimore, Md., 1960), 98, n. 75; Fulton (n. 8 above), 311–12; Catlett (n. 17 above), 541, and Bacon and Hamor (n. 8 above) 1: 133–34. For a contemporary discussions of beehive coke oven wastes, see Catlett, "Increasing the Coke Yield From Bee-Hive Ovens," 534–42; William Gilbert Irwin, "Coke-Making in the United States," *Cassier's Magazine* 19 (January 1901): 205; Heinrich J. Freyn, "The Wastefulness of Coke Ovens," *Scientific American Supplement No. 1998* (April 1914), 246–47; and Floyd W. Parsons, "Everybody's Business: Needless Coal Wastes," *Saturday Evening Post* (Mar. 13, 1920), 36–38.

20. M. Camp and C. B. Francis, *The Making, Shaping and Treating of Steel*, 4th ed. (Pittsburgh, Pa., 1919,), 101–16.

21. Camp and Francis (n. 20 above), 101–02; F. H. Wagner (n. 11 above), 303–06; and Camp and Francis (n. 20 above), 116–41. See W. W. Davis, "The Semet-Solvay By-Product Oven," in

Journal of the Engineers' Society of Pennsylvania 2 (Oct., 1910): 401. For the conservation movement, see Samuel P. Hays, *Conservation and the Gospel of Efficiency* (Cambridge, Mass., 1959).

22. Quivik (in n. 14 above), 15. William H. Blauvelt, "The By-Product Coke Oven," *Engineering News* 60 (Aug. 27, 1908): 218–23; and Aaron J. Ihde, *The Development of Modern Chemistry* (New York, 1964), 454–61.

23. C. A. Meissner, "The Modern By-Product Coke Oven," in James T. McCleary (ed.), *Year Book of the American Iron and Steel Institute* (New York, 1913), 118–78; Hyde, *The Development of Modern Chemistry,* 614–17, 671–94.

24. Parsons (n. 19 above), 38; Wilbert G. Fritz and Theodore A. Veenstra, *Regional Shifts in the Bituminous Coal Industry With Special Reference to Pennsylvania* (Pittsburgh, Pa., 1935), 116–20. Beehive ovens continued to be constructed in situations where the expense of a by-product plant could not be justified. See Warren (n. 9 above), 112–15; and Eavenson (n. 12 above), 582–85. During World War II, many beehive ovens were fired-up to meet wartime demands. See G. S. Scott, J. A. Kelley, E. L. Fish, and L. D. Schmidt, *Modern Beehive Coke-Oven Practice, Preliminary Report,* U.S. Bureau of Mines R. I. 3738 (Washington, D.C., Dec., 1943).

25. Richard G. Luthy, "Problems of Water Reuse in Coke Production," in E. Joe Middlebrooks (ed.), *Water Reuse* (Ann Arbor, Mich., 1982), 501–20. Phenols is the broad term applied to the mono-hydroxy derivatives of the benzene ring, and includes phenol, cresols, and xylenols. Coke oven waste includes a combination of all the phenols. See Ohio River Valley Water Sanitation Commission, *Phenol Wastes: Treatment by Chemical Oxidation* (Cincinnati, Oh., 1951), 9. R. D. Leitch, "Stream Pollution by Wastes from By-Product Coke Ovens: A Review, With Special Reference to Methods of Disposal," *Public Health Reports* 40 (Sept. 25, 1925): 2022.

26. Stuart Galishoff, "Triumph and Failure: The American Response to the Urban Water Supply Problem, 1860–1923," in Martin V. Melosi, ed., *Pollution and Reform in American Cities, 1870–1930* (Austin, Tx., 1980), 50–54; H. Bohmann, "Find Causes of Obnoxious Tastes in Milwaukee Water," *Engineering News-Record* 82 (Jan. 23, 1919): 181–82. H. R. Crohurst, a sanitary engineer with the USPHS, estimated in 1924 that 1 part of phenol waste to 75–100 million parts of chlorinated water would bring detectable tastes and odors. See E. S. Tisdale, "Cooperative State Control of Phenol Wastes on the Ohio River Watershed," *Journal of American Water Works Association* 18 (1927): 575; and Wellington Donaldson, "Industrial Wastes in Relation to Water Supplies," *American Journal of Public Health* 11 (March 1921): 195. Interestingly, very high concentrations of chlorine would neutralize the phenol. See Ohio River Valley Water Sanitation Commission, *Phenol Wastes,* 9.

27. "Discussion of Report of Committee No. 6 on Industrial Wastes in Relation to Water Supply," *JAWWA* 12 (1924): 411–15; F. Holman Waring, "Results Obtained in Phenolic Wastes Disposal Under the Ohio River Basin Interstate Stream Conservation Agreement," *AJPH* 19 (1929): 758–70; and "Progress Report of Committee on Industrial Wastes in Relation to Water Supply," *JAWWA* 10 (1923): 415–30; 16 (1926): 302–03.

28. Waring (n . 10 above), 758–70; "Progress Report on Recent Developments in the Field of Industrial Wastes in Relation to Water Supply," *JAWWA* 16 (1926): 310–11.

29. In 1912, the city of New Castle, Pennsylvania sued the Carnegie Steel Company of Farrel, Pennsylvania to stop it from polluting New Castle's water supply with phenolic wastes from its by-product coke oven. The company settled the case out of court and agreed to prevent further difficulties. See Waring, (n . 27 above), "Results Obtained in Phenolic Wastes Disposal under the Ohio River Basin Interstate Stream Conservation Agreement," *AJPH* 19 (1929): 759–61.

30. The Youghiogheny River waters were plagued by extreme hardness, acidity, and turbidity. See E .C. Trax, "A New Raw Water Supply for the City of McKeesport, Pennsylvania," *Journal of the American Water Works Association* 3 (1916): 947–53.

31. See Camp and Francis (n. 20 above), 103–04, for a description of the Clairton by-product coke plant.

32. McKeesport v Carnegie Steel Company 66 *Pittsburgh Legal Journal* 695 (1918); 66 *Pittsburgh Legal Journal* 696.

33. E. B. Besselievre, "Statutory Regulation of Stream Pollution and the Common Law," *Transactions of the American Institute of Chemical Engineers* 16 (1924): 217–29; and John H. Fertig, "The Legal Aspects of the Stream Pollution Problem," *AJPH* 16 (August 1926): 782–88.

34. Joel A. Tarr, "Industrial Wastes and Public Health: Some Historical Notes, Part 1, 1876–1932," *AJPH* 75 (September 1985): 1059–1067; W. L. Stevenson, "Pennsylvania Sanitary Water Board," *Engineering News-Record* 91 (Oct. 25, 1923): 684–85; and Almon L. Fales, "Progress in the Control of Pollution by Industrial Wastes," *AJPH* 18 (1928): 715–21. The Board established three classes of streams: those that were relatively pure; those in which pollution needed to be controlled; and those that were already so polluted that they could not be used for public water supplies or recreational purposes, and therefore were not worthy of pollution removal. Some conservationists criticized the Pennsylvania law as focusing too strongly on drinking water quality and not paying enough attention to larger environmental questions. See Tarr, "Industrial Wastes and Public Health," 1062; Philip Wells, Deputy Attorney General, to Charles E. Miner, M.D., Secretary of Health, April 12, 1924, printed in 72 *Pittsburgh Legal Journal* 648; "Discussion of Report of Committee No. 6 on Industrial Wastes in Relation to Water Supply," *JAWWA* 12 (1924): 411, comments of W. L. Stevenson; and John E. Monger, "Administrative Phases of Stream Pollution Control," *AJPH* 16 (August 1926): 790–91. Existing plants, however, were exempt from the requirement.

35. Ellis Hawley, "Herbert Hoover, the Commerce Secreatariat, and the Vision of an Associative State, 1921–28," *Journal of American History* 61 (1974): 116–40; Monger (n. 34 above), 16: 792–94.

36. W. L. Stevenson, "The Sanitary Conservation of Streams by Cooperation," *Transactions of the American Institute of Chemical Engineers* 27 (1931): 13–21. The agreement also required the firms to notify state authorities immediately in case of an accident that threatened water supplies. John E. Monger to Homer S. Cumming, Nov. 6, 1923, U.S. Public Health Service Archives, Record Group 90, National Archives, Washington, D.C. Cooperation was sometimes slow in forthcoming, and the threat of legal action was always in the background. See "Discussion of Report of Committee No. 6 on Industrial Wastes in Regard to Water Supply," *JAWWA* 12 (May 22, 1924): 411–12, comments of J. W. Ellms; Tisdale (n. 26 above), 18:577–78.

37. Monger to Cummings, Nov. 6, 1923; Tarr (n. 34 above), 1064. See a series of letters from various state departments of health to Surgeon General Homer S. Cummings in USPHS Records, Record Group 90, National Archives, Washington, D.C.

38. John E. Monger to Surgeon General Homer S. Cummings, Feb. 7, 1923, H. H. Streeter to Cummings, Dec. 24, 1923, and Wade Frost to Cummings, Nov. 19, 1923, U.S. Public Health Service Records, Record Group 90, National Archives, Washington, D.C.; Tisdale (n. 26 above), 256–57. The state health officials officially requested that the USPHS conduct an investigation of the health effects of phenols and that the Bureau of Mines investigate its "industrial" aspects. Only the Bureau of Mines study was published. See Leitch (n. 25 above), 2021–26. Leitch was a chemical engineer employed by the Bureau of Mines.

39. See Tisdale (n. 26 above), 256–57.

40. Waring (n. 29 above), 764–66. The agreement was eventually signed by the states of Kentucky, New York, Maryland, Illinois, Indiana, Tennessee, North Carolina, and Virginia. This compact was an early example of a single-purpose basin-wide agreement. For a discussion of such agreements, see Roscoe C. Martin, et. al., *River Basin Administration and the Delaware* (Syracuse, N.Y., 1960), 227–37. For the growing importance of engineers within the field of public health and their conflict with physicians, see Joel A. Tarr, et. al., "Disputes Over Water Quality Policy: Professional Cultures in Conflict," *AJPH* 70 (April 1980): 427–35, and Joel A. Tarr et. al., "Water and Wastes: A Retrospective Assessment of Wastewater Technology in the United States, 1800–1932," *Technology & Culture* 25 (April 1984): 246–50.

41. Willard W. Hodge, "Waste Disposal Problems in the Coal Mining Industry," in Willem Rudolfs (ed.), *Industrial Wastes: Their Disposal and Treatment* (New York, 1953), 312–418; and Luthy (n. 25 above), 501–20. The quenching ratio was about 500 gallons of water to quench the coke produced by one ton of coal. Other methods such as biological treatment were also used but primarily for discharge to sewers. Monger to Cummings, Aug. 6, 1923, and enclosures.

42. Waring (n. 29 above), 761, 761–63; Leitch (n. 25 above), 40:2021–2026. See also Luthy (n. 25 above), 511.

43. Sanitary Engineering Division, Chicago Board of Health, "Water Contamination With Phenol Wastes," *Annual Reports, 1926–30* (Chicago, 1930), 470–76, 492–504; Arthur E. Gorman, "Pollution of Lake Michigan: Survey of Sources of Pollution," *Civil Engineering* 3 (1933): 519–22; and Craig E. Colten, *Industrial Wastes in the Calumet Areas, 1869–1970* (Champaign, Ill., 1985).

44. Norman F. Prince, chemist and engineer from the Rochester Gas and Electric Corporation, in "Discussion," *JAWWA* 18 (1927): 582–83. See also F. W. Sperr, Jr., "Disposal of Phenol Wastes from By-Product Coke Plants," *AJPH* 19 (1929): 907.

45. Mellon Institute, (n. 18 above), 72. By-product ovens did produce fumes and steam but not necessarily heavy smoke pollution.

46. In 1969, Allegheny County, Pennsylvania, the home of U.S. Steel's Clairton coke works, approved new Air Pollution Rules and Regulations requiring that "The water utilized for the quenching of coke, prior to use as a quenching agent, shall be of a quality as may be discharged into the nearest stream or river, in accordance with the Acts of the Commonwealth of Pennsylvania." Compliance with this act would have required water completely free of phenols, an almost impossible task, and the code was amended in 1970 to require that 99 percent of the phenol be removed. Since that time, the coke industry and the regulatory authorities have engaged in constant conflict and negotiation about control of coke plant emissions to the environment, with the primary concern focused on health damages to workers and county residents. For a discussion of these acts and of subsequent events, see Charles O. Jones, *Clean Air: The Policies and Politics of Pollution Control* (Pittsburgh, Pa., 1975). In 1929, for instance, a case was filed in Allegheny County by a group of plantiffs trying to secure compensation for personal and property damage from coke oven chimney wastes. The case was not disposed of until 1940, having been to the U.S. Supreme Court, the Pennsylvania Supreme Court, and to the U.S. Court of Appeals a number of times on procedural grounds. The merits of the case were never dealt with and the plantiffs secured no relief. See J. Philip Bromberg, *Clean Air Act Handbook*, 2nd. ed. (Rockville, Md., 1985), 16.

47. "Enforcement through persuasion and cooperation" was also characteristic of enforcement of the Allegheny County Air Pollution statutes from 1949 through the 1960s. See Jones (n. 46 above), 94. In 1924, the congress had passed the Oil Pollution Act, but it only applied to coastal waters and made no reference to water supplies. Quoted in Tisdale (n. 26 above), 580. The preferred vehicle for dealing with interstate river pollution problems was the Ohio River Valley Sanitation Commission (ORSANCO), an interstate compact founded by authority of Congress in 1936. See Edward J. Cleary, *The ORSANCO Story: Water Quality Management in the Ohio Valley under an Interstate Compact* (Baltimore, Md., 1967), 249–83.

Name Index

This index is alphabetized letter-by-letter. Page numbers in italics refer to illustrations or their captions.

Place Index

This index is alphabetized letter-by-letter. Page numbers in italics indicate illustrations or their captions. Entries for states are for general references to states, not for specific agencies or locations within states such as health boards or cities; cities, agencies (in Name Index*), etc. are listed individually. "Passim" refers to passing references to the subject entry over a long series of pages.*

ABOUT THE AUTHOR

Joel A. Tarr is the Richard S. Caliguiri Professor of Urban and Environmental History and Policy at Carnegie Mellon University. He received his B.S. and M.A. degrees at Rutgers University and his Ph.D. from Northwestern University. Among his many publications is the co-edited collection, *Technology and the Rise of the Networked City in Europe and America*, 1988 winner of the Abel Wolman Award from the Public Works Historical Society. He has been awarded fellowships and grants from several institutions, including the National Science Foundation and the National Endowment for the Humanities.